The Quadratic Assignment Problem

COMBINATORIAL OPTIMIZATION

VOLUME 1

Through monographs and contributed works the objective of the series is to publish state of the art expository research covering all topics in the field of combinatorial optimization. In addition, the series will include books which are suitable for graduate level courses in computer science, engineering, business, applied mathematics, and operations research.

Combinatorial (or discrete) optimization problems arise in various applications, including communications network design, VLSI design, machine vision, airline crew scheduling, corporate planning, computer-aided design and manufacturing, database query design, cellular telephone frequency assignment, constraint directed reasoning, and computational biology. The topics of the books will cover complexity analysis and algorithm design (parallel and serial), computational experiments and applications in science and engineering.

The Quadratic Assignment Problem

Theory and Algorithms

by

Eranda Çela
Institute of Mathematics,
Technical University Graz,
Graz, Austria

Springer-Science+Business Media, B.V.

A C.I.P. Catalogue record for this book is available from the Library of Congress.

DOI 10.1007/978-1-4757-2787-6

Printed on acid-free paper

CONTENTS

PREFACE

The quadratic assignment problem (QAP) was introduced in 1957 by Koopmans and Beckmann to model a plant location problem. Since then the QAP has been object of numerous investigations by mathematicians, computers scientists, operations researchers and practitioners. Nowadays the QAP is widely considered as a classical combinatorial optimization problem which is (still) attractive from many points of view. In our opinion there are at last three main reasons which make the QAP a popular problem in combinatorial optimization. First, the number of real-life problems which are mathematically modeled by QAPs has been continuously increasing and the variety of the fields they belong to is astonishing. To recall just a restricted number among the applications of the QAP let us mention placement problems, scheduling, manufacturing, VLSI design, statistical data analysis, and parallel and distributed computing. Secondly, a number of other well known combinatorial optimization problems can be formulated as QAPs. Typical examples are the traveling salesman problem and a large number of optimization problems in graphs such as the maximum clique problem, the graph partitioning problem and the minimum feedback arc set problem. Finally, from a computational point of view the QAP is a very difficult problem. The QAP is not only NP-hard and NP-hard to approximate, but it is also practically intractable: it is generally considered as impossible to solve (to optimality) QAP instances of size larger than 20 within reasonable time limits. Notice that the remarkable progress in data structures and algorithmic developments, as well as major advances in computer hardware, have enabled a tremendous increase in the size of NP-hard problems which can be solved in practice. Recall for example large-scale instances of combinatorial problems such as the maximum clique problem on graphs with thousands of vertices and millions of edges, or the traveling salesman problem with thousands of cities which can be solved in practice. In this context, the quadratic assignment problem remains a challenging exception.

Despite the toughness of the QAP and also because of it, the literature on this problems abounds in results on almost all usually considered aspects of combinatorial optimization problems. This monograph aims at giving a general overview on the most studied aspects of the problem and the obtained results, as well as at outlining a number of research directions on the QAP which currently seem to be promising. In this way this monograph might probably serve as a relevant reference to researchers and students interested in the QAP, but also to practitioners who face the QAP and wish to better understand this problem in its inherent complexity.

The monograph is organized in 7 chapters. In Chapter 1, after introducing the problem and briefly describing some of its numerous applications, we present a number of linearizations of the QAP together with the first significant results concerning the QAP polytope and the investigation of the QAP from a polyhedral point of view. Further, a picture on the computational complexity of the QAP is given by summarizing a number of results concerning this aspect of the problem. In Chapter 2 algorithms to solve the QAP to optimality are considered. We focus on bounding procedures as a crucial component of such algorithms. (Recall that the QAP is NP-hard and therefore, enumeration methods are the only approach towards solving it to optimality.) The first part of Chapter 3 gives an overview on heuristic approaches which have been used to approximately solve the QAP. Further, the asymptotic behavior of the QAP is described in the second part of this chapter. In Chapters 4, 5 and 6 restricted versions of the QAP are considered. We point out polynomially solvable special cases of the QAP, as well as provably hard versions of the problem, aiming at drawing a border-line between polynomially solvable versions of the problem and NP-hard ones. The restricted versions of the QAP investigated in these three chapter arise either by imposing specific combinatorial properties on the coefficient matrices or as QAP formulations of optimization problems in graphs. The special coefficient matrices may posses Monge-like properties (Monge, Anti-Monge or Kalmanson matrices), may be diagonally structured (Toeplitz or circulant matrices), may be graded on the rows, on the columns or on both rows and columns etc. Restricted versions of the QAP where the coefficient matrices have properties as those mentioned above are discussed and analyzed in Chapter 4 and 5. The matrix classes which are characterized by this properties, and a number of restricted versions of the QAP with coefficient matrices taken from this classes are discussed in Chapter 4. Another restricted version of the QAP with special coefficient matrices, the so-called Anti-Monge–Toeplitz QAP, together with some of its applications is discussed in Chapter 5. This problem is considered in a separate chapter for two reasons. First, because of its interesting applications this restricted version of the QAP seems to be more relevant than the others. Secondly, in our opinion the methods used to deal with this problem which basically are quite simple, are however somewhat specific from the mathematical point of view. Chapter 5 is concluded by a brief discussion on permuted polynomially solvable cases of the QAP. Chapter 6 reviews and reformulates in the language of the QAP some results on optimization problems in graphs which can be formulated as QAPs. Among others, packing problems, the minimum feedback arc set problem and the linear arrangement problem are considered. This chapter is concluded by a theorem on the complexity of the so called pyramidal QAP. Finally, we present a generalization of the QAP, the so called biquadratic assignment problem (BiQAP), which is the mathematical formulation of a problem arising in VLSI design. This problem which seems to be interesting because of its above mentioned application, is apparently very difficult. Some results obtained for QAPs, e.g. results concerning linearizations, the Gilmore-Lawler bound or the asymptotic behavior, can also be generalized for the BiQAP. However, such results are of no particular relevance in terms of the solvability of BiQAPs which arise in real-life applications and hence, a lot remains to be done in this direction.

My most special thanks are due to Prof. Rainer Burkard at the Technical University Graz who introduced the Quadratic Assignment Problem to me, supervised my graduate studies, and encouraged me to write this monograph. His guidance and support during the preparation of my Ph. D. thesis of which this monograph is an extension and generalization, and his suggestions and remarks during the preparation of this monograph gave me an invaluable help.

I would like to express my deepest gratitude to Gerhard Woeginger for his tutelage and guidance during my graduate studies, when I was mainly working on the QAP. Most of the results on special cases of the QAP presented in Chapter 4, 5, and 6 are result of our joint work.

I am gratefully indebted to Stefan Karisch, Bettina Klinz, Ulrich Pferschy, Rüdiger Rudolf, and Marc Wennink, for carefully reading parts of the manuscript and pointing out many errors in earlier versions, as well as for their helpful suggestions which have been indispensable for improving the presentation. Of course, I am the only responsible for the remaining mistakes and errors: I apologize in advance.

I would also like to thank Prof. Panos Pardalos for encouraging me to extend my thesis into a monograph and for his willingness to propose the inclusion of this monograph in the series "Combinatorial Optimization" edited by P.Pardalos and D.-Zh. Du.

Last but not least, I gratefully acknowledge the financial support that I received from the "Fonds zur Förderung der wissenschaftlichen Forschung (FWF)", research project "Spezialforschungsbereich F003 Optimierung und Kontrolle, Projektbereich Diskrete Optimierung" which made possible the writing of this monograph.

Graz
August 1997

Eranda Çela

LIST OF FIGURES

LIST OF TABLES

1

PROBLEM STATEMENT AND COMPLEXITY ASPECTS

In this chapter we introduce the quadratic assignment problem (QAP). We start with the definition of the QAP and some frequently used formulations for it. Then, we consider the applications of the QAP and discuss in more details the most celebrated ones: applications in facility location and applications in wiring problems. Further, three mixed integer linear programming (MILP) formulations of the QAP are introduced: the Kaufman and Broeckx linearization [141], the Frieze and Yadegar linearization [83], and the linearization of Padberg and Rijal [177]. These formulations are chosen among numerous MILP formulation proposed in the literature in the hope that they can better help for a deep understanding of the QAP and its combinatorial structure. Consider that the linearization of Kaufman and Broeckx is perhaps the smallest QAP linearizations in terms of the number of variables and constraints, whereas some of the best existing bounding procedures for the QAP are obtained by building upon the linearization of Frieze and Yadegar. Furthermore, the linearization of Padberg and Rijal is the basis of recent significant results on the QAP polytope. Based on this linearization, the affine hull and the dimension of the QAP polytope (symmetric QAP polytope) have been computed. Moreover, some valid inequalities for the QAP polytope and some facet defining equalities for the symmetric QAP polytope have been identified. Finally, we consider computational complexity aspects of the QAP, discussing among others the complexity of approximating the problem, and the complexity of the local search.

We hope this chapter will fulfill its goal to provide the reader with a clear understanding of the problem and the extent of its applications and difficulty.

1.1 PROBLEM STATEMENT

Consider the set $\{1, 2, \ldots, n\}$ and two $n \times n$ matrices $A = (a_{ij})$, $B = (b_{ij})$. The quadratic assignment problem with coefficient matrices A and B, shortly denoted by QAP(A,B), can be stated as follows

$$\min_{\pi \in \mathcal{S}_n} \sum_{i=1}^{n} \sum_{j=1}^{n} a_{\pi(i)\pi(j)} b_{ij}, \tag{1.1}$$

where $\mathcal{S}_n$ is the set of permutations of $\{1, 2, \ldots, n\}$[1]. That is, QAP(A,B) is the problem of finding a permutation $\pi \in \mathcal{S}_n$ which minimizes the double sum in the above formulation. Obviously, the value of this sum depends on the matrices A and B and on the permutation π. To formalize these dependencies we denote:

$$Z(A, B, \pi) = \sum_{i=1}^{n} \sum_{j=1}^{n} a_{\pi(i)\pi(j)} b_{ij}.$$

The function $Z(A, B, \pi)$ is called the *objective function* of QAP(A,B) and a permutation π_0 which minimizes it over $\mathcal{S}_n$ is called an *optimal solution* to QAP(A,B). The corresponding value of the objective function, $Z(A, B, \pi_0)$, is called *the optimal value* of QAP(A,B). The size n of the coefficient matrices A and B is *the size* of QAP(A,B). Given an $n \times n$ matrix $A = (a_{ij})$ and a permutation $\pi \in \mathcal{S}_n$, we denote by $A^{\pi} = (a^{\pi}_{ij}$ the matrix obtained from A by permuting its rows and columns according to permutation π, i.e., $a^{\pi}_{ij} = a_{\pi(i)\pi(j)}$, for $1 \le i, j \le n$. A similar notation is adopted for an n-dimensional vector $V = (v_i)$, where $V^{\pi} = (v^{\pi}_i)$ is the vector obtained from V by permuting its elements according to permutation π, i.e., $v^{\pi}_i = v_{\pi(i)}$, for $1 \le i \le n$.

If any of the coefficient matrices A, B is symmetric, QAP(A,B) is termed *symmetric QAP*. Otherwise, QAP(A,B) is said to be *asymmetric*. A matrix $A = (a_{ij})$ is said to *fulfill the triangle inequality* if for each triple of indices (i, j, k) the following inequality holds:

$$a_{ij} \le a_{ik} + a_{kj}$$

A matrix A is said to be *Euclidean* if it is the distance matrix of a set of points in the Euclidean space $\mathbb{R}^d$ with some l_p norm, where d and p are two natural numbers. If any of the matrices A, B fulfills the triangle inequality (is Euclidean), we say that *QAP(A,B) fulfills the triangle inequality* (*is Euclidean*).

[1]The adjective "quadratic" in the name of the problem is related to the formulation of the problem as an integer program with quadratic cost function.

The QAP defined in (1.1) is often called *Koopmans-Beckmann QAP*[141] in order to distinguish it from a more general problem due to Lawler [152]

$$\min_{\pi \in \mathcal{S}_n} \sum_{i=1}^{n} \sum_{j=1}^{n} d_{\pi(i) i \pi(j) j}, \tag{1.2}$$

where $D = d_{kilj}$ is a 4-dimensional array of reals,$i, j, k, l = 1, 2, \ldots, n$. The QAP (1.2) will be also called *Lawler QAP* throughout the rest of this monograph. Obviously, in the case that the numbers d_{kilj} fulfill the equalities $d_{kilj} = a_{kl} b_{ij}$, for $1 \leq i, j, k, l \leq n$, the problem given in (1.2) is equivalent to QAP(A,B) defined in (1.1). In this monograph we mainly focus on the Koopmans-Beckmann version of the problem as defined in (1.1). However, most of the results presented in the first three chapters extend also to the more general QAP (1.2).

A slightly different problem also addressed as a QAP, and investigated by several authors is the following. Besides the two coefficient matrices A and B we are given a third matrix $C = (c_{ij})$, whose entries are the coefficients of a linear term in the objective function:

$$\min_{\pi \in \mathcal{S}_n} \left(\sum_{i=1}^{n} \sum_{j=1}^{n} a_{\pi(i)\pi(j)} b_{ij} + \sum_{i=1}^{n} c_{\pi(i) i} \right), \text{ or} \tag{1.3}$$

$$\min_{\pi \in \mathcal{S}_n} \left(\sum_{i=1}^{n} \sum_{j=1}^{n} d_{\pi(i)\pi(j) ij} + \sum_{i=1}^{n} c_{\pi(i) i} \right). \tag{1.4}$$

These problems will be called *generalized Koopmans-Beckmann QAP* and *generalized Lawler QAP*, respectively. In the case that $c_{ij} = 0$, for all $1 \leq i, j \leq n$, we get the problems formulated in (1.1) and (1.2), respectively. Again, most of the results presented in the first three chapters extend also to problems (1.3) and (1.4).

1.2 APPLICATIONS

The first occurrence of the QAP dates back to 1957, when Koopmans and Beckmann [141] derived it as a mathematical model of assigning a set of economic activities to a set of locations. Thus, the QAP occurred at first in the context of *facility location problems* which still remain one of its major applications. Nowadays, a large variety of other applications of the QAP is known including such areas as scheduling [91], wiring problems in electronics [215], parallel and distributed computing [22], statistical data analysis [41, 124, 223], design of control panels and typewriter keyboards [39, 186], sports [118], chemistry [224, 80], archeology [107, 143], balancing of turbine runners [149, 210], and computer manufacturing [104, 127]. Recently, Malucelli [161] has proposed some applications of the QAP and its relatives in the field of transportation.

In the following, we shortly describe applications of the QAP in the field of *facility location* and in *wiring problems*.

Facility location. In this context n facilities are to be assigned to n locations. $A = (a_{ij})$ is the flow matrix, i.e. a_{ij} is the flow of materials moving from facility i to facility j, per time unit, and $B = (b_{ij})$ is the distance matrix, i.e. b_{ij} represents the distance from location i to location j. The cost of simultaneously locating facility $\pi(i)$ to location i and facility $\pi(j)$ to location j is $a_{\pi(i)\pi(j)}b_{ij}$. Obviously, an assignment of all facilities to locations can be represented mathematically by a permutation $\pi \in \mathcal{S}_n$. In this model, the total cost of an assignment π of all facilities to locations is equal to $Z(A, B, \pi)$. The objective is to find an assignment π of locations to facilities such that the total cost $Z(A, B, \pi)$ is minimized. This amounts to solving QAP(A,B). Concrete applications of the QAP in a facility location context are described by Dickey and Hopkins in [68] and by Elshafei [72]. In [68] a campus planning model is presented, whereas in [72] the design of a hospital layout is modeled as a QAP.

In the following we will often refer to the QAP in the facility-location context. The terms "facility" and "location" will be used even if there is no concrete occurrence of a facility location problem.

Wiring problems. In such problems a number of modules have to be placed on a board. The modules are pairwise connected by a number of wires. We wish to find a placement of the modules on the board, such that the total length of the connecting wires is minimized. Again, an assignment of n modules to n places provided for them on the board can be represented mathematically by a permutation π of $\{1, 2 \ldots, n\}$. Assume that the number of wires connecting two modules i and j is given by a_{ij}, and the distance between two places i and j on the board is given by d_{ij}, $1 \leq i, j \leq n$. Then, the length of the wires needed for connecting the modules $\pi(i)$ and $\pi(j)$ which are assigned to the places i and j, respectively, is given by $a_{\pi(i)\pi(j)}b_{ij}$, and the overall length of the wires needed for connecting all pairs of modules is equal to $Z(A, B, \pi)$, where $A = (a_{ij})$ and $B = b_{ij}$. Hence, looking for an assignment π which minimizes the overall length of the connecting wires amounts to solving QAP(A,B).

Steinberg [215] describes in detail a concrete application of the QAP in backboard wiring in electronics.

1.3 TWO ALTERNATIVE FORMULATIONS OF THE QAP

We usually refer to QAP(A,B) as defined in (1.1). The main reason for this is that the problem formulation (1.1) expresses the combinatorial structure of the QAP better than the alternative equivalent formulations. However, when powerful techniques such as subgradient optimization, integer or mixed integer programming, and semidefinite programming, are applied, or when the facial structure of the QAP polytope is investigated [177], the alternative formulations turn out to be useful.

Koopmans and Beckmann formulation. The following formulation of the QAP used initially by Koopmans and Beckmann [141] relies basically on the one-to-one correspondence between the set of permutations $\mathcal{S}_n$ and the set of *permutation matrices* defined as follows.

Definition 1.1 *Let $X = (x_{ij})$ be an $n \times n$ matrix. If the entries x_{ij} fulfill the following conditions*

$$\begin{aligned} &\sum_{i=1}^{n} x_{ij} = 1, && 1 \leq j \leq n \\ &\sum_{j=1}^{n} x_{ij} = 1, && 1 \leq i \leq n \\ &x_{ij} \in \{0,1\}, && 1 \leq i,j \leq n \end{aligned} \tag{1.5}$$

then X is called a permutation matrix. *The set of all $n \times n$ permutation matrices is denoted by Π_n.*

The one-to-one correspondence mentioned above is realized by associating a permutation $\pi_X \in \mathcal{S}_n$ to each permutation matrix $X = (x_{ij}) \in \Pi_n$, where $\pi_X(i) = j$ if and only if $x_{ij} = 1$. Then, it is easy to see that QAP(A,B) is equivalent to the following minimization problem on the set of permutation matrices.

$$\begin{aligned} \min \quad & \sum_{i=1}^{n} \sum_{j=1}^{n} \sum_{k=1}^{n} \sum_{l=1}^{n} a_{ij} b_{kl} x_{ik} x_{jl} \\ \text{subject to} \quad & \end{aligned}$$

$$
\begin{aligned}
&\sum_{i=1}^{n} x_{ij} = 1 && 1 \le j \le n \\
&\sum_{j=1}^{n} x_{ij} = 1 && 1 \le i \le n \\
&x_{ij} \in \{0,1\} && 1 \le i,j \le n
\end{aligned}
\tag{1.6}
$$

The problem formulation given in (1.6) is called *Koopmans-Beckmann formulation* of the QAP.

The equivalence of the problems (1.1) and (1.6) becomes more clear in the facility location context. Namely, $x_{ij} = 1$ if facility i is placed at location j and $x_{ij} = 0$ otherwise. The other restrictions in (1.6) formalize the constraints that each facility should be assigned to exactly one location and that to each location exactly one facility should be assigned. The term $a_{ij}b_{kl}x_{ik}x_{jl}$ contributes to the objective function with a value equal to $a_{ij}b_{kl}$ if and only if $x_{ik} = x_{jl} = 1$, that is, if and only if facility i is assigned to location k and facility j is assigned to location l.

In the case that we wish to minimize a linear function over variables x_{ij}, $1 \le i,j \le n$, fulfilling the so-called *assignment constraints* (1.5), we obtain the *linear assignment problem (LAP)*

$$
\begin{aligned}
\min \quad & \sum_{i=1}^{n}\sum_{j=1}^{n} c_{ij}x_{ij} \\
\text{subject to} \quad & \\
& \sum_{i=1}^{n} x_{ij} = 1 && 1 \le j \le n \\
& \sum_{j=1}^{n} x_{ij} = 1 && 1 \le i \le n \\
& x_{ij} \in \{0,1\} && 1 \le i,j \le n,
\end{aligned}
\tag{1.7}
$$

where $C = (c_{ij})$ an $n \times n$ *cost matrix*. LAP is a fundamental combinatorial optimization problem which is very well studied. It is solvable in polynomial time ($O(n^3)$) for example by the *Hungarian method*. LAP can be seen as a special case of the general QAP (1.2). Indeed, a LAP with an $n \times n$ cost matrix (c_{ij}) is equivalent to the QAP (1.2) with d-coefficients defined as follows: $d_{ijkl} = nc_{ik}$, $1 \le i,j,k,l \le n$. For bibliographical pointers to this problem the reader is referred to [66].

Trace formulation. Another equivalent formulation of the QAP can be derived by using the notion of permutation matrices. Namely, for a QAP instance QAP(A,B) of size n, a function $f_{A,B}$ can be defined on the set Π_n of permutation matrices:

$$f_{A,B} \colon \Pi_n \to \mathbb{R}$$

$$X \mapsto \mathrm{tr}(AXBX^t),$$

where the superscript ^{t} denotes the transposed of the corresponding matrix, and $tr(A)$ is the trace of matrix A, i.e. $\mathrm{tr}(A) = \sum_{i=1}^n a_{ii}$, for any $n \times n$ matrix A. Now, QAP(A,B) is equivalent to the following minimization problem on the set of permutation matrices Π_n:

$$\min_{X \in \Pi_n} f_{A,B}(X) = \min_{X \in \Pi_n} \mathrm{tr}(AXBX^t) \tag{1.8}$$

This formulation of the QAP, called *trace formulation*, was introduced by Edwards in [70, 71]. It may be used for a flexible algebraic manipulation of the problem data. In the next chapter we will see how this formulation is used for deriving the so-called *eigenvalue related lower bounds*.

1.4 LINEARIZATIONS

When dealing with QAPs, it seems that the quadratic form in its objective function destroys every hope of finding efficient solution methods. One of the first ideas to cope with the quadratic form was the so-called *linearization* of the QAP. This is an equivalent transformation of the quadratic form in the QAP objective function into a linear one. Numerous QAP linearizations have been proposed by different authors. Most of them are mixed integer linear programs (MILP) with a large number of variables and equations and are hardly exploited in numerical computations. For real life QAPs even solving the related relaxed linear programs to derive lower bounds cannot be done within reasonable time limits. An extensive list of literature pointers to various QAP linearizations can be found in [184] and some linearizations are described in detail in [28]. We present two of them here: The linearization of Kaufman and Broeckx [138], which is one of the "smallest" linearizations ever proposed for the QAP, and the linearization of Frieze and Yadegar [83], which has turned out to be the ground stone of many other linearizations leading to the best known lower bounds for the QAP.

Although, as mentioned above, several authors have proposed linearizations of the QAP, almost nothing was known about the structure of the associated integer QAP polytopes. A lot of efforts were paid to obtain *compact* linearizations, i.e. linearizations with relatively few variables and constraints. Of course, the "size" of the linearization matters in the case that pure enumeration procedures (implicit or not) are applied to the problem at hand. However, as pointed out by Padberg and Rijal in [177], "... these considerations do not matter at all if the overall problem is embedded into a *continuum*, such as it is done when we use linear programming, assignment problem-type relaxations and the like...".

We describe in the following the *ideal linear description* of the QAP proposed in [177]. The approach of Padberg et al. is based on the concept of a *locally ideal linearization* which is an *ideal*, i.e. *minimal* and *complete*, linearization for the pairs of variables which appear as interacting in the quadratic objective function. (In our case these are the pairs (x_{ij}, x_{kl}), for indices i, j, k, l fulfilling specific conditions due to the structure of the set of feasible solutions .) Moreover, the first steps towards understanding the facial structure of the resulting QAP polytope are done and some valid inequalities have been identified. Such efforts may finally lead to the construction of efficient polyhedral cutting plane algorithms for the QAP. For a detailed information on this topic and on the related theoretical background the reader is addressed to [128, 129, 177] and [176], respectively.

1.4.1 Kaufman and Broeckx linearization

Kaufman and Broeckx [138] derived a formulation of the QAP as an MILP with $O(n^2)$ Boolean variables, $O(n^2)$ real variables and $O(n^2)$ constraints. This linearization, which is probably the smallest one in terms of the number of variables and constraints, works also for the more general QAP defined in (1.2).

Consider the Koopmans and Beckmann formulation (1.6) of QAP(A,B). For $1 \leq i, k \leq n$, let us introduce real variables y_{ik} by

$$y_{ik} := x_{ik} \sum_{j=1}^{n} \sum_{l=1}^{n} a_{ij} b_{kl} x_{jl} \tag{1.9}$$

Using these new variables the objective function of QAP(A,B) in (1.6) can be linearized:

$$\sum_{i=1}^{n} \sum_{j=1}^{n} \sum_{k=1}^{n} \sum_{l=1}^{n} a_{ij} b_{kl} x_{ik} x_{jl} = \sum_{i=1}^{n} \sum_{k=1}^{n} y_{ik} \tag{1.10}$$

Moreover, define new constants d_{ik}, $1 \leq i, k \leq n$, by

$$d_{ik} = \sum_{j=1}^{n} \sum_{l=1}^{n} a_{ij} b_{kl} \tag{1.11}$$

The following theorem due to Kaufman and Broeckx, whose proof can also be found in [28], gives a linearization of QAP(A,B).

Theorem 1.1 (Kaufman and Broeckx [138], 1978)
QAP(A,B) given in (1.1) is equivalent to the following MILP with n^2 Boolean variables, n^2 real variables, and $n^2 + 2n$ constraints:

$$
\begin{array}{lll}
\min & \sum_{i=1}^{n}\sum_{k=1}^{n} y_{ik} & \\
\textit{subject to} & & \\
& \sum_{i=1}^{n} x_{ik} = 1 \qquad 1 \le k \le n & \\
& \sum_{k=1}^{n} x_{ik} = 1 \qquad 1 \le i \le n & \\
& d_{ik}x_{ik} + \sum_{j=1}^{n}\sum_{l=1}^{n} a_{ij}b_{kl}x_{jl} - y_{ik} \le d_{ik} & 1 \le i,k \le n \\
& x_{ik} \in \{0,1\},\ y_{ik} \ge 0, \qquad 1 \le i,k \le n &
\end{array}
\tag{1.12}
$$

where d_{ik} are defined in (1.11). □

The idea of the proof relies on showing that each feasible solution (x_{ij}, y_{ik}), $1 \le i,j,k \le n$, of (1.12) fulfills (1.9). Then it is proven that there exists a one-to-one correspondence between feasible solutions of (1.12) and feasible solutions of the Koopmans-Beckmann formulation (1.6), which preserves the objective function values.

Notice that for large n even this linearization which perhaps the smallest one, has a large number of variables and constraints. Under these conditions, even powerful tools to cope with linear integer programs such as Benders' decomposition [18] or cutting planes [19] do not help a lot. It turns out that for QAPs arising in practical applications even solving the relaxed linear program is computationally a hard job.

1.4.2 Frieze and Yadegar linearization

Again, consider the Koopmans and Beckmann formulation of QAP(A,B) of size n in (1.6). Introduce n^4 new Boolean variables y_{ijkl} by:

$$y_{ijkl} := x_{ik}x_{jl} \quad \text{for} \quad 1 \le i,j,k,l \le n$$

By using these variables Frieze and Yadegar derive a linearization of QAP as shown by the following theorem.

Theorem 1.2 (Frieze and Yadegar [83], 1983)
QAP(A,B) in (1.6) is equivalent to the following MILP with n^4 real variables, n^2 Boolean variables, and $n^4 + 4n^3 + n^2 + 2n$ constraints:

$$
\begin{array}{lll}
\min & \sum_{i=1}^{n}\sum_{j=1}^{n}\sum_{k=1}^{n}\sum_{l=1}^{n} a_{ij}b_{kl}y_{ijkl} & \\
\textit{subject to} & & \\
\sum_{i=1}^{n} x_{ik} = 1 & 1 \le k \le n & \\
\sum_{k=1}^{n} x_{ik} = 1 & 1 \le i \le n & \\
\sum_{i=1}^{n} y_{ijkl} = x_{jl} & 1 \le j,k,l \le n & \\
\sum_{j=1}^{n} y_{ijkl} = x_{ik} & 1 \le i,k,l \le n & \\
\sum_{k=1}^{n} y_{ijkl} = x_{jl} & 1 \le i,j,l \le n & \\
\sum_{l=1}^{n} y_{ijkl} = x_{ik} & 1 \le i,j,k \le n & \\
y_{iikk} = x_{ik} & 1 \le i,k \le n & \\
x_{ik} \in \{0,1\} & 1 \le i,k \le n & \\
0 \le y_{ijkl} \le 1 & 1 \le i,j,k,l \le n &
\end{array}
\tag{1.13}
$$

□

Frieze and Yadegar use this mixed integer programming formulation to derive lower bounds for the QAP by solving a Lagrangean relaxation of it. Queyranne observes that only half of the variables y_{ijkl} are really needed in (1.13), since $y_{ijkl} = y_{jilk}$, for all $1 \le i,j,k,l \le n$ (see [161]). Considering this formulation of the QAP, he proposes some valid inequalities for the QAP polytope. These inequalities have not been proved to be facet defining. However, they can be used for improving the performance of lower bound computations when solving relaxations of MILP formulations for the QAP, as done by Malucelli in [161].

1.4.3 Padberg and Rijal linearization

The general QAP. This linearization involves again real variables of the form $y_{ijkl} = x_{ik}x_{jl}$, but over a restricted set of indices $1 \le i < j \le n$, $1 \le k \ne l \le n$. Indeed, exploiting the equalities $y_{ijkl} = x_{ik}x_{jl} = x_{jl}x_{ik} = y_{jilk}$, we can restrict our attention to variables y_{ijkl} with $i < j$. Moreover, simple properties of the set of feasible solutions such as the implications

$$\forall 1 \le k \ne l \le n, \quad (x_{ik} = 1) \Rightarrow (x_{il} = 0)$$

$$\forall 1 \le i \ne j \le n, \quad (x_{ik} = 1) \Rightarrow (x_{jk} = 0)$$

allow us to further restrict the set of the considered y_{ijkl} variables to those with $i < k$ and $j \ne l$.

The set of constraints is built as the union of subsets of "locally ideal" constraints. More concretely, let us introduce the following notations.

$$D = \left\{ \begin{array}{lr} (x,y) \in \mathbb{R}^{n^2+n^2(n-1)^2/2}: & \\ \sum_{i=1}^{n} x_{ik} = 1, & 1 \le k \le n \\ \sum_{k=1}^{n} x_{ik} = 1, & 1 \le i \le n \\ y_{ijkl} = x_{ik}x_{jl}, & 1 \le i < j \le n, 1 \le k \ne l \le n \\ x_{ik} \in \{0,1\}, & 1 \le i,k \le n \end{array} \right\}$$

$$D^{(1,1)} = \left\{ \begin{array}{rcll} (x,y) \in \mathbb{R}^{n^2-1}: & & & \\ \sum_{k=1}^{n} x_{1k} & = & 1, & \\ \sum_{i=1}^{n} x_{i1} & = & 1, & \\ y_{1jk1} & = & x_{1k}x_{j1}, & 1 < k \le n, 1 < j \le n \\ x_{1k} & \in & \{0,1\}, & 1 \le k \le n \end{array} \right\}$$

Furthermore, let us introduce the so-called *local polytope* P as the convex hull of all points in $D^{(1,1)}$, $P = conv(D^{(1,1)})$, and the QAP polytope of size n, QAP_n, as the convex hull of all points in D, $QAP_n = conv(D)$.

Padberg and Rijal [177] give an ideal linear description of the local polytope P.

Lemma 1.3 (Padberg and Rijal [177], 1996)
The following system of equalities and inequalities is an ideal linear description of the local polytope P

$$\begin{array}{rcll} \sum_{k=1}^{n} x_{1k} & = & 1 & \\ \sum_{i=1}^{n} x_{i1} & = & 1 & \\ -x_{1k} + \sum_{i=2}^{n} y_{1ik1} & = & 0 & 2 \le k \le n \\ -x_{i1} + \sum_{k=2}^{n} y_{1ik1} & = & 0 & 1 \le i \le n-1 \\ x_{11} & \ge & 0 & \\ y_{1ik1} & \ge & 0 & 2 \le i,k \le n, \end{array} \tag{1.14}$$

That is, if P_L is the polytope consisting of all $(x,y) \in \mathbb{R}^{n^2-1}$ which fulfill (1.14), then $P = P_L$. □

First, Padberg and Rijal consider all equalities and inequalities which result from the locally ideal linearization of all variables giving rise to quadratic terms in objective function of the QAP. Then, by removing redundant equations so that the remaining system of constraints has full rank, the following linearization of the QAP (1.6) is obtained:

$$\min \quad \sum_{1\le i<j\le n} \sum_{1\le k\neq l\le n} q_{ijkl} y_{ijkl} \qquad \text{subject to} \tag{1.15}$$

$$\sum_{i=1}^{n} x_{ik} = 1 \quad 1 \le k \le n \tag{1.16}$$

$$\sum_{k=1}^{n} x_{ik} = 1 \quad 1 \le i \le n \tag{1.17}$$

$$-x_{jl} + \sum_{i=1}^{j-1} y_{ijkl} + \sum_{i=j+1}^{n} y_{jilk} = 0 \quad \begin{array}{l} 1 \le k \neq l \le n, 1 \le j \le n-1, \\ \text{or} \quad 1 \le l < k \le n, j = n \end{array} \tag{1.18}$$

$$-x_{ik} + \sum_{l=1}^{k-1} y_{ijkl} + \sum_{l=k+1}^{n} y_{ijkl} = 0 \quad \begin{array}{l} 1 \le k \le n, 1 \le i \le n-3, \\ i < j \le n-1 \quad \text{or} \\ 1 \le k \le n-1, i = n-2, \\ j = n-1 \end{array} \tag{1.19}$$

$$-x_{jk} + \sum_{l=1}^{k-1} y_{ijlk} + \sum_{l=k+1}^{n} y_{ijlk} = 0 \quad \begin{array}{l} 1 \le k \le n-1, 1 \le i \le n-3, \\ i < j \le n-1 \end{array} \tag{1.20}$$

$$y_{ijkl} \ge 0 \quad 1 \le i < j \le n, 1 \le k \neq l \le n \tag{1.21}$$

$$x_{ik} \in \{0,1\} \quad 1 \le i,k \le n\ , \tag{1.22}$$

where $q_{ijkl} = a_{ij}b_{kl} + a_{ji}b_{lk}$, for $1 \le i,j,k,l \le n$. Concerning this linearization Padberg and Rijal prove the following theorem.

Theorem 1.4 (Padberg and Rijal [177], 1996)

(i) The mixed 0-1 linear program (1.15), ..., (1.22) is equivalent to the QAP(A,B) in (1.6). It has $n^2 + n^2(n-1)^2/2$ variables and $2n(n-1)^2 - (n-1)(n-2)$ equations.

(ii) The rank of (1.16), ..., (1.20) equals $2n(n-1)^2 - (n-1)(n-2)$.

(ii) The system of equalities (1.16), ..., (1.20) is an ideal linear description of the affine hull of the set D.

(ii) The dimension of QAP_n *equals* $1 + (n-1)^2 + n(n-1)(n-2)(n-3)/2$, *for all* $n \geq 3$, *and the inequalities (1.21) define distinct facets of* QAP_n, *for all* $n \geq 4$.

□

Padberg and Rijal [177] identified two classes of valid inequalities for QAP_n, the so-called *clique inequalities* and *cut* inequalities. This terminology is due to the interpretation of these inequalities in an undirected graph $G = (V, E)$ with n^2 vertices and $n^2(n-1)^2/2$ edges associated to QAP_n and constructed as described below.

The vertex set V consists of all pairs of indices (i, k) with $1 \leq i, k \leq n$. The edge set E consists of pairs of vertices $((i,k),(j,l))$ for which $i \neq j$ and $k \neq l$. Hence, the Boolean variables x_{ik} are in one-to-one correspondence with the vertices (i,k) of G, and the real variables y_{ijkl}, $1 \leq i < j \leq n$, $1 \leq k \neq l \leq n$, are in one-to-one correspondence with the edges $((i,k),(j,l))$ of the graph. A clique C in this graph is a set of pairs (i,k) such that each two elements of C are connected by an edge, or equivalently, for each two elements (i_1,k_1), (i_2,k_2) in C, the inequalities $i_1 \neq i_2$ and $k_1 \neq k_2$ hold. If $(C, E(C))$ is a maximal clique in G, where $E(C)$ is the set of all edges from E which have both endpoints in C, the variables x_{ik} defined by the equalities $x_{ik} = 1$ for $(i,k) \in E(C)$ and $x_{ik} = 0$ otherwise, fulfill the assignment constraints (1.16), (1.17), (1.22). Vice-versa, every permutation matrix $X = (x_{ij}) \in \Pi_n$ gives rise to a maximal clique $C = \{(i,k): x_{ik} = 1\}$ in G. Thus, there is a one-to-one correspondence between permutation matrices $X \in \Pi_n$ and maximal cliques in G. Consequently, there are exactly $n!$ maximal cliques in G and each of those has cardinality equal to n.

For $S \subseteq V$, $T \subseteq V \setminus S$, we denote

$$(S:T) = \{((i,k),(j,l)) \in E: (i,k) \in S, (j,l) \in T\}, \qquad x(S) = \sum_{(i,k)\in S} x_{ik}$$

$$y(E(S)) = \sum_{((i,k),(j,l))\in E(S)} y_{ijkl}, \qquad y(S:T) = \sum_{(i,k)\in S} \sum_{(j,l)\in T} y_{ijkl}$$

Then, it is not difficult to see that the following proposition holds.

Proposition 1.5 (Padberg and Rijal [177], 1996)

(i) For any $C \subseteq V$ and any integer α, the clique inequality

$$\alpha x(C) - y(E(C)) \leq \alpha(\alpha+1)/2$$

is satisfied by all $(x,y) \in QAP_n$.

(ii) For any $S \subset V$ with $|S| \geq 1$ and $T \subseteq V \setminus S$, $|T| \geq 2$, the cut inequality

$$-x(S) - y(E(S)) + y(S:T) - y(E(T)) \leq 0$$

is satisfied by all $(x,y) \in QAP_n$.

□

Clearly, not all of these inequalities are facet defining. Today, the identification of the facet defining inequalities among them is an open question. However, the first efforts in solving this problem have led to the identification of a set of conditions under which the cut inequalities are not facet defining (see Proposition 7.3 in [177]). Moreover, the authors conjecture that at least one of these conditions is necessarily satisfied by some cut inequality which is not facet defining.

The symmetric QAP. For the symmetric QAP Padberg and Rijal [177] propose a special linearization which is "smaller" than that for the general QAP. The proposed linearization is obtained by introducing $n^2(n-1)^2/4$ variables $y_{ijkl} = x_{ik}x_{jl} + x_{il}x_{jk}$, for $1 \leq i < j \leq n$, $1 \leq k < l \leq n$.

The symmetric QAP polytope of size n, denoted by $SQAP_n$, is then the convex hull of all points in SD:

$$SD = \left\{ \begin{array}{ll} (x,y) \in \mathbb{R}^{n^2+n^2(n-1)^2/4}: & \\ \quad \sum_{i=1}^n x_{ik} = 1, & 1 \leq k \leq n \\ \quad \sum_{k=1}^n x_{ik} = 1, & 1 \leq i \leq n \\ y_{ijkl} = x_{ik}x_{jl} + x_{il}x_{jk}, & 1 \leq i < j \leq n, 1 \leq k < l \leq n \\ \quad x_{ik} \in \{0,1\}, & 1 \leq i,k \leq n \end{array} \right\}$$

For obtaining their linearization the authors proceed analogously as for the general case, introducing the so-called *local symmetric polytope* (SP) as the convex hull of all points in $SD^{(1,2)}$:

$$SD^{(1,2)} = \left\{ \begin{array}{rcll} (x,y) \in \mathbb{R}^{3n-1}: & & & \\ \sum_{k=1}^{n} x_{ik} & = & 1, & 1 \leq i \leq 2 \\ \sum_{i=1}^{2} x_{ik} & \leq & 1, & 1 \leq k \leq n \\ y_{121k} & = & x_{11}x_{2k} + x_{1k}x_{21}, & 2 \leq k \leq n \\ x_{ik} & \in & \{0,1\}, & 1 \leq i \leq 2, , 1 \leq k \leq n \end{array} \right\}$$

First, linear description for the symmetric local polytope SP is given, and it is conjectured that this linearization is (locally) *ideal*. Then, the authors put together all equations and inequalities resulting from the local linearizations for all variables which give rise to quadratic terms in the objective function of the symmetric QAP. The following linearization for the symmetric QAP is obtained from these equalities and inequalities by keeping only a maximal set of independent constraints.

$$\min \quad \sum_{1 \leq i < j \leq n} \sum_{1 \leq k < l \leq n} q_{ijkl} y_{ijkl} \tag{1.23}$$

subject to

$$\sum_{k=1}^{n} x_{ik} = 1 \qquad 1 \leq i \leq n \tag{1.24}$$

$$\sum_{i=1}^{n} x_{ik} = 1 \qquad 1 \leq k \leq n-1 \tag{1.25}$$

$$-x_{ik} - x_{jk} + \sum_{l=1}^{k-1} y_{ijlk} + \sum_{l=k+1}^{n} y_{ijkl} = 0 \qquad \begin{array}{l} 1 \leq i < j \leq n \\ 1 \leq k \leq n, \end{array} \tag{1.26}$$

$$-x_{jk} - x_{jl} + \sum_{i=1}^{j-1} y_{ijkl} + \sum_{i=j+1}^{n} y_{jikl} = 0 \qquad \begin{array}{l} 1 \leq j \leq n \\ 1 \leq k \leq n-3, \\ 1 \leq k < l \leq n-1 \end{array} \tag{1.27}$$

$$y_{ijkl} \geq 0 \qquad \begin{array}{l} 1 \leq i < j \leq n \\ 1 \leq k < l \leq n \end{array} \tag{1.28}$$

$$x_{ik} \geq 0 \qquad 1 \leq i,k \leq n \tag{1.29}$$

$$x_{ik} \in \{0,1\} \qquad 1 \leq i,k \leq n \tag{1.30}$$

where $q_{ijkl} = a_{ij}b_{kl} + a_{ji}b_{lk}$, for $1 \leq i,j,k,l \leq n$.

The following results have been proven partially by Padberg and Rijal [177] and partially by Jünger and Kaibel [129].

Theorem 1.6 (Padberg and Rijal [177], 1996, Jünger and Kaibel [129], 1996)

(i) The mixed 0-1 linear program (1.23), ..., (1.30) is equivalent to the symmetric QAP(A,B) in (1.6). It has $n^2 + n^2(n-1)^2/4$ *variables and* $n^2(n-2) + 2n - 1$ *equations.*

(ii) The rank of (1.24), ..., (1.27) equals $n^2(n-2) + 2n - 1$.

(ii) The system of equalities (1.24), ..., (1.27) is a linear description of the affine hull of the set SD.

(ii) The dimension of the $SQAP_n$ *equals* $(n-1)^2 + n^2(n-3)^2/4$, *for all* $n \geq 4$, *and the inequalities (1.28), (1.29) define facets of* $SQAP_n$, *for all* $n \geq 3$. □

Recently, Jünger and Kaibel [129] have identified the first non-trivial facet defining inequalities for the $SQAP_n$, the so-called *curtain inequalities*. The curtain inequalities are divided into *row* and *column* curtain inequalities. The names arise from the interpretation of these inequalities and the linearization (1.23), ..., (1.30) in a hypergraph with vertex set V as defined in the case of the general QAP.

The row curtain inequalities are

$$-\sum_{k \in S} x_{ik} + \sum_{\substack{k,l \in S \\ k<l}} y_{ijkl} + \sum_{\substack{k,l \in S \\ k>l}} y_{iljk} \leq 0$$

for $1 \leq i < j \leq n$, $S \subseteq \{1, 2, \ldots, n\}$.

Analogously, the column curtain inequalities are

$$-\sum_{i \in S} x_{ik} + \sum_{\substack{i,j \in S \\ i<j}} y_{ijkl} + \sum_{\substack{i,j \in S \\ i>j}} y_{jikl} \leq 0$$

for $1 \leq k < l \leq n$, $S \subseteq \{1, 2, \ldots, n\}$.

Theorem 1.7 (Jünger and Kaibel [129], 1996)
All curtain inequalities with $3 \leq |S| \leq \lfloor \frac{n}{2} \rfloor$ *define facets of the* $SQAP_n$. *(Clearly,* $3 \leq \lfloor \frac{n}{2} \rfloor$ *implies* $n \geq 6$*).* □

In [129] the curtain inequalities are used in a cutting planes approach for strengthening the LP relaxation of the MILP formulation (1.23),...,(1.30) for the symmetric QAP. The numerical results, which are preliminary, as stated by the authors, show that the curtain inequalities are not very attractive from the computational point of view. As the complexity of the separation problem for these inequalities is still open, a heuristic separation approach is used.

1.5 COMPUTATIONAL COMPLEXITY ASPECTS

This section is dedicated to computational complexity aspects of the QAP. All results in this section bring evidence to the fact that the QAP is a "very hard" problem from the theoretical point of view. Other results presented in the coming chapters will show that the QAP is a big challenge also from the practical point of view.

We first show that QAP belongs to the class of NP-hard problems. Then we discuss the complexity of approximately solving the QAP in general and QAPs with an arbitrarily large optimal value in particular, showing that these approximation problems are hard to solve. In the second part of this section the complexity of finding a locally optimal solution of the QAP is considered. Two underlying neighborhood structures which lead to PLS-complete problems are described.

1.5.1 The complexity of optimally and approximately solving QAPs

The following two theorems reformulate two early results obtained by Sahni and Gonzalez [206] in 1976. Throughout this monograph we use basic concepts from complexity theory following in the main lines the seminal work of Garey and Johnson [88]. Therefore, the formulations given below differ slightly from those given in [206].

Theorem 1.8 (Sahni and Gonzalez [206], 1976)
The quadratic assignment problem is strongly NP-hard.

Proof. The proof consists of showing that the existence of a polynomial time algorithm for solving QAPs with the entries of the coefficient matrices belonging to $\{0, 1, 2\}$ implies the existence of a polynomial time algorithm for an NP-complete decision problem. This would imply then by definition that the QAP is strongly NP-hard. The above mentioned NP-complete decision problem is the Hamiltonian cycle problem (for short HC) (see [136, 88]).

(HC) Given a graph $G = (V, E)$ with vertex set V and edge set E. Does G contain a Hamiltonian cycle?

Assume that there exists a polynomial time algorithm for solving QAPs with coefficients in $\{0,1,2\}$. Consider an arbitrary instance of the Hamiltonian cycle problem, i.e. we seek a Hamiltonian cycle in an arbitrary graph $G=(V,E)$. Let $|V|=n$, $V=\{v_1,v_2,\ldots,v_n\}$. Define two $n\times n$ matrices $A=(a_{ij})$ and $B=(b_{ij})$ as follows:

$$a_{ij}=\begin{cases}1 & \text{if } (v_i,v_j)\in E\\ 2 & \text{otherwise}\end{cases}\qquad b_{ij}=\begin{cases}1 & \text{if } \begin{array}{c} j=i+1, 1\le i\le n-1\\ \text{or } \; i=n, j=1\end{array}\\ 0 & \text{otherwise}\end{cases}$$

and consider QAP(A,B). It is easily seen that the optimal value of QAP(A,B) is equal to n if and only if G contains a Hamiltonian cycle. Thus, we translate the given instance of HC to an instance of QAP as above and apply the polynomial algorithm which is assumed to exist. Then ,we check whether the solution provided by this algorithm is equal to n. All this can be done in polynomial time with respect to the size of the HC instance. Therefore, we would have a polynomial time algorithm for the Hamiltonian cycle problem. □

In addition, Sahni and Gonzalez proved that even finding an ϵ-approximate solution for QAPs is a hard problem, in the sense that the existence of a polynomial ϵ-approximation algorithm implies $\mathcal{P}=\mathcal{NP}$.

Let us first introduce the notion of an ϵ-approximation algorithm for the QAP (or, similarly, for an optimization problem in general).

Definition 1.2 *Given a real number $\epsilon>0$, an algorithm Υ for the QAP is said to be an* ϵ-approximation algorithm *if and only if for every instance QAP(A,B) the following holds:*

$$\left|\frac{Z(A,B,\pi_\Upsilon)-Z(A,B,\pi_{opt})}{Z(A,B,\pi_{opt})}\right|\le\epsilon\,, \tag{1.31}$$

where π_Υ is the solution to QAP(A,B) computed by algorithm Υ and π_{opt} is an optimal solution to QAP(A,B). The solution of QAP(A,B) produced by an ϵ-approximation algorithm is called an ϵ-approximate solution.

Theorem 1.9 (Sahni and Gonzalez [206], 1976)
For an arbitrary $\epsilon>0$, the existence of a polynomial time ϵ-approximation algorithm for the QAP implies $\mathcal{P}=\mathcal{NP}$.

Proof. We prove the theorem by showing that the existence of a polynomial time ϵ-approximation algorithm for the QAP implies the existence of a polynomial time algorithm for the Hamiltonian cycle problem, which is an NP-complete decision problem.

Indeed, assume that Υ is a polynomial time ϵ-approximation algorithm for the QAP, for some fixed ϵ. Making use of the algorithm Υ, we can design a polynomial time algorithm for HC. Consider an instance of HC in an arbitrary graph $G = (V, E)$ with vertex set $V = \{v_1, \ldots, v_n\}$. Construct (in polynomial time) a QAP instance QAP(A,B) by defining two $n \times n$ coefficient matrices $A = (a_{ij})$ and $B = (b_{ij})$ as follows:

$$a_{ij} = \begin{cases} 1 & \text{if } (v_i, v_j) \in E \\ \omega & \text{otherwise} \end{cases} \qquad b_{ij} = \begin{cases} 1 & \text{if } \begin{array}{c} j = i+1, 1 \le i \le n-1 \\ \text{or } \; i = n, j = 1 \end{array} \\ 0 & \text{otherwise} \end{cases},$$

where $\omega > 1 + n\epsilon$. Apply algorithm Υ to QAP(A,B). Denote by π_Υ the solution produced by Υ and by π_{opt} an optimal solution to QAP(A,B). Inequality (1.31) implies

$$Z(A, B, \pi_{\text{opt}}) \ge \frac{Z(A, B, \pi_\Upsilon)}{1+\epsilon}$$

This inequality shows that if $Z(A, B, \pi_\Upsilon) > n(1+\epsilon)$, then $Z(A, B, \pi_{\text{opt}}) > n$, and therefore G does not contain a Hamiltonian cycle. Vice versa, if G does not contain a Hamiltonian cycle, then $Z(A, B, \pi) \ge n - 1 + \omega > n(1+\epsilon)$, for all $\pi \in \mathcal{S}_n$. Thus, $Z(A, b, \pi_\Upsilon) > n(1+\epsilon)$. Hence, HC in G has an answer "yes" if and only if $Z(A, B, \pi_\Upsilon) > n(1+\epsilon)$. Then, our polynomial time algorithm for solving HC consists of three steps.

(1) Translate the HC instance into an appropriate QAP instance as above.

(2) Apply the algorithm Υ to the QAP instance.

(3) Check the objective function value $Z(A, B, \pi_\Upsilon)$ of the solution π_Υ provided by algorithm Υ.

Noticing that these three steps can be performed in polynomial time completes the proof. □

Considering the general belief that $\mathcal{P} \ne \mathcal{NP}$ and Theorem 1.9, it seems very unlikely to find a polynomial time ϵ-approximation algorithm for the QAP, for some $\epsilon > 0$. Thus, solving the QAP to optimality or even finding an ϵ-approximate solution to it are considered to be hard problems.

Queyranne [188] derives a stronger result which further confirms the widely spread belief on the inherent difficulty of the QAP even in comparison with other difficult combinatorial optimization problems. It it well known and very easy to see that the *traveling salesman problem (TSP)* is a special case of the QAP. Namely, the TSP on n cities can be formulated as a QAP(A,B) where A is the distance matrix of the TSP and B is the adjacence matrix of a Hamiltonian cycle on n vertices (matrix B in the proof of Theorem 1.9). In the case that the distance matrix is symmetric and satisfies the triangle inequality, the TSP is approximable in polynomial time within 3/2 [51]. Queyranne [188] showed that, unless $\mathcal{P} = \mathcal{NP}$, QAP(A,B) is not approximable in polynomial time within some finite approximation ratio, even if A is the distance matrix of some set of points on the Euclidean line and B is a block diagonal symmetric matrix.

In order to formulate Queyranne's result we need one more definition related to the theoretical analysis of the performance of heuristics (cf. [88]).

Definition 1.3 *Let Υ be a heuristic for the QAP. The performance ratio $R_\Upsilon(A,B)$ for heuristic Υ with respect to an instance QAP(A,B) is given as*

$$R_\Upsilon(A,B) = \frac{Z(A,B,\pi_\Upsilon)}{Z(A,B,\pi_{opt})},$$

where π_Υ is the solution produced by Υ when applied to QAP(A,B) and π_{opt} is an optimal solution to QAP(A,B).
The absolute performance ratio R_Υ for heuristic Υ is given as

$$R_\Upsilon = \inf\{r \geq 1 \colon R_\Upsilon(A,B) \leq r \;\; \textit{for all matrices } A,\; B\}$$

The asymptotic performance ratio R_Υ^∞ for heuristic Υ is given by the following equality:

$$R_\Upsilon^\infty = \inf\{r \geq 1 \colon \exists m \in \mathbb{N},\; R_\Upsilon(A,B) \leq r \;\forall A,B \textit{ with } Z(A,B,\pi_{opt}) \geq m\}$$

Theorem 1.10 (Queyranne [188]), 1988)
The existence of a polynomial time heuristic Υ with a bounded asymptotic performance ratio for the QAP(A,B) fulfilling the triangle inequality, that is, the existence of a constant $K \geq 1$ such that $R_\Upsilon^\infty \leq K$, implies $\mathcal{P} = \mathcal{NP}$. □

In other words, unless $\mathcal{P} = \mathcal{NP}$, for each polynomial time algorithm Υ and for each $K_1, K_2 \in \mathbb{N}$, there exist a matrix A fulfilling the triangle inequality and an arbitrary matrix B, such that the optimal value of QAP(A,B) is at least K_1, $Z(A,B,\pi_{opt}) \geq K_1$, and

$$\frac{Z(A,B,\pi_\Upsilon)}{Z(A,B,\pi_{opt})} \geq K_2 ,$$

where π_Υ is the permutation produced by Υ when applied to QAP(A,B). The proof consists of a reduction from the 3-Partition Problem (see [88]). As mentioned above the instance QAP(A,B) involved in the reduction has a very restrictive structure. A is the distance matrix of a set of points on the Euclidean line and B is a symmetric matrix with block diagonal structure. Namely, for a prespecified sequence $s_0 = 0 < s_1 < \ldots < s_l$, $l < n$, the $n \times n$ matrix $B = (b_{ij})$ is symmetric with zeroes on the diagonal and the entries b_{ij}, $i < j$ are given as follows

$$b_{ij} = \begin{cases} \gamma & s_{k-1} < i < j \leq s_k, \text{for some } 1 \leq k \leq l \\ \omega & \text{otherwise} \end{cases} \tag{1.32}$$

where $\gamma > \omega \geq 0$. It is easy to see that $QAP(A,B)$ is equivalent to a $QAP(A,B')$, where the matrix B' is a symmetric 0-1 matrix with zeroes on the diagonal, whose entries are obtained by (1.32) for $\gamma = 1$ and $\omega = 0$. From Queyranne's result follows that such a $QAP(A,B')$ is NP-hard to approximate within a finite asymptotic performance ratio.

A recent result of Arora, Frieze and Kaplan [6] answers partially one of the open questions stated by Queyranne in [188]: What happens if matrix A is the distance matrix of n points which are regularly spaced on the Euclidean line, i.e. points with abscissae given by $x_p = p$, $p = 1, \ldots, n$? This special case of the QAP is termed *linear arrangement problem* and it is a well studied NP-hard problem. In the linear arrangement problem the matrix B is not restricted to have the block diagonal structure described above, but is simply a symmetric $0-1$ matrix. This problem will be reconsidered in Chapter 6 where some polynomially solvable special cases of it are described. Arora et al. give a *polynomial time approximation scheme (PTAS)* for the linear arrangement problem in the case that the $0-1$ matrix B is *dense*, i.e. the number of 1-entries in B is in $\Omega(n^2)$. This restricted version of the problem is termed *dense linear arrangement problem*. Arora et al. show that for each $\epsilon > 0$ there exists an ϵ-approximation algorithm for the dense linear arrangement problem with time complexity depending polynomially on n and exponentially on $1/\epsilon$, and hence being polynomial for each *fixed* $\epsilon > 0$.

Proposition 1.11 (Arora, Frieze and Kaplan [6], 1996)
For each $\epsilon > 0$, there exists an ϵ-approximation algorithm for the dense linear arrangement problem with time complexity $O\left(n^{(1/\epsilon)}\right)$. □

Proposition 1.11 is a corollary of a more general result presented in [6] which concerns the approximability of optimizing (minimizing or maximizing) polynomial of degree d in the variables x_{ij}, $1 \leq i,j \leq n$, where x_{ij} fulfill the assignment constraints (1.5). The proof exploits a new procedure to round the fractional solutions of linear programs which represent continuous relaxations of the linear assignment problem (1.7) with additional linear constraints.

1.5.2 The complexity of local search

In combinatorial optimization we often seek an optimal solution to an NP-hard problem, a task which can be prohibitively difficult even for problem instances of relatively small size. In the case that we are not able to find an optimal solution of the problem we are dealing with, we often content ourselves with a solution which is provably "good" in some sense, although it is probably not optimal. For example, instead of looking for the best solution among all feasible ones, we may seek a solution which is the best out of a subset of feasible solutions. Intuitively, finding such a solution could be easier than finding the optimal one.

Formalizing this kind of compromise leads to so-called *local search* approaches. Consider an instance of an optimization problem P obtained by specifying a ground set $\mathcal{E}$, a feasible solution set $\mathcal{F} \subseteq 2^{\mathcal{E}}$ and a cost function $c: \mathcal{E} \to \mathbb{R}$. This cost function c implies an objective function $f: \mathcal{F} \to \mathbb{R}$ defined by $f(S) = \sum_{x \in S} c(x)$, for all $S \in \mathcal{F}$. The goal is to find a feasible solution which minimizes or maximizes the objective function. Assume for concreteness that P is a minimization problem and let $S \in \mathcal{F}$ be a feasible solution. Define a set $\mathcal{N}(S) \subset \mathcal{F}$ consisting of feasible solutions which are somehow "close" to S. Such a set $\mathcal{N}(S)$ is called a *neighborhood* of S. By defining neighborhoods $\mathcal{N}(S)$ for all $S \in \mathcal{F}$ we obtain a so-called neighborhood structure $\{\mathcal{N}(S): S \in \mathcal{F}\}$. Now, instead of looking for an optimal solution of the problem P, that is an $S^* \in \mathcal{F}$ such that

$$f(S^*) = \min_{S \in \mathcal{F}} f(S)$$

we look for a *locally optimal solution* or (in our case) *local minimum* of P, that is an $\bar{S} \in \mathcal{F}$ such that

$$f(\bar{S}) = \min_{S \in \mathcal{N}(\bar{S})} f(S)$$

A local minimum is not necessarily the optimal solution, or in other words *the global minimum* of P. However, the global minimum is a local minimum. The local optimality of a solution is strongly related with the neighborhood structure we are considering. If the neighborhoods $\mathcal{N}(S)$ are replaced by new neighborhoods $\mathcal{N}'(S)$, one would generally expect changes in the local optimality status of a solution.

As mentioned above, sometimes it is reasonable and convenient to look for local optima rather than for global optima. The algorithms that do this job, i.e. the algorithms which produce locally optimal solutions, are frequently called *local search* algorithms. Basically, every local search algorithm makes use of an *improving* routine which can be generally defined as follows:

$$improve(S) = \begin{cases} \text{any } S' \in \mathcal{N}(S) \text{ with } f(S') < f(S), & \text{if such an } S' \text{ exists} \\ \\ \text{output "}S\text{ is local optimum"}, & \text{otherwise} \end{cases}$$

In general, a local search algorithm starts with some initial feasible solution $S \in \mathcal{F}$ and uses the subroutine *improve* to search for a better solution in the neighborhood of the current solution. As long as an improvement exists, the current solution is updated and the neighborhood search is repeated. The algorithm stops when it reaches a locally optimal solution. The implementation of a local search algorithms involves a number of important choices. Two very important decisions are the choice of a feasible starting solution, and the choice of a "good" neighborhood structure for the problem at hand together with an appropriate method for searching through it. It is beyond the scope of this monograph to provide general criteria for distinguishing a "good" neighborhood structure from a "bad" one, in particular, and to present a detailed theory of local search, in general. The reader interested in theoretical aspects of local search is referred to [181, 208]. Some results concerning applications of local search algorithms to QAPs are presented in Chapter 3. Moreover, in Section 7.6 some local search approaches for the biquadratic assignment problem (BiQAP) are described.

Obviously, everything we said about an arbitrary optimization problem P applies also for the QAP. In this subsection we consider the intriguing question "Is it easy to find a locally optimal solution for the QAP?", whose answer clearly depends on the definition of the neighborhood structure that is considered.

The theoretical basis for facing this kind of problems was introduced by Johnson, Papadimitriou and Yannakakis in [130]. They define the so-called *polynomial-time local search problems*, shortly *PLS problems*. A pair $(P, \mathcal{N})$, where P is a (combinatorial) optimization problem P and $\mathcal{N}$ is an associated neighborhood structure, defines a *local search problem* which consists of finding a locally optimal solution of P with respect to the neighborhood structure $\mathcal{N}$. Neglecting the details, for which the interested reader is addressed to the original paper of Johnson et al., we can consider a PLS problem as a local search problem for which local optimality can be checked in polynomial time. The class of PLS problems is denoted by $\mathcal{PLS}$. In analogy with decision problems, the existence of complete problems in this class can be shown. In analogy with the NP-complete decision problems, the PLS-complete problems, are – in the usual complexity sense – the most difficult problems in the class $\mathcal{PLS}$. Again, the notion of reduction, here called PLS-reduction, is crucial for the whole theory.

Murthy, Pardalos and Li [171] introduce a neighborhood structure for the QAP and show that the corresponding local search problem is PLS-complete. The proposed neighborhood structure is similar to the neighborhood structure proposed by Kernighan and Lin [139] for the graph partitioning problem. For this reason we will call it a *K-L type neighborhood structure for the QAP.*

A K-L type neighborhood structure for the QAP. Consider QAP(A,B) and an arbitrary permutation $\pi_0 \in \mathcal{S}_n$. A *swap* of π_0 is a permutation $\pi \in \mathcal{S}_n$ obtained from π_0 by applying a transposition (i,j) to it, $\pi = \pi_0 \circ (i,j)$. A transposition (i,j) is defined as a permutation which maps i to j, j to i, and k to k for all $k \notin \{i,j\}$. In the facility location context a swap is obtained by interchanging the facilities assigned to two locations i and j. A *greedy swap* of permutation π_0 is a swap π_1 which minimizes the difference $Z(A,B,\pi) - Z(A,B,\pi_0)$ over all swaps π of π_0. Let $\pi_0, \pi_1, \ldots, \pi_l$ be a set of permutations in $\mathcal{S}_n$, each of them being a greedy swap of the preceding one. Such a sequence is called *monotone* if for each pair of permutations π_k, π_t in the sequence, $\{i_k, j_k\} \cap \{i_t, j_t\} = \emptyset$, where π_k (π_t) is obtained by applying transposition (i_k, j_k) ((i_t, j_t)) to the preceding permutation in the sequence. The *neighborhood* of π_0 consists of all permutations which occur in the (unique) maximal monotone sequence of greedy swaps starting with permutation π_0. Let us denote this neighborhood structure for the QAP by $\mathcal{N}_{\text{K-L}}$. It is not difficult to see that, given a QAP(A,B) of size n and a permutation $\pi \in \mathcal{S}_n$, the cardinality of $\mathcal{N}_{\text{K-L}}(\pi)$ does not exceed $\lfloor n/2 \rfloor + 1$.

It is easily seen that the local search problem $(\text{QAP}, \mathcal{N}_{\text{K-L}})$ is a member of $\mathcal{PLS}$. Indeed, generating a feasible solution to the QAP and computing the corresponding value of the objective function can be obviously done in polynomial time. Moreover, we can determine in polynomial time whether a given permutation π is a local optimum and, if it is not, generate a better permutation among its $\lceil n/2 \rceil$ neighbors. In [184] it is shown that a PLS-complete problem, namely the graph partitioning problem with the neighborhood structure defined by Kernighan and Lin [139] is PLS-reducible to $(\text{QAP}, \mathcal{N}_{\text{K-L}})$. This implies the following theorem.

Theorem 1.12 (Pardalos, Rendl and Wolkowicz [184], 1994)
The local search problem $(\text{QAP}, \mathcal{N}_{\text{K-L}})$, *where* $\mathcal{N}_{\text{K-L}}$ *is the Kernighan-Lin type neighborhood structure for QAP, is PLS-complete.* □

The PLS-completeness of (QAP, $\mathcal{N}_{\text{K-L}}$) implies that, in the worst case, a general local search algorithm as described above involving the Kernighan-Lin type neighborhood finds a local minimum only after a time which is exponential in the size of the problem. However, numerical results show that such local search algorithms perform quite well when applied on QAP test instances, as reported in [171].

Another simple and frequently used neighborhood structure in $\mathcal{S}_n$ is the so-called *pair-exchange* (or *2-opt*) neighborhood $\mathcal{N}_2$. The pair-exchange neighborhood of a permutation $\pi_0 \in \mathcal{S}_n$ consists of all permutations $\pi \in \mathcal{S}_n$ obtained by π_0 when applying some transposition (i,j) to it. Thus, $\mathcal{N}_2(\pi) = \{(i,j) \circ \pi : i \neq j,\ 1 \leq i,j \leq n\}$.

It can be shown that $(QAP, \mathcal{N}_2)$ is PLS-complete. Namely, Schäffer and Yannakakis [209] have proven that the graph partitioning problem with a neighborhood structure analogous to $\mathcal{N}_2$ is PLS-complete. A similar PLS-reduction as in [184] implies the following result.

Theorem 1.13 *The local search problem* (QAP, $\mathcal{N}_2$), *where* $\mathcal{N}_2$ *is the pair-exchange neighborhood, is PLS-complete.* □

Again, Theorem 1.13 implies that the time complexity of a general local search algorithm for the QAP involving the pair-exchange neighborhood is exponential in the worst case.

At this point, it should be mentioned that no local criteria are known for deciding how good a locally optimal solution is as compared to a global one. From the complexity point of view, deciding whether a given permutation is an optimal solution to a given instance of QAP, is a hard problem.

Theorem 1.14 (Papadimitriou and Wolfe [182], 1985)
Let Υ *be an algorithm which outputs whether an input permutation* $\pi \in \mathcal{S}_n$ *is a (globally) optimal solution to QAP(A,B), where A and B are part of the input* Υ *is not a polynomial time algorithm, unless* $\mathcal{P} = \mathcal{NP}$. □

2

EXACT ALGORITHMS AND LOWER BOUNDS

This chapter gives a summary on approaches and methods used for solving QAPs to optimality. Since the QAP is an NP-hard problem, only explicit (simple and straightforward) and implicit enumeration methods are known for solving it exactly. For a long time, branch and bound algorithms have been the most successful optimization approaches to QAPs, outperforming cutting plane algorithms whose convergence running time is simply unfeasible. Recently, theoretical results obtained on the combinatorial structure of the QAP polytope have raised new hopes that, in the future, polyhedral cutting planes can be successfully used for solving reasonably sized QAPs. Clearly, the design of efficient branch and cut methods in the vein of those already developed for the TSP (see for example [178]) is conditioned by the identification of new valid and possibly facet defining inequalities for the QAP polytope, and the development of the corresponding separation algorithms. Hence, quite a lot of efforts are required before the current size limits of solvable QAPs can be significantly improved.

Nowadays, the results achieved by applying the best existing exact algorithms (branch and bound) are modest: generally problems of size larger than 20 cannot be solved to optimality in reasonable time [58, 156] and problems of size larger than 15 are considered to be difficult. Recently, two "stubborn" test instances from QAPLIB [37], the Nugent et al. instances of size 21 and 22, were solved by a parallel implementation of a branch and bound algorithm [26].

The low efficiency of branch and bound algorithms for the QAP is mainly due to a lack of efficient bounding approaches for problems of relatively large size, which confirms implicitly the inherent difficulty of the QAP. As the performance of branch and bound methods depends very much on the quality of the involved lower bounds, quite a lot of efforts have been spent for designing efficient lower bounding schemes.

In the following we describe in some detail the most interesting and successful techniques developed to date for the generation of lower bounds for the QAP. Further, we focus on *branch and bound* and *cutting plane* approaches, which are essentially the types of algorithms used for solving QAPs to optimality. We review the application of the principal fundamental ideas for the design of such algorithms in the case of the QAP.

2.1 LOWER BOUNDS

The computation of lower bounds is the most studied topic on the QAP. Although a lot of efforts have been done to derive tight and computationally efficient lower bounds, such bounds have not been found yet. Lower bounds are a component of crucial importance in branch and bound techniques. Moreover, they are also a basic tool for testing the quality of the solutions produced by heuristics. When considering branch and bound methods, both the tightness of the bounds and the computation time requirements are important. In the case that the lower bounds are used for evaluation of heuristics, the emphasis is on their tightness rather than on the computation time requirements. In this section we shortly describe the major bounding techniques for QAPs. The lower bounds for the QAP known to date can be classified in five main groups: *Gilmore-Lawler and related lower bounds*, *eigenvalue related lower bounds*, *reformulation based bounds*, *lower bounds based on LP relaxations* and *lower bounds based on semidefinite relaxations*.

2.1.1 The bound of Gilmore-Lawler and other Gilmore-Lawler-like bounds

The Gilmore-Lawler bound (GL). One of the first lower bounds for QAPs proposed in the literature [92, 152] is the so-called *Gilmore-Lawler bound.* Consider a QAP(A,B) of size n. Define a new $n \times n$ matrix $C = (c_{ij})$ by

$$c_{ij} := \min_{\substack{\pi \in S_n: \\ \pi(j)=i}} \sum_{k=1}^{n} a_{i\pi(k)} b_{jk}, \qquad 1 \leq i, j \leq n$$

It is well known that the entries c_{ij} can be easily computed by sorting the vectors $A_{(i,\cdot)}$ and $B_{(j,\cdot)}$ in increasing and decreasing order, respectively, where $A_{(i,\cdot)}$ is the

i-th row of matrix A. This follows from an early result due to Hardy, Littlewood, and Pòlya in [115] (pp. 260–270), which will turn out to be useful in many proofs throughout the rest of this monograph. Before formulating these results we introduce some notations. The *scalar product* of two n dimensional vectors $U = (u_i)$ and $V = (v_i)$, denoted by $\langle U, V \rangle$, is defined by

$$\langle U, V \rangle := \sum_{i=1}^{n} u_i v_i$$

Consider three permutations $\phi, \psi, \rho \in \mathcal{S}_n$ such that $u_{\phi(1)} \le u_{\phi(2)} \le \ldots \le u_{\phi(n)}$, $u_{\rho(1)} \ge u_{\rho(2)} \ge \ldots \ge u_{\rho(n)}$ and $v_{\psi(1)} \ge v_{\psi(2)} \ge \ldots \ge v_{\psi(n)}$. As we will see more than once in this chapter, the scalar products $\langle U^\phi, V^\psi \rangle$ and $\langle U^\rho, V^\psi \rangle$ play a special role and therefore we introduce special notations for them.

$$\langle U, V \rangle_- := \langle U^\phi, V^\psi \rangle \quad \text{and} \quad \langle U, V \rangle_+ := \langle U^\rho, V^\psi \rangle$$

With these notations the above mentioned result of Hardy et al. reads as follows.

Proposition 2.1 (Hardy, Littlewood and Pòlya [115], 1952)
Let $U = (u_i)$ and $V = (v_i)$ be two n dimensional vectors. Then, the following inequalities hold, for any $\pi \in \mathcal{S}_n$,

$$\langle U, V \rangle_- \le \langle U^\pi, V \rangle \le \langle U, V \rangle_+$$ □

Now let us return to the Gilmore-Lawler bound. Assume that a preprocessing step has been performed for sorting the rows of both matrices A and B in increasing and decreasing order, respectively. Then, it takes $O(n^3)$ elementary operations to compute all c_{ij} and the preprocessing step takes $O(n^2 \log n)$ elementary operations. It is easily seen that the following equalities and inequalities hold, for each $\pi \in \mathcal{S}_n$:

$$Z(A, B, \pi) = \sum_{i=1}^{n} \sum_{j=1}^{n} a_{\pi(i)\pi(j)} b_{ij} \ge \sum_{i=1}^{n} \langle A^\pi_{(\pi(i),.)}, B_{(i,.)} \rangle \ge \sum_{i=1}^{n} c_{\pi(i)i}$$

The last inequality implies

$$\min_{\pi \in \mathcal{S}_n} Z(A, B, \pi) \ge \min_{\pi \in \mathcal{S}_n} \sum_{i=1}^{n} c_{\pi(i)i} =: GL$$

Thus, GL is a lower bound for the optimal value of QAP(A,B). This is called the Gilmore-Lawler bound for QAP(A,B). Once the entries of matrix C have been computed, it takes $O(n^3)$ time to compute GL by solving a linear assignment problem. Hence, the overall time complexity for the computation of the Gilmore-Lawler bound amounts to $O(n^3)$ for the Koopmans-Beckmann QAP. Analogously, the GL bound can be also computed for the Lawler QAP (1.2). In this case, the computation of the entries c_{ij} is done by solving a linear assignment problem for each $1 \le i, j \le n$.

$$c_{ij} := \min_{\substack{\pi \in \mathcal{S}_n: \\ \pi(j)=i}} \sum_{k=1}^{n} d_{i\pi(k)jk}, \qquad 1 \le i, j \le n$$

Hence, the computation of the matrix (c_{ij}) can be done in $O(n^5)$ time, and this is also the time complexity for the computation of the GL bound in this case.

The Gilmore-Lawler bound is one of the simplest and cheapest bounds to compute, but unfortunately this bound is not tight. Moreover, in general, the gap between the Gilmore-Lawler bound and the optimal solution increases quite fast as the size of the problem increases.

Reduction methods. One would expect that the quality of the GL bound could be improved by transforming the coefficient matrices A and B to new matrices $\bar{A}$ and $\bar{B}$ with smaller entries. Then, the quadratic term $Z(\bar{A}, \bar{B}, \pi)$ will generally have a smaller value than $Z(A, B, \pi)$, $\pi \in \mathcal{S}_n$, and also the gap between the lower bound and the optimal value of the objective function is expected to be smaller. The transformation of the coefficient matrices yields an additional term in the objective function. This term should also be minimized over $\mathcal{S}_n$. The crucial point is to use only such transformations of A, B to $\bar{A}$, $\bar{B}$, that lead to a *linear* additional term. In this case, the minimal value of the additional term over $\mathcal{S}_n$ could be computed exactly. This kind of approaches are generally termed *reduction methods*. They were originally introduced by Conrad [60] and further studied by several authors [27, 28, 71, 78, 83, 199].

Generally, a reduction approach transforms the coefficient matrices $A = (a_{ij})$ and $B = (b_{ij})$ in new matrices $\bar{A} = (\bar{a}_{ij})$ and $\bar{B} = (\bar{b}_{ij})$ by using formulas of the form

$$\bar{a}_{ij} = a_{ij} - \bar{f}_i - f_j \quad 1 \le i, j \le n, \ i \ne j$$

$$\bar{b}_{ij} = b_{ij} - \bar{e}_i - e_j \quad 1 \le i, j \le n, \ i \ne j$$

After this transformation one gets the so-called *reduced* QAP with a linear term in its objective function as in (1.3). For the quadratic term in the objective function of the reduced QAP the costs c_{ij}, $1 \le i, j \le n$, are computed. These coefficients c_{ij} are updated by adding to them the coefficients of the additional linear term in the objective function resulting from the reduction. Further, a linear assignment problem (LAP) with the new matrix C is solved. The optimal value of the LAP is an obvious lower bound for the reduced QAP and therefore, also for the original QAP. Clearly, different reduction schemes, i.e. different choices for the vectors (f_i), $(\bar{f}_i)$, (e_i), $(\bar{e}_i)$ above, yield GL bounds of different quality.

There is no clear evidence showing that some reduction scheme is better than the others. Frieze and Yadegar [83] have shown that the reduction along the rows is redundant, that is we can set $\bar{f}_i = 0$, $\bar{e}_i = 0$, for $1 \leq i \leq n$. Further, the same authors have shown that the best among the GL bounds obtained by applying reductions of the form

$$\begin{array}{rcll} \bar{a}_{ij} & = & a_{ij} - f_j & 1 \leq i,j \leq n, \quad i \neq j \\ \bar{b}_{ij} & = & b_{ij} - e_j & 1 \leq i,j \leq n, \quad i \neq j \end{array} \tag{2.1}$$

equals the optimal value of a Lagrangean relaxation of the MILP (1.13) with best possible values for the Lagrangean multipliers. Frieze et al. use two different subgradient approaches for finding approximate values for the best Lagrangean multipliers and get two new lower bounds, denoted by FY1 and FY2, which are generally better than the GL bound (even that obtained after reduction). As we will see in more details in the coming paragraphs, Resende, Ramakrishnan and Drezner [196] tried to exploit the theoretical result achieved by Frieze and Yadegar for deriving better lower bounds. Namely, Frieze and Yadegar notice that for each QAP(A,B) the lower bound obtained as solution of an LP relaxation of (1.13) is at least as good as each GL bound obtained after some reduction of the form 2.1. In [196] an interior point method is used for solving the LP relaxation proposed by Frieze and Yadegar.

Finally, note that, given a QAP(A,B), checking whether the corresponding Gilmore-Lawler bound equals the optimal solution or not is an NP-complete problem (cf. [156]).

Christofides and Gerrard bound (CG). Another, more general, but essentially similar bounding procedure is proposed by Christofides and Gerrard in 1981 [54]. This bounding procedure is based on the following graph theoretical interpretation of the QAP. Consider a QAP(A,B) of size n. A and B are considered to be the weighted adjacency matrices of two complete graphs G_1 and G_2 on n vertices, respectively. Obviously, each $\pi \in \mathcal{S}_n$ is an isomorphism between G_1 and G_2. $Z(A, B, \pi)$ is considered to be the *cost* of this isomorphism. Solving QAP(A,B) means finding an isomorphism between G_1 and G_2 which has minimum cost. The basic idea of Christofides and Gerrard consists of decomposing the graphs G_1 and G_2 into isomorphic subgraphs $G_1^{(1)}, G_1^{(2)}, \ldots, G_1^{(k)}$ and $G_2^{(1)}, G_2^{(2)}, \ldots, G_2^{(k)}$, respectively. The subgraphs $G_1^{(i)}$ ($G_2^{(i)}$) should have the same vertex set as G_1 (G_2) and each edge of G_1 (G_2) should occur in at least one of the corresponding subgraphs. Moreover, the number of occurrences of an edge e of G_1 (G_2) in the subgraphs $G_1^{(i)}$ ($G_2^{(i)}$) is a constant, say d, for all edges e. An isomorphism between G_1 and G_2 maps each subgraph $G_1^{(i)}$ to exactly one subgraph $G_2^{(j)}$, $1 \leq i, j \leq k$. Let us denote the cost of such a mapping, which is an isomorphism between the corresponding subgraphs, by c_{ij}. Clearly, the optimal value of the linear assignment problem with cost matrix $C = (c_{ij})$ multiplied by $1/d$ is a lower bound for the optimal value of QAP(A,B). This bound is denoted by CG.

Obviously, this approach makes sense only in the case that the costs c_{ij} can be computed in polynomial time. For the sake of efficiency, the size of the linear assignment problem to be solved at the end of the bounding procedure should preferable not be larger than $O(n^2)$. These two requirements restrict the choice for the subgraphs used for the decomposition. Examples of good subgraphs used for the decomposition are *stars*, *double stars* and graphs consisting of a single edge [54]. A *star* has a special vertex, called *central vertex* which is connected by an edge to all other vertices in G and is incident with all edges of G. A *double star* has *two* central vertices which are connected by edges to all other vertices. Moreover, each edge of a double star is incident with only one of the central vertices. Figure 2.1 represents a star and a double star with 7 vertices, respectively.

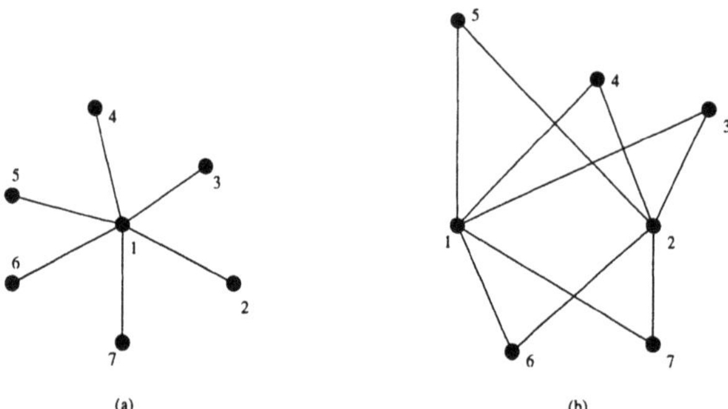

Figure 2.1 (a) A star with 7 vertices and central vertex labelled by 1. (b) A double star with 7 vertices and central vertices labelled by 1 and 2.

The time complexity for the computation of the CG bounds depends on the structure of the decomposing subgraphs. Their structure determines both the number of the entries c_{ij} to be computed, and the time needed for the computation of each of them. For example, in the case of decomposition into a star we have n

decomposing subgraphs, that is n^2 entries c_{ij} to compute, and the computation of each of these entries requires $O(n)$ elementary operations, after a preprocessing step which sorts the columns of the coefficient matrices A, B. Summarizing, the time complexity of computation of the CG bound amounts to $O(n^3)$, in this case. Notice that the CG bound for a decomposition into stars is exactly the GL bound.

The application of the CG bound for a decomposition into graphs consisting of single edges leads to a bound proposed originally by Land [148] and by Gavett and Plyter [89]. In this case, there are n^2 decomposing subgraphs, hence n^4 entries c_{ij} to be computed. Each of these entries can be computed in constant time. At the end, an LAP of size n^2 has to be solved, and this can be done in $O(n^6)$ time which is also the time complexity of this CG bound computation.

A decomposition into double stars leads to bounds which are tighter than those arising from other decompositions. However, the computation of the this CG bound is expensive. There are $O(n^2)$ decomposing subgraphs, and hence $O(n^4)$ entries c_{ij} to be computed. Since the computation of each of these entries requires the solution of an LAP, the computation of the whole matrix (c_{ij}) amounts to $O(n^7)$ which is also the time complexity for the computation of this CG bound.

Variance reduction based lower bounds. This class of lower bounds, proposed by Li, Pardalos, Ramakrishnan and Resende [156], is based on optimal reduction schemes for the QAP. Consider a QAP(A,B) of size n and a partition of A and B of the form $A = A_1 + A_2$, $B = B_1 + B_2$, where $A_1 = (a_{ij}^{(1)})$, $A_2 = (a_{ij}^{(2)})$, $B_1 = (b_{ij}^{(1)})$ and $B_2 = (b_{ij}^{(2)})$. For each pair (i, j), $1 \le i, j \le n$, denote by l_{ji} the solution of the following minimization problem:

$$l_{ji} = \min_{\substack{\pi \in \mathcal{S}_n \\ \pi(i)=j}} \left\{ \sum_{k=1}^{n} a_{j\pi(k)}^{(1)} b_{ik}^{(1)} + \sum_{k=1}^{n} a_{\pi(k)j}^{(2)} b_{ki} + \sum_{k=1}^{n} a_{\pi(k)j} b_{ki}^{(2)} - \sum_{k=1}^{n} a_{\pi(k)j}^{(2)} b_{ki}^{(2)} \right\} \quad (2.2)$$

The key observation is that the solution of the following linear assignment problem

$$\min_{\pi \in \mathcal{S}_n} \sum_{i=1}^{n} l_{\pi(i)i}$$

is a lower bound for QAP(A,B) (cf. [156]):

$$\min_{\pi \in \mathcal{S}_n} \sum_{i=1}^{n} l_{\pi(i)i} \le \sum_{i=1}^{n} \sum_{j=1}^{n} a_{\pi(i)\pi(j)} b_{ij} \quad (2.3)$$

Obviously, different choices for A_1, A_2, B_1 and B_2 yield different lower bounds. The GL bound is obtained from (2.3) for $A_1 = A$, $A_2 = (0)$ and $B_1 = B$ and

$B_2 = (0)$, i.e. by applying the new bounding approach without partitioning the matrices A and B. The partitions considered in [156] are such that the *variances* of the matrices A_1, A_2, B_1, B_2 and the average of their *row variances* are minimized. For an $n \times m$ matrix M, its *average* $\gamma(M)$ and its *variance* $V(M)$ are defined as follows:

$$\gamma(M) := \frac{1}{mn}\sum_{i=1}^{n}\sum_{j=1}^{m} m_{ij} \quad V(M) := \sum_{i=1}^{n}\sum_{j=1}^{m}(\gamma(M) - m_{ij})^2$$

The *total variance* $T(M, \lambda)$ of matrix M depends on the parameter λ, $0 \le \lambda \le 1$, and is defined by

$$T(M, \lambda) := \lambda \sum_{i=1}^{n} V(M_{(i,.)}) + (1 - \lambda)V(M),$$

where $M_{(i,.)}$ is the ith row vector of M considered as a $1 \times m$ matrix. Thus, the total variance of M is a convex combination of the sum of the row variances of M and the variance of M. Considering partitions of the type $A_1 = A + \Delta$, $A_2 = -\Delta$ and trying to minimize the total variances of matrices A_1 and A_2, we get the following minimization problem:

$$\min_{\Delta \in \mathbb{R}^{n\times n}} \theta T(A + \Delta, \lambda) + (1 - \theta)T(-\Delta, \lambda), \qquad (2.4)$$

for fixed values of the control parameters $0 \le \theta \le 1$ and $0 \le \lambda \le 1$. It turns out that the minimization of each of the summands in the above sum is independent from λ. Two approximate solutions are proposed for (2.4) giving the entries δ_{ij} of matrix Δ.

(R_1) $\delta_{ij} = \theta(a_{nn} - a_{ij}) + \delta_{nn}, \qquad i, j = 1, \ldots, n.$

(R_2) $\delta_{ij} = \theta\left(\gamma(A_{(.,n)}) - \gamma(A_{(.,j)}))\right) + \delta_{nn}, \qquad i, j = 1, \ldots, n.$

Obviously, the quality of the bounds which arise when (R_1), (R_2) are applied, depends on the control parameter θ. When applying reduction scheme (R_1) with $\theta = 0$ one gets the GL bound. Experimental results have shown that the best choices for the parameter θ are: $\theta = 0.5$, if (R_1) is applied, and $\theta = 1$, if (R_2) is applied. Matrix B is partitioned in an analogous way.

The partition of each of the matrices A, B according to R_1 or R_2 involves $O(n^2)$ elementary operations. Then, the computation of each l_{ji} entry requires the solution of an LAP which can be done in $O(n^3)$ time. Hence the computation of the whole matrix (l_{ji}) requires $O(n^5)$ elementary operations, and this is also the time complexity of the computation of the variance reduction based lower bound.

In order to speed up the computation of these bounds, the entries l_{ji} are replaced by lower bounds for the problems (2.2). These lower bounds are obtained as the sum of the optimal objective function values of the linear assignment problems arising when all summands in (2.2) are minimized separately. With this modification, the computation of the variance reduction lower bound for a given partitioning A_1,A_2, B_1,B_2 takes $O(n^3)$ elementary steps, assuming that a preprocessing step has been applied for sorting the rows of all involved matrices. It has been proven experimentally that this bound, which is fast to compute, is in general much better than the GL bound, and competes successfully with the other "cheap" bounds[1]. For problems with high variances of the coefficient matrices, the new bounding procedure often yields the best existing Gilmore-Lawler-like bounds.

Lower bounds for the QAP based on a dual formulation. Recently, a new bounding procedure has been proposed by P. Hahn and T. Grant [108, 109]. This procedure combines Gilmore-Lawler bound ideas with reduction steps in a general framework which works also for the Lawler QAP (1.2). The resulting bound will be denoted by DP through the rest of this chapter. Without loss of generality we may assume that all d_{ijkl} in (1.2) are nonnegative [2]. The four dimensional array $D = (d_{kilj})$ is thought as being an $n^2 \times n^2$ matrix composed of n^2 submatrices $D^{(ki)}$, $1 \leq i, k \leq n$, where each $D^{(ki)}$ is an $n \times n$ matrix given by $D^{(ki)} = (d_{kilj})$. This structure of the array (d_{kilj}) complies with the structure of the Kronecker product $X \otimes X$, where X is an $n \times n$ permutation matrix. The entries d_{kiki} are called *leaders*. Clearly, there is only one leader in each submatrix. The objective function value corresponding to permutation π consists of the sum of those entries d_{kilj} which correspond to 1-entries in the Kronecker product $X_\pi \otimes X_\pi$, where X_π is the permutation matrix corresponding to permutation π. Hence, entries of the form d_{kili}, $k \neq l$, or d_{kikj}, $i \neq j$, do not contribute to the value of the objective function. Such entries are called *disallowed entries*. Entries which are not disallowed are said to be *allowed*.

The bounding procedure uses the following classes of operations acting on the array (d_{ijkl}):

(R1) Add a constant to all allowed entries of some row (column) of some submatrix $D^{(ik)}$ and either subtract the same constant from the allowed entries of another

[1] Bounds whose computation involves $O(n^3)$ elementary operations are considered as "cheap".

[2] If this is not the case, a large (enough) constant M can be added to all entries d_{ijkl}, so as to obtain a new QAP which is equivalent to the original one.

row (column) of the same submatrix, or subtract it from the leader in that submatrix.

(R2) Add a constant to all allowed entries of some row (column) of the $n^2 \times n^2$ matrix (d_{ijkl}).

Let D' be the four dimensional array obtained from D after applying to it some of the above defined operations. If the applied operation belongs to class R1 the objective function values corresponding to permutations of $\{1, 2, \ldots, n\}$ remain unchanged. Thus, operations of class R1 maintain the values of the objective function but redistribute the entries of the submatrices $D^{(ik)}$. If the applied operation belongs to class R2, the amount added to the row or column of D is added to the objective function value corresponding to each permutation. Hence, operations of class R2 maintain the ordering of the permutations with respect to the corresponding value of the objective function. The main idea of the method is the following. If the operations applied on D decrease the objective function value corresponding to each permutation by an amount R, and are performed in such a way that the entries of the transformed array D' remain nonnegative, then, clearly, R is a lower bound for the optimal solution of the given QAP. If, moreover, the 0-entries in the transformed matrix D' comply with the pattern of zeros in the Kronecker product $X_\pi \otimes X_\pi$ for some permutation matrix X_π, then R is the optimal objective function value of the original QAP and permutation π is an optimal solution.

The procedure developed to find such a lower bound R, or possibly, to optimally solve the problem, is essentially similar to the Hungarian method for the linear assignment problem. It uses operations of classes R1 and R2 to redistribute the entries of D so as to obtain a pattern of zeros which complies with the pattern of zeros of the Kronecker product $X \otimes X$ for some permutation matrix X. It is an iterative approach which starts by applying the Hungarian method to each of the submatrices $D^{(ik)}$, $1 \leq i, k \leq n$. Then, the optimal values are added to the corresponding leaders and the Hungarian method is applied to the matrix of the leaders. The resulting optimal value R is added to some dummy "superleader" element which is initially set to 0. As argued above, R is a lower bound for the original problem. After the first iteration this lower bound coincides with the Gilmore-Lawler bound. Then, it is checked whether the pattern of zeroes of the transformed array D, that is the array resulting after the above $n^2 + 1$ applications of the Hungarian method, coincides with the pattern of zeros of $X \otimes X$, for some permutation matrix X. If this is the case, the procedure terminates with an optimal solution. If this is not the case, it is checked whether the matrix of leaders contains some non-zero elements. If it does not, the procedure terminates with a lower bound for the original problem. Otherwise, the non-zero leaders are distributed (uniformly) over the elements of the corresponding submatrices by applying operations of class R1. Again, the Hungarian method is applied to each of the resulting submatrices $D^{(ik)}$ and the whole process is repeated until the value of the superleader, i.e. the lower bound R cannot be improved any more or some other termination criterion is fulfilled.

In summary, the whole process is a repeated computation of Gilmore-Lawler bounds on iteratively transformed problem data, where the transformations generalize the ideas of reduction methods, described before. The time complexity of each iteration is basically that of the GL bound computation for a Lawler QAP, i.e. $O(n^5)$. The additional computational load for the redistribution of the non-zero leaders over the elements of the corresponding submatrices amounts to $O(n)$ elementary operations for each of the leaders, hence $O(n^3)$ elementary operations altogether, which is irrelevant in comparison with $O(n^5)$.

A deeper investigation of this bounding procedure reveals that it is an iterative approach in which the dual of some LP relaxation of the original problem is solved and reformulated iteratively. The reformulation step makes use of the information furnished by the preceding solution step.

As reported in [108] this bounding procedure has been tested on small and middle sized QAP instances from QAPLIB. The computational results show that the new bounds are competitive with the best existing lower bounds in terms of quality and superior in terms of computation times. In another paper of Hahn et al.[109] the DP bounds are used in a branch and bound approach for the QAP. The proposed algorithm is tested on instances from QAPLIB and, among others, on the very difficult Nugent et al. problem of size 20. The computation time needed for solving this instance is about three times less than that needed by the branch and bound algorithm of Clausen et al. [58] for solving the same instance[3]. As pointed out in [109], the good performance of the proposed branch and bound algorithms seems to be mainly due to the good quality of the DP bounds and the efficiency of their computation, as well as to the iterative nature of this bounding procedure, which permits stopping the computation as soon as the lower bound has reached the best existing upper bound. Other favorable features of the DP bounds are their reasonable memory requirements, their being potentially parallelizable, and the fact that they allow to produce "good" feasible solutions in early stages of search, far from the leaves of the branch and bound tree.

[3] Considering that the algorithm of Clausen et al. is a parallel one, in order to have a fair comparison, the running time reported in [58] is multiplied by an appropriate speed-up coefficient based on the number of processors involved in the computation.

2.1.2 Eigenvalue related lower bounds

The eigenvalue related bounding techniques for the QAP are based on the relationship between the objective function value of the QAP in the trace formulation and the eigenvalues of its coefficient matrices. These bounding techniques work for the Koopmans-Beckmann QAP, but cannot be applied to the Lawler QAP (1.2). When designed and implemented carefully, these techniques produce bounds of good quality in comparison with Gilmore-Lawler type lower bounds or, more generally, with bounds based on linear relaxations (to be discussed in more detail in the coming sections). However, these bounds are quite expensive in terms of computation time requirements and hence, are not really appropriate to be used as basic bounding techniques within branch and bound algorithms. Eigenvalue based bounds for QAP(A,B), with A and B being symmetric matrices, are proposed by several authors in a series of papers [78, 111, 113, 194].

Bound EV. Consider the trace formulation of the QAP (1.8). The matrices A and B are assumed to be symmetric in order to guarantee their eigenvalues to be reals. Let $\lambda_1 \le \lambda_2 \le \ldots \le \lambda_n$ be the eigenvalues of matrix A and $\mu_1 \ge \mu_2 \ge \ldots \ge \mu_n$ be the eigenvalues of matrix B. There exist two orthogonal matrices P_1 and P_2 and two *diagonal matrices*, ie. matrices whose off-diagonal elements are equal to zero, $\Lambda_1 = Diag(\lambda_1, \lambda_2, \ldots, \lambda_n)$and $\Lambda_2 = Diag(\mu_1, \mu_2, \ldots, \mu_n)$, such that $A = P_1 \Lambda_1 P_1^t$, $B = P_2 \Lambda_2 P_2^t$. As shown in [78], the following equality holds:

$$tr(AXBX^t) = \lambda^t S(X) \mu \ , \quad \forall X \in \Pi \ , \tag{2.5}$$

where $S(X) = (\langle P_{1_{(.,i)}}, X P_{2_{(.,j)}} \rangle^2)$ and $P_{1_{(.,i)}}$, $P_{2_{(.,i)}}$ are the column vectors of matrices P_1 and P_2[4]. λ and μ are the vectors of the eigenvalues: $\lambda = (\lambda_i)$ and $\mu = (\mu_i)$. The following inequality establishes the first eigenvalue related bound for the QAP(A,B), proposed by Finke et al. in 1987 [78].

$$\sum_{i=1}^{n} \lambda_i \mu_i \le tr(AXBX^t) \le \sum_{i=1}^{n} \lambda_{n-i+1} \mu_i \ , \quad \text{for all } X \in \Pi \tag{2.6}$$

The bound $\sum_{i=1}^{n} \lambda_i \mu_i$, denoted by EV, is usually very poor because of the large variance of the eigenvalues λ_i, μ_i. In most cases the EV bound is found to be negative. The time complexity for computing this bounds is determined by the computation of the eigenvalues of the symmetric matrices A and B, and hence amounts to $O(n^3)$ (by using standard algorithms, e.g. the tridiagonalization and the symmetric QR algorithm [99]), where n is the size of the problem.

[4] $P_{1_{(.,i)}}$, $P_{2_{(.,i)}}$ are also the eigenvectors of A and B, respectively.

Bound EV1. One possibility to sharpen the EV bound is to try to decrease the difference $\sum_{i=1} \lambda_{n-i+1}\mu_i - \sum_{i=1} \lambda_i \mu_i$. This can be done by reducing the *spreads* of the matrices A and B, where the spread $sd(A)$ is defined as follows

$$sd(A) = \max\{\lambda_i : 1 \leq i \leq n\} - \min\{\lambda_i : 1 \leq i \leq n\}.$$

According to this definition, in our case we have

$$sd(A) = \lambda_n - \lambda_1 \quad \text{and} \quad sd(B) = \mu_1 - \mu_n$$

There exist no simple formulas for computing the spread of a given matrix, but there exists a simple formula which gives an upper bound for it. Making use of these formulas one tries to replace the coefficient matrices A, B by new symmetric matrices $\bar{A}$ and $\bar{B}$ with smaller spreads. These ideas lead to the application of *reduction* methods which turn out to be successful for improving the EV bound, whereas their role in improving the GL bound is disputable. In [78] the following reduction formulas are used

$$\bar{A} = A - F_1 - F_1^t - D_1, \qquad \bar{B} = B - F_2 - F_2^t - D_2, \tag{2.7}$$

where F_1, F_2 are matrices with constant columns, F_1^t, F_2^t are their transposed matrices, and D_1, D_2 are diagonal matrices. F_1, D_1, F_2 and D_2 are appropriately chosen in order to minimize the upper bounds on the spreads $sd(\bar{A})$ and $sd(\bar{B})$. F_1, D_1, F_2 and D_2 can be easily computed by using explicit formulas (see [78]). Similarly as in the case of the GL bound, the reduction gives rise to an additional linear term in the objective function. Then, for the quadratic term the bound EV can be applied, whereas the linear term can be minimized efficiently. The sum of the lower bound for the quadratic term and the minimum of the linear term is a new lower bound for the QAP denoted by EV1. The bound EV1 is generally better than EV, especially for QAPs of size larger than 20.
The reduction, i.e. the computation of the new matrices $\bar{A}$, $\bar{B}$ and of the coefficients of the additional linear term in the objective function involves $O(n^2)$ elementary operations. Thus, the time complexity for the computation of the bound EV1 is dominated by the time for the computation of the eigenvalues of the matrices $\bar{A}$, $\bar{B}$, which amounts to $O(n^3)$, for the problem of size n.

Bound EV2. If the symmetric matrices A and B have zero diagonal elements, the EV1 bound can be further improved. For the sake of simplicity let us assume that we reduce only one of the coefficient matrices, say A[5]. After the reduction, the objective function looks as follows:

$$Z(A, B, \pi) = \sum_{i=1}^{n} \sum_{j=1}^{n} \bar{a}_{\pi(i)\pi(j)} b_{ij} + \sum_{i=1}^{n} c_{\pi(i)} d_i \tag{2.8}$$

[5]Everything goes through analogously in the case that both matrices A, B will be reduced.

where $c_i = 2f_{ki}$ for any $k \in \{1, 2, \ldots, n\}$ and $d_i = \sum_{j=1}^{n} b_{ij}$. We denote $C = (c_i)$, $D = (d_i)$, and have $EV1 = \langle \bar{\lambda}, \mu \rangle_- + \langle C, D \rangle_-$, where $\bar{\lambda}$ and μ are the vectors of the eigenvalues of matrices $\bar{A}$ and B sorted in increasing and decreasing order, respectively. Separately bounding the quadratic and the linear part of the objective function as done by EV1, does not have to yield always good bounds. An idea for a compromising lower bound is given by Rendl [191]. He introduces the so-called *measure of linearity* L for the QAP(A,B) as follows:

$$L = \left(\langle \bar{\lambda}, \mu \rangle_+ - \langle \bar{\lambda}, \mu \rangle_- \right) / \left(\langle C, D \rangle_+ - \langle C, D \rangle_- \right)$$

A small L indicates a small influence of the quadratic term in (2.8) as compared to the linear term. In this case it is suggested to order the scalar products $\langle C^\pi, D \rangle$, $\pi \in \mathcal{S}_n$, in increasing order and compute the objective function values corresponding to the respective permutations. Assume that the permutations π_i, $1 \le i \le k$, yield the k best values of the scalar product $\langle C^\pi, D \rangle$ over $\mathcal{S}_n$ Rendl [191] proves that the following inequalities hold:

$$Z(\bar{A}, B, \pi_1) \ge Z(\bar{A}, B, \pi_2) \ge \ldots \ge Z(\bar{A}, B, \pi_k) \ge \langle \bar{\lambda}, \mu \rangle_- + \langle C^{\pi_k}, D \rangle \ge \ldots \ge \langle \bar{\lambda}, \mu \rangle_- + \langle C^{\pi_1}, D \rangle$$

Moreover, if $Z(\bar{A}, B, \pi_i) > \langle \bar{\lambda}, \mu \rangle_- + \langle C^{\pi_i}, D \rangle$ then $\langle \bar{\lambda}, \mu \rangle_- + \langle C^{\pi_i}, D \rangle$ is a lower bound for QAP(A,B) (obviously, at least as good as EV1), otherwise π_i is an optimal solution to QAP(A,B). The bound obtained in this manner is often denoted by EV2. Finding k permutations which yield the k smallest values for the scalar product $\langle C^\pi, D \rangle$ over $\mathcal{S}_n$ takes $O(n \log n + (n + \log k)k)$ time [191]. Hence, in this case, the overall time complexity for the computation of the bound EV2 amounts to $O(n^3 + (n + k) \log k)$. The bounds produced by this approach are much better than EV1 for matrices A and B whose row sums cover a relatively large interval of values.

The remaining case is when the value of the ratio L is large. A large value of L indicates a negligible linear term in (2.8). In this case, an improvement for the eigenvalue bound $\langle \bar{\lambda}, \bar{\mu} \rangle_-$ of the quadratic term is obtained as follows. The quadratic term of the objective function in (2.8) can be written as

$$\sum_{i=1}^{n} \sum_{j=1}^{n} \bar{a}_{\pi(i)\pi(j)} b_{ij} = tr(\bar{A} X_\pi B X_\pi^t) = \sum_{i=1}^{n} \sum_{j=1}^{n} \bar{\lambda}_i \mu_j \langle U_i, X_\pi V_j \rangle^2 ,$$

where X_π is the permutation matrix corresponding to π, $\bar{\lambda}_i$, μ_i, $1 \le i \le n$, are the eigenvalues of matrices $\bar{A}$, B (appropriately sorted), and U_i, V_i, $1 \le i \le n$, are their eigenvectors, respectively. The entries $\langle U_i, X_\pi V_j \rangle^2$ are bounded as

$$\ell_{ij} \le \langle U_i, X_\pi V_j \rangle^2 \le u_{ij},$$

for all π, where $u_{ij} = \max\left\{\langle U_i, V_j\rangle_-^2, \langle U_i, V_j\rangle_+^2\right\}$ and

$$\ell_{ij} = \begin{cases} 0 & \text{if } \langle U_i, V_j\rangle_-, \langle U_i, V_j\rangle_+ \text{ have different signs} \\ \min\left\{\langle U_i, V_j\rangle_-^2, \langle U_i, V_j\rangle_+^2\right\} & \text{otherwise} \end{cases}$$

Replacing $\langle\bar{\lambda}, \mu\rangle$ in EV1 by the optimal value of the following capacitated transportation problem, yields the improved lower bound EV2:

$$\begin{array}{ll} \min & \sum_{i=1}^{n}\sum_{j=1}^{n} \bar{\lambda}_i \mu_j x_{ij} \\ \text{subject to} & \\ & \sum_{i=1}^{n} x_{ij} = 1 \\ & \sum_{j=1}^{n} x_{ij} = 1 \\ & \ell_{ij} \le x_{ij} \le u_{ij} \end{array}$$

In this case, the time complexity for the computation of the bound EV2 amounts to $O(n^3)$, being determined by the computation of the eigenvalues for the matrices $\bar{A}$, B, and the solution of a capacitated transportation problem.

Other eigenvalue related bounds. The reduction used to improve bound EV in order to get bound EV1 is performed independently for the matrices A, B, and the linear term of the reduced QAP is then bounded separately from the quadratic term. A way to obtain better bounds is to consider all these factors jointly. This idea is developed in Rendl and Wolkowicz [194].

Consider QAP(A,B) and the generic reduction formula in (2.7). Obviously, the matrices $\bar{A}$, $\bar{B}$ and also the matrix C, which defines the additional linear term of the reduced QAP, depend on the matrices F_1, F_2, D_1 and D_2. As the matrices F_1, F_2 have constant columns, and the matrices D_1, D_2 are diagonal matrices, the above generic reduction is completely defined by a $4n$-dimensional parameter vector. Let us denote it by Γ. Further, let us denote the minimal value of the scalar products of the eigenvalues $\bar{\lambda}$, $\bar{\mu}$ of $\bar{A}$, $\bar{B}$ obtained by the reduction with parameter Γ, by $msp(\Gamma) := \langle\bar{\lambda}, \bar{\mu}\rangle_-$. Moreover, denote

$$lap(\Gamma) := \min_{\pi\in\mathcal{S}_n} \sum_{i=1}^{n}\sum_{j=1}^{n} c_{\pi(i)i}, \tag{2.9}$$

for C obtained by the reduction with parameter Γ. Clearly, for each parameter vector Γ, $msp(\Gamma) + lap(\Gamma)$ is a lower bound for the QAP(A,B). The natural goal is to find a parameter vector Γ which maximizes this bound, i.e.

$$\max\left\{msp(\Gamma) + lap(\Gamma) \colon \Gamma \in R^{4n\times 1}\right\} \tag{2.10}$$

$msp(\Gamma)$ is a differentiable function (at least when the eigenvalues have multiplicity 1), whereas $lap(\Gamma)$ is a piecewise linear concave function. In [194], problem (2.10) is solved by using a steepest ascent algorithm. This procedure produces sometimes the best existing bounds for QAPs with symmetric coefficient matrices. However, it is quite time-consuming as an EV bound must be computed in each iteration (which amounts to a time complexity of $O(n^3)$ per iteration), and a large number of iterations is needed to solve (2.10).

Another bounding idea related to the eigenvalues of the coefficient matrices arises when considering relaxations of the trace formulation of the QAP (1.8). It is well known that the class of permutation matrices is the intersection of three matrix classes: the class of *orthogonal matrices*, the class of matrices with *columns and rows sums equal to* 1 and the class of matrices with *nonnegative entries*. (The intersection of the last two classes yields the *doubly stochastic* matrices.) Minimizing the objective function of the problem (1.8) over one of these three classes of matrices, or over intersections of pairs of them, would yield relaxations of the QAP. The relaxation of (1.8) over the class of orthogonal matrices has been proven to be useful. Rendl and Wolkowicz [194] show that the EV bound is actually the optimal solution of the relaxation of the QAP (1.8) on the class of orthogonal matrices. The relaxation of (1.8) on the class of orthogonal matrices which additionally have row and column sums equal to 1 is studied by Hadley et al. [113]. The constraints on the sums over the rows and columns of matrix X are reformulated to obtain a projected problem. The latter is an $n-1$ dimensional minimization problem which is reformulated as a QAP relaxation on the set of $(n-1) \times (n-1)$ orthogonal matrices. But this relaxation can be solved to optimality by an eigenvalue approach as described above. In the literature the bounds obtained by this procedure are often termed *elimination bounds (ELI)*. More recently, the relaxation of QAP (1.8) on the class of orthogonal matrices with row and column sums equal to one has further been studied in relationship with trust region problems [134].

All eigenvalue related bounds described up to now work only for symmetric QAPs. If only one matrix is symmetric, the other one can be symmetricized yielding an equivalent QAP. Hadley, Rendl and Wolkowicz [112] propose also a technique to transform an asymmetric QAP to a QAP with (complex) Hermitian coefficient matrices. Then, an eigenvalue type bound analogous to EV is obtained for the transformed QAP (and equivalently for the given QAP) by applying the Hoffman-Wielandt inequality for Hermitian matrices.

All eigenvalue related approaches produce in most of the cases high quality bounds, but their computation is too expensive. Another reason why these bounds are not well suited for use within branch and bound algorithms concerns the quick deterioration of their quality when searching lower levels of the branch and bound tree (see [57]).

2.1.3 Bounding techniques based on reformulations

First of all, let us remark that lower bounds of this class can be applied to the general QAPs formulated as in (1.2) and (1.4). Obviously, they can be also applied to the Koopmans-Beckmann QAP. However, in this case, they cannot take advantage of the specific structure of the objective function coefficients . Consider the generalized Lawler QAP, below denoted by P.

$$(P)\quad \begin{array}{ll} \text{minimize} & \left\{ \sum\limits_{i,k=1}^{n} \sum\limits_{j,l=1}^{n} d_{ijkl} x_{ik} x_{jl} + \sum\limits_{i,k=1}^{n} c_{ik} x_{ik} \right\} \\ \text{subject to} & \\ & \sum\limits_{i=1}^{n} x_{ik} = 1 \quad 1 \le k \le n \\ & \sum\limits_{k=1}^{n} x_{ik} = 1 \quad 1 \le i \le n \\ & x_{ik} \in \{0,1\} \quad 1 \le i,k \le n \end{array}$$

As already mentioned above, we may assume without loss of generality that the coefficients d_{ijkl}, $1 \le i,j,k,l \le n$ are nonnegative.

A *reformulation* of the QAP P is another QAP P' of the same form as P, but with new coefficients d'_{ijkl}, $1 \le i,j,k,l \le n$, and c'_{ik}, $1 \le i,k \le n$, such that for all (x_{ij}) fulfilling the assignment constraints

$$\sum_{i,k=1}^{n} \sum_{j,l=1}^{n} d_{ijkl} x_{ik} x_{jl} + \sum_{i,k=1}^{n} c_{ik} x_{ik} = \sum_{i,k=1}^{n} \sum_{j,l=1}^{n} d'_{ijkl} x_{ik} x_{jl} + \sum_{i,k=1}^{n} c'_{ik} x_{ik}$$

The basic idea of these methods is to derive a sequence of reformulations $P_0 = P, P_1, \ldots, P_k$ for a given problem P by applying some "appropriate" reformulation rule. As pointed out by Carraresi and Malucelli in [44], the new coefficients can be derived by the following generic equalities:

$$d'_{ijkl} = d_{ijkl} + \tau_{ijkl} - \alpha_{ijl} - \beta_{jkl} + \theta_{ik} \quad 1 \le i,j,k,l \le n$$

$$c'_{ik} = c_{ik} + \sum_{j=1}^{n} \alpha_{ijk} + \sum_{l=1}^{n} \beta_{ikl} - (n-1)\theta_{ik} \quad 1 \le i,k \le n$$

Clearly, not all choices of the parameters τ, α, β and θ in the above formulas produce a reformulation of P, but there are settings of those parameters which do, as shown in [7, 43].

To each of the reformulations P_i, $0 \le i \le k$, a bounding procedure is applied, for example the GL bound, and a lower bound, say B_i, is derived. The reformulation rule is "appropriate" if the sequence of the lower bounds B_i is monotonically nondecreasing: $B_0 \le B_1 \le \ldots \le B_k$. Usually, the construction of a new reformulation exploits the previous reformulations and the bounds obtained for them. The procedure is repeated iteratively until a stop criterion is fulfilled. The lower bound for the last reformulation is a valid lower bound for the original problem. In principle, any bounding technique can be involved in such a reformulation scheme. However, as this approach performs one lower bound computation per iteration, it is reasonable to choose cheap and easy-to-compute bounds (for example the GL bound) to be applied in each reformulation.

This type of bounding strategies has been proposed by Carraresi and Malucelli [43] and Assad and Xu [7]. Both procedures compute GL bounds for each reformulation of the original problem. The parameters α, β, τ and θ are updated in each iteration (reformulation step). Their values are determined by making use of the lower bound obtained for the last reformulation, the optimal values and the optimal dual variables of the linear assignment problems solved during the last GL computation.

More concretely, the bounds proposed by Assad and Xu [7] can be embedded into the above reformulation scheme by setting $\tau_{ijkl} = 0$, $\alpha_{ijk} = 0$, $\beta_{ijk} = 0$, for $1 \le i,j,k,l \le n$ $\theta^0_{ik} = 0$, $1 \le i,k \le n$, and

$$\theta^{(t+1)}_{ik} = \theta^{(t)}_{ik} + z^{(t)}_{ik}/(n-1)$$

$\theta^{(t)}_{ik}$ are the values of the parameter θ used for the t-th reformulation and $z^{(t)}_{ik}$ is the optimal value of the following linear assignment problem

$$\left(P^{(t)}_{ik}\right) \qquad \min\left\{ c^{(t)}_{ik} + \sum_{\substack{j,l=1\\(j,l)\ne(i,k)}}^{n} d^{(t)}_{ijkl} x_{jl} \colon (x_{jl}) \text{ fulfill (1.5)} \right\},$$

where $d^{(t)}_{ijkl}$ and $c^{(t)}_{ik}$ in the last equality are the coefficients of the t-th reformulation P_t. The termination criterion is

$$\max_{1\le i,k\le n} z^{(t)}_{ik} - \min_{1\le i,k\le n} z^{(t)}_{ik} \le \epsilon$$

for some prespecified small accuracy measure ϵ. The bounds computed in this way are usually denoted by AX. The time complexity amounts to $O(n^5)$ per iteration. This is due to the computation of the coefficients of the corresponding reformulation which requires the solution of n^2 LAPs. Numerical experiments on the Nugent et al. test instances have shown that this algorithm converges quickly in practice. The produced bounds are comparable to the bounds FY1 and FY2 of Frieze and Yadegar [83] in terms of quality, while requiring less computation time. However, to the best of our knowledge, no theoretical proof for the convergence of this algorithm has been given yet.

The bounds proposed by Carraresi and Malucelli [43] can be embedded into the above reformulation scheme by setting $\theta_{ik}^{(t)} = 0$, $1 \le i, k \le n$, for all t, and

$$\tau_{ijkl}^{(t+1)} = d_{ijkl}^{(t)} - d_{jilk}^{(t)} \tag{2.11}$$

$$\alpha_{ijl}^{(t+1)} = u_{ijl}^{(t)} \tag{2.12}$$

$$\beta_{jkl}^{(t+1)} = v_{jkl}^{(t)} \tag{2.13}$$

for $1 \le i, j, k, l \le n$, where $u_{ijl}^{(t)}$, $v_{jkl}^{(t)}$ are the optimal values of the dual variables of the linear assignment $P_{jl}^{(t)}$ defined analogously to $P_{ik}^{(t)}$ above. The termination criterion is $z_{ik}^{(t+1)} = z_{ik}^{(t)}$, for all $1 \le i, k \le n$. The authors show that this stop criterion is fulfilled after finitely many reformulations steps. The bounds computed in this way are usually denoted by CM1.

Slightly different reformulation formulas have been proposed by Carraresi and Malucelli in [44]. The setting for the τ, α and β parameters are those given in (2.11), (2.12), (2.13), whereas the coefficients $\theta^{(t)}$ are computed as follows:

$$\theta_{ik}^{(t+1)} = \frac{1}{n-1}\left(c_{ik}^{(t)} + u_i^{(t)} + v_k^{(t)}\right), \quad 1 \le i, k \le n,$$

where $u_i^{(t)}$, $v_k^{(t)}$ are the optimal dual variables of the following linear assignment problem

$$\min\left\{\sum_{i,k=1}^{n} (c_{ik}^{(t)} + z_{ik}^{(t)})x_{ik} \colon (x_{ik}) \text{ fulfill } (1.5)\right\}$$

and $z_{ik}^{(t)}$ are defined as previously. The bound produced with these settings is often denoted by CM2 in the literature.

Analogously as for the bound AX, it can per shown that the time complexity for the bounds CM1, CM2, amounts to $O(n^5)$ elementary operations per iteration. Computational results show that the reformulation schemes generally produce bounds of good quality. However, these bounding techniques are quite time-consuming, as n^2+1 linear assignment problems per iteration have to be solved. Moreover, when computing bounds CM1, CM2 and AX for a Koopmans-Beckmann QAP(A,B), the nice structure of the objective function coefficients is not preserved by the reformulation rules. That is, even if in the above sequence of reformulations P_0 is a Koopmans-Beckmann QAP, the problems P_i, $i \geq 1$, are (generalized) Lawler QAPs of the form (1.2) or (1.4). Notice that the computation of the GL bound for a Lawler QAP takes $O(n^5)$ time, whereas for a Koopmans-Beckmann QAP it takes only $O(n^3)$. Thus, CM1, CM2, AX are computed in $O(kn^5)$ time, where k is the number of iterations (reformulations). Finally, notice that, in the case that $d_{ijkl} = d_{jilk}$, for all $1 \leq i,j,k,l \leq n$, the general reformulation scheme cannot produce lower bounds which are better than the optimal value of the LP relaxation of the MILP formulation (1.13) of Frieze and Yadegar, as shown in [43].

2.1.4 Bounds based on linear programming relaxations

Several authors have proposed mixed integer linear programming (MILP) formulations for the QAP. In Chapter 1 we have described the most significant among them, in terms of their size, their impact for the computation of the lower bounds, and the analysis and understanding of the QAP polytope. In the context of lower bound computation the MILP formulation of Frieze and Yadegar (1.13) is of essential importance. (Bounds FY1 and FY2 are obtained in terms of this MILP.)

Clearly, the optimal solution of the LP relaxation of an MILP formulation is a lower bound for the optimal value of the corresponding QAP. Moreover, each feasible solution of the dual LP is also a lower bound. As already mentioned in this chapter, good lower bounds should be tight, in the sense that they should be as close as possible to the optimal value of the QAP, and also efficient to compute. In order to derive bounds which match this requirements several efforts have been made concerning the identification of appropriate LP relaxations of MILP formulations, and also the development of solution methods for solving these LP relaxations or the corresponding duals.

Among the MILP formulations for QAPs, the one proposed by Adams and Johnson [2] and obtained by building upon the MILP formulation of Frieze and Yadegar (1.13) (see Chapter 1), is of particular significance. The optimal value of the corresponding LP relaxation is provably at least as good as most of the Gilmore-Lawler-like or reformulation lower bounds known to date for the QAP. Moreover, the solution method for the corresponding LP relaxation generalizes and unifies most of the bounding techniques which rely on reduction and reformulation ideas [2]. The MILP formulation of Adams and Johnson denoted by AJ through the rest of this chapter, is presented below. The authors make use of the variables $y_{ijkl} = x_{ik}x_{jl}$, $1 \le i, j, k, l \le n$, introduced by Frieze and Yadegar, and show that two groups of constraints of the MILP (1.13) are redundant under the constraint $y_{ijkl} = y_{jlik}$, $1 \le i, j, k, l \le n$. In this way, they obtain the following MILP formulation:

$$
\text{(AJ)} \qquad \begin{array}{lll}
\text{minimize} & \sum_{i=1}^{n}\sum_{j=1}^{n}\sum_{k=1}^{n}\sum_{l=1}^{n} a_{ij}b_{kl}y_{ikjl} & \\
\text{subject to} & & \\
& \sum_{i=1}^{n} x_{ik} = 1 & 1 \le k \le n \\
& \sum_{k=1}^{n} x_{ik} = 1 & 1 \le i \le n \\
& \sum_{j=1}^{n} y_{ikjl} = x_{ik} & 1 \le i, k, l \le n \\
& \sum_{l=1}^{n} y_{ikjl} = x_{ik} & 1 \le i, j, k \le n \\
& y_{jlik} = y_{ikjl} & 1 \le i, j, k, l \le n \qquad (*) \\
& x_{ik} \in \{0, 1\} & 1 \le i, k \le n \\
& 0 \le y_{ikjl} \le 1 & 1 \le i, j, k, l \le n
\end{array}
$$

In [2] it is shown that the constraints of the continuous relaxation of (1.13) can be obtained as a linear combination of the constraints of the continuous relaxation of AJ. Moreover, in the case that both coefficient matrices $A = (a_{ij})$ and $B = (b_{ij})$ are symmetric, the optimal value of CAJ is equal to the optimal value of the continuous relaxation of (1.13). The continuous relaxation of AJ is denoted by CAJ throughout the rest of this chapter. Among the constraints of AJ the equalities $(*)$ play a special role. For solving CAJ approximately, a Lagrangean dual is considered, where the constraints $(*)$ are added to the objective function multiplied by the Lagrangean multipliers u_{ikjl}, $1 \le i, j, k, l \le n$. So we get

$$
\begin{array}{ll}
\text{minimize} & \sum_{i=1}^{n}\sum_{\substack{j=1\\ j>i}}^{n}\sum_{k=1}^{n}\sum_{\substack{l=1\\ l\neq k}}^{n}(a_{ij}b_{kl} - u_{ikjl})y_{ikjl} - \\
& \sum_{i=1}^{n}\sum_{\substack{j=1\\ j<i}}^{n}\sum_{k=1}^{n}\sum_{\substack{l=1\\ l\neq k}}^{n}(a_{ij}b_{kl} - u_{jlik})y_{ikjl} + \sum_{i=1}^{n}\sum_{k=1}^{n} a_{ik}b_{ik}x_{ik}
\end{array}
$$

subject to

$$(\mathrm{AJ(u)}) \qquad \begin{array}{ll} \sum_{i=1}^{n} x_{ik} = 1 & 1 \le k \le n \\ \sum_{k=1}^{n} x_{ik} = 1 & 1 \le i \le n \\ \sum_{j=1}^{n} y_{ijkl} = x_{ik} & 1 \le i,k,l \le n \\ \sum_{l=1}^{n} y_{ijkl} = x_{ik} & 1 \le i,j,k \le n \\ x_{ik} \in \{0,1\} & 1 \le i,k \le n \\ 0 \le y_{ijkl} \le 1 & 1 \le i,j,k,l \le n \end{array}$$

Let $\theta(u)$ be the optimal value of AJ(u). It is a well known result in dual linear programming that $\max_u \theta(u)$ equals the optimal objective function value of CAJ. An iterative dual ascent procedure is applied for solving CAJ approximately. In each iteration the u-multipliers are first kept fixed and the problem AJ(u) is solved to optimality. Its optimal value $\theta(u)$ is a lower bound for the optimal value of CAJ. Notice that for any fixed set of u-multipliers, the x and y variables in the problem AJ(u) can be separated and AJ(u) can be solved to optimality by solving n^2+1 linear assignment problems (LAPs). Then, the values of the multipliers u_{ijkl} are updated, using the information contained in the dual variables of the LAPs solved during the current iteration, and the procedure is repeated. The procedure stops after having performed a prespecified number of iterations. The proposed updating rule leads to a non-decreasing sequence of lower bounds $\theta(u)$ with the increasing number of the iterations. The time complexity of this method is dominated by the solution of n^2+1 LAPs in each iteration and hence, amounts to $O(n^5)$ per iteration.

In [2] it is shown that $\theta(0)$ equals the Gilmore-Lawler bound for the QAP, whereas the other Gilmore-Lawler-like bounds which are obtained after some reduction (see Subsection 2.1.1) equal $\theta(u)$ for special settings of the Lagrangean multipliers u_{ijkl}. The theoretical relationship between the lower bounds proposed lately by Hahn et al. [108] and the bounds related to the AJ formulation of the QAP, is not clear yet. According to the numerical results reported in [2, 108], the bounds proposed by Hahn et al. outperform those obtained by the dual ascent procedure proposed by Adams et al., in terms of quality, whereas the latter procedure seems to be considerably faster. As for the reformulation based bounds (Section 2.1.3) in [2] it is argued that the reformulation bounds can be obtained by applying a dual ascent procedure as described above with specific update rules for the Lagrangean multipliers.

Hence, these bounds cannot exceed the optimal value of the CAJ. The dual procedure proposed in [2] was tested on the Nugent et al. test examples from QAPLIB.

The resulting bounds are, as expected, better than most of the other LP-based bounds described in the literature.

As for other solution methods for the CAJ, Adams et al. point out that this problem is highly degenerated and degeneracy poses a problem for primal approaches. As reported in [2], preliminary computational experiments show that primal methods indeed perform poorly. Another effort for solving the LP relaxation of the MILP formulation AJ is done by Resende, Ramakrishnan and Drezner [196]. They use an interior point approach, the so-called *approximate dual projective* algorithm of Karmarkar and Ramakrishnan [135], and obtain bounds which beat those obtained by Adams et al. in [2], in terms of quality. The algorithm of Resende at al. does not solve the CAJ exactly. In general it computes a "good" solution for the dual, this being obviously a lower bound for the primal. The linear systems arising in each iteration of the algorithm are solved by a preconditioned conjugate gradient approach.

Because of memory requirements, the bounds of Resende et al., frequently denoted by IPLP, are tested on QAP instances from QAPLIB of size smaller than or equal to 30. The IPLP bounds turn out to be the best existing bounds for a large number of QAP test instances from QAPLIB. In terms of quality, they are beaten only by the so-called triangle decomposition bounds (to be considered below)[6]. However, the computation of the IPLP bounds requires very high computation times, which is mainly due to a huge number of calls of the preconditioned conjugate gradient routine, cf. Resende et al. [196].

Due to these high computation time requirements, the IPLP bounds cannot be used for practical purposes in bounding subroutines within branch and bound algorithms, despite the good quality.
The DP bound of Hahn et al. [108] is the only bound which can compete with the IPLP, in terms of tightness. Moreover, generally, the DP bounds can be computed much faster than the IPLP bounds. As an illustrative example consider the Nugent at al. test problem od size 20 from QAPLIB, where the gap between the IPLP and the DP bound is about 0.3%, and the CPU time requirements amount to 3611.4 and 382.7 seconds for the computation of the IPLP and the DP bounds, respectively. The computations were performed on a Silicon Graphics Challenge computer with a 150 MHz IP19 processor and a SPARC 10 workstation with a 75 MHz SuperSparc processor, for IPLP and DP, respectively.

[6]The triangle decomposition bounds can be computed only for specially structured QAPs, the so-called *grid QAPs*.

Finally, let us mention an application of dynamic programming to compute lower bounds for the QAP based on MILP formulations. In 1989, Christofides and Benavent [52] considered a special case of the Koopmans-Beckmann QAP(A,B), the so-called *tree-QAP*. In the tree QAP A is an arbitrary matrix, and B is the weighted adjacency matrix of a tree. It is easily seen that this special case of the QAP is NP-hard, since it contains the shortest Hamiltonian path problem as a special case. Christofides and Benavent give a MILP formulation for this special QAP, and consider several relaxations which can be solved in polynomial time, producing in this way lower bounds for the original problem. Among others, they distinguish a Lagrangean type relaxation which, due to the special structure of the problem, can be solved in polynomial time by a fast dynamic programming approach. In general, the bounds produced by solving this relaxation are of very good quality and even match the optimal solution of the MILP formulation in many cases. When involved in a branch and bound algorithm these bounds produce a relatively small branch and bound tree. This allows the authors to solve to optimality test instances of size up to 25, within reasonable time limits (less than 600 seconds of CPU time on a UNIVA 1100/60, in almost all cases). Recall that nowadays QAPs of size larger than 22 are considered intractable.

2.1.5 Bounds based on semidefinite relaxations

Semidefinite programming (SDP) is a generalization of linear programming where the variables are taken from the Euclidean space of matrices with the trace operator acting as an inner product (see below). The non-negativity constraints are replaced by semidefiniteness constraints and the linear constraints are formulated in terms of linear operators on the above mentioned Euclidean space of matrices. From the early 90's on, semidefinite programming was a subject of renewed interest in discrete optimization. This is due to successful applications of semidefinite programming in discrete optimization as presented in [96, 146].

Recently, semidefinite programming relaxations for the QAP were considered by Karisch [132], Zhao [231], and Zhao, Karisch, Rendl and Wolkowicz [232]. The SDP relaxations considered in these papers are solved by interior point methods or cutting plane methods, and the obtained solutions are valid lower bounds for the QAP. The solution methods described in [132, 231, 232] require at least $O(n^6)$ time per iteration, where n is the size of the problem. In terms of quality the bounds obtained in this way are competitive with the best existing lower bounds for the QAP. For many test instances from QAPLIB, such as some instances of Hadley, Roucairol, Nugent et al. and Taillard, they are the best existing bounds. However, due to prohibitively high computation time requirements, the use of such approaches as basic bounding procedures within branch and bound algorithms is impossible.

We refer to [132, 232] for a detailed description of SDP approaches to the QAP. The semidefinite programming relaxations used for the lower bound computations in these papers, are presented below.

As already mentioned in this chapter, the set of $n \times n$ permutation matrices Π_n is the intersection of the set of $n \times n$ 0-1 matrices, denoted by $\mathcal{Z}_n$, and the set of $n \times n$ matrices with row and column sums equal to 1, denoted by $\mathcal{H}_n$ throughout the rest of this chapter. Moreover, Π_n is also the intersection of $\mathcal{Z}_n$ with the set of $n \times n$ orthogonal matrices, denoted by $\mathcal{O}_n$. Hence

$$\Pi_n = \mathcal{Z}_n \cap \mathcal{H}_n = \mathcal{Z}_n \cap \mathcal{O}_n$$

Recall that

$$\mathcal{O}_n = \left\{X \in \mathbb{R}^{n\times n} : XX^t = X^tX = I\right\} \quad \text{and}$$
$$\mathcal{H}_n = \left\{X \in \mathbb{R}^{n\times n} : XE = X^tE = E\right\},$$

where I is the $n \times n$ identity matrix and E is the n-dimensional vector of all ones. Then, the trace formulation of QAP(A,B) (1.8) with the additional linear term

$$-2\sum_{i=1}^{n}\sum_{j=1}^{n} c_{ij}x_{ij}$$

(see Section 1.3) can be represented equivalently as follows:

$$(QAP_{\mathcal{H}}) \qquad \begin{array}{ll} \min & tr(AXBX^t - 2CX^t) \\ \text{subject to} & \\ & XX^t = X^tX = I \\ & XE = X^tE = E \\ & x_{ij}^2 - x_{ij} = 0 \end{array}$$

In order to obtain a semidefinite relaxation for the QAP from the formulation $QAP_{\mathcal{H}}$ above, we introduce first an n^2-dimensional vector $vec(X)$. $vec(X)$ is obtained as a column-wise ordering of the entries of matrix X. Then the vector $vec(X)$ is *lifted* into the space of $(n^2+1) \times (n^2+1)$ matrices by introducing a matrix Y_X

$$Y_X = \begin{pmatrix} x_0 & vec(X)^t \\ vec(X) & vec(X)vec(X)^t \end{pmatrix}$$

Thus, Y_X has some entry x_0 in the left-upper corner followed by the vector $vec(X)$ in its first row (column). The remaining terms are those of the matrix

$$vec(X)vec(X)^t$$

sitting on the right lower $n^2 \times n^2$ block of Y_X.

Secondly, the coefficients of the problem are collected in an $(n^2+1) \times (n^2+1)$ matrix K given as

$$K = \begin{pmatrix} 0 & -vec(C)^t \\ vec(C) & B \otimes A \end{pmatrix},$$

where the operator vec is defined as above and $B \otimes A$ is the Kronecker product of B and A.

It is easy to see that with these notations the objective function of $QAP_{\mathcal{H}}$ equals $tr(KY_X)$. By setting $y_{00} := x_0 = 1$ as done in Zhao et al. [232], one obtaines two additional constraints to be fulfilled by the matrix Y_X: Y_X is positive semidefinite and matrix Y_X is a rank-one matrix. Whereas the semidefiniteness and the equality $y_{00} = 1$ can be immediately included in an SDP relaxation, the rank-one condition is hard to handle and is discarded in an SDP relaxation. In order to assure that the rank-one positive semidefinite matrix Y_X is obtained by an $n \times n$ permutation matrix as described above, other constraints are to be imposed on Y_X. Such conditions can be formulated as valid constraints of an SDP formulation for the QAP by means of some new operators, acting on matrices or vectors as introduced below. $diag(A)$ produces a vector containing the diagonal entries of matrix A in their natural order, i.e. from top-left to bottom-right. The adjoint operator $Diag$ acts on a vector V and produces a square matrix $Diag(V)$ with off-diagonal entries equal to 0 and the components of V on the main diagonal. Clearly, for an n dimensional vector V, $Diag(V)$ is an $n \times n$ matrix.
$arrow$ acts on an $(n^2+1) \times (n^2+1)$ matrix Y and produces an n^2+1 dimensional vector $arrow(Y) = diag(Y) - (0, Y_{0,1:n^2})$, where $(0, Y_{(0,1:n^2)})$ is an n^2+1 dimensional vector with first entry equal to 0 and other entries coinciding with the entries of Y lying on the 0-th row and in columns between 1 and n^2, in their natural order[7]. The adjoint operator $Arrow$ acts on an n^2+1 dimensional vector W and produces an $(n^2+1) \times (n^2+1)$ matrix $Arrow(W)$

$$Arrow(W) = \begin{pmatrix} w_0 & 1/2W^t_{1:n^2} \\ 1/2W_{(1:n^2)} & Diag(W_{1:n^2}) \end{pmatrix},$$

where $W_{(1:n^2)}$ is the n^2 dimensional vector obtained from W by removing its first entry w_0.

[7] Note here that the rows and columns of an $(n^2+1) \times (n^2+1)$ matrix are indexed by $0, 1, \ldots, n^2$.

Further, we are going to consider an $(n^2+1) \times (n^2+1)$ matrix Y as composed of its first row $Y_{(0,.)}$, of its first column $Y_{(.,0)}$, and of n^2 submatrices of size $n \times n$ each, which are arranged in an $n \times n$ array of $n \times n$ matrices, and produce its remaining $n^2 \times n^2$ block. The entry $y_{\alpha\beta}$, $1 \leq \alpha, \beta \leq n^2$, will be also denoted by $y_{(ij)(kl)}$, with $1 \leq i,j,k,l \leq n$, where $\alpha = (i-1)n+j$ and $\beta = (k-1)n+l$. Hence, $y_{(ij)(kl)}$ is the element with coordinates (j,l) within the $n \times n$ block with coordinates (i,k). These ideas are illustratited in Figure 2.2. In this figure the entries $y_{\alpha\beta}$ of the matrix Y are denoted by y_α^β. This is done for technical reasons and ease of presentation[8].

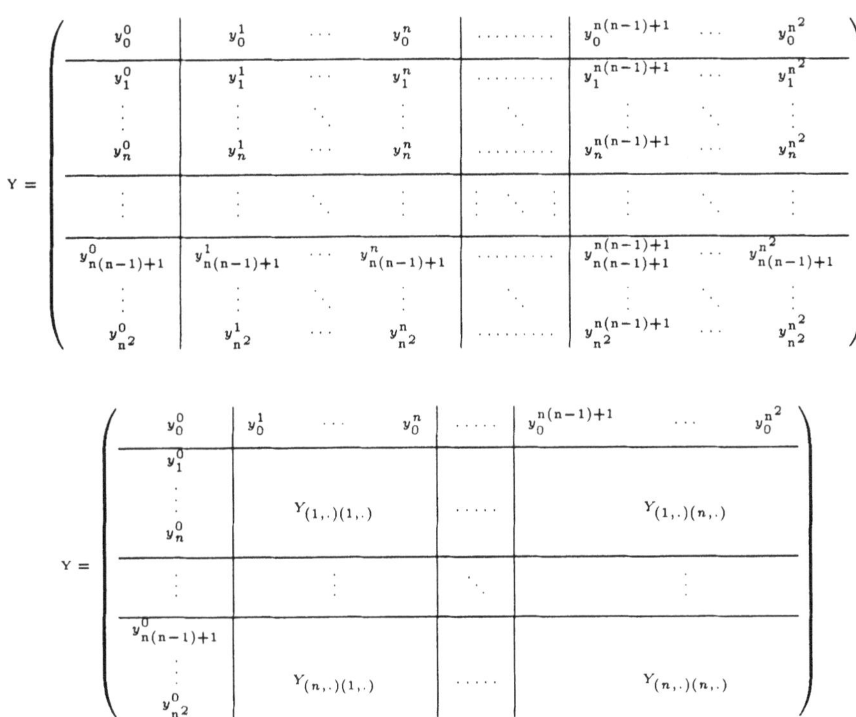

Figure 2.2 An $(n^2+1) \times (n^2+1)$ matrix Y considered as an $n \times n$ array of $n \times n$ matrices $Y_{(i,.)(j,.)}$, for $1 \leq i,j \leq n$.

With these formal conventions let us define now the so-called *block-0-diagonal* and *off-0-diagonal* operators, acting on an $(n^2+1) \times (n^2+1)$ matrix Y, and denoted

[8]If the usual notations $y_{\alpha\beta}$ would have been used, the width of the figure would exceed the pagewidth.

by b^0diag and o^0diag, respectively. $b^0diag(Y)$ and $o^0diag(Y)$ are $n \times n$ matrices given as follows:

$$b^0diag(Y) = \sum_{k=1}^{n} Y_{(k,\cdot)(k,\cdot)} \quad o^0diag(Y) = \sum_{k=1}^{n} Y_{(\cdot,k),(\cdot,k)},$$

where, for $1 \leq k \leq n$, $Y_{(k,\cdot)(k,\cdot)}$ is the k-th $n \times n$ matrix on the diagonal of the $n \times n$ array of matrices, defined as described above. Analogously, $Y_{(\cdot,k),(\cdot,k)}$ is an $n \times n$ matrix consisting of the diagonal elements sitting on the position (k, k) of the $n \times n$ matrices (n^2 matrices altogether) which form the $n^2 \times n^2$ lower right block of matrix Y. The corresponding adjoint operators B^0Diag and O^0Diag act on an $n \times n$ matrix S and produce $(n^2+1) \times (n^2+1)$ matrices as follows:

$$B^0Diag = \begin{pmatrix} 0 & 0 \\ 0 & I \otimes S \end{pmatrix} \quad O^0Diag = \begin{pmatrix} 0 & 0 \\ 0 & S \otimes I \end{pmatrix}$$

Finally, let us denote by E_0 the n^2+1 dimensional unit vector with first component equal to 1 and all other components equal to 0, and let D be the $(n^2+1) \times (n^2+1)$ matrix given by

$$D = \begin{pmatrix} n & -E^t \otimes E^t \\ -E \otimes E & I \otimes \Xi \end{pmatrix} + \begin{pmatrix} n & -E^t \otimes E^t \\ -E \otimes E & \Xi \otimes I \end{pmatrix},$$

where Ξ is the $n \times n$ matrix of all ones.

With these notations, a semidefinite relaxation for $QAP_{\mathcal{H}}$ is given as follows

$$(QAP_{R0}) \qquad \begin{array}{ll} \min & tr(KY) \\ \text{subject to} & \\ & b^0diag(Y) = I \\ & o^0diag(Y) = I \\ & arrow(Y) = E_0 \\ & tr(DY) = 0 \\ & Y \succeq 0 \end{array}$$

where $\preceq$ is the so-called *Löwner partial order*, i.e. $A \preceq B$ if and only if $B - A \succeq 0$, that is $B - A$ is positive semidefinite.

Zhao et al. [232] have shown that an equivalent formulation for the considered QAP is obtained from QAP_{R0} by imposing one additional condition on the matrix Y, namely, the rank-one condition. On the other hand, it is not very difficult to see that any feasible solution of QAP_{R0} is singular, and hence the set of the feasible solutions of QAP_{R0} has empty interior. This may be inconvenient for the application of interior point methods to solve this problem. In order to avoid such difficulties, the authors project the problem onto the so-called *minimal face* of the semidefinite cone. The minimal face is characterized by an $(n^2+1) \times \left[(n-1)^2+1\right]$ full-rank matrix $\hat{V}$ given as

$$\hat{V} = \begin{pmatrix} 1 & 0 \\ \frac{1}{n}(E \otimes E) & V \otimes V \end{pmatrix}$$

where V is an $n \times (n-1)$ matrix whose columns form a basis of the orthogonal complement of E, i.e. $V^t E = 0$. In [132, 232] $V = \begin{pmatrix} I_{n-1} \\ -e_{n-1}^t \end{pmatrix}$ is chosen. The minimal face of the semidefinite cone contains only matrices of the form $\hat{V} R \hat{V}^t$, for some arbitrary $\left[(n-1)^2+1\right] \times \left[(n-1)^2+1\right]$ matrix R. The feasible solutions of the QAP are then required in the form $Y = \hat{V} R \hat{V}^t$ and the following semidefinite relaxation for the QAP, called *basic semidefinite relaxation* is obtained:

$$(QAP_{R1}) \qquad \begin{array}{ll} \min & tr((\hat{V}^t K \hat{V}) R) \\ \text{subject to} & \\ & b^0 diag(\hat{V} R \hat{V}^t) = I \\ & o^0 diag(\hat{V} R \hat{V}^t) = I \\ & arrow(\hat{V} R \hat{V}^t) = E_0 \\ & R \succeq 0 \end{array}$$

It can be shown that the operators block-0-diagonal and off-0-diagonal produce many redundant constraints in QAP_{R1}. Moreover, the dimensions of the image spaces of both operators are $(n^2-3n)/2$ (see [232]). Therefore, only parts of those operators which cover systems of linearly independent inequalities, need to be considered.

The set of the feasible solutions of the relaxation QAP_{R0} is convex but not polyhedral. This set may contain, among others, matrices with non-zero elements in places where all affine combinations of the matrices Y_X, $X \in \Pi$, have zeroes. Hence, the relaxation QAP_{R1} can be strengthened by requiring that the feasible solutions have zeroes in the above mentioned places. To this end, the so-called *gangster operator* is introduced. Let J be the set of pairs (i,j), $0 \leq i,j \leq n^2$, such that $Y_{ij} = 0$, for every $(n^2+1) \times (n^2+1)$ matrix Y generated by some permutation matrix X[9]. The gangster operator $\mathcal{G}$ operates on an $(n^2+1) \times (n^2+1)$ matrix

[9] See [232] for an explicit description of J.

Y and produces a new $(n^2+1) \times (n^2+1)$ matrix $\mathcal{G}(Y)$, with components $\mathcal{G}(Y)_{ij}$ given by

$$\mathcal{G}(Y)_{ij} = \begin{cases} 0 & \text{if } (i,j) \in J \\ Y_{ij} & \text{otherwise} \end{cases}$$

By means of the gangster operator the following simple SDP relaxation of the QAP is obtained:

$$QAP_{R2} \qquad \begin{array}{ll} \min & tr((\hat{V}^t K \hat{V})R) \\ \text{subject to} & \\ & \mathcal{G}(\hat{V}R\hat{V}^t) = 0 \\ & R \succeq 0 \end{array}$$

We add to this relaxation the simple inequalities $y_{(ij),(kl)} \geq 0$, which are facet defining for the QAP polytope, as shown in [128, 177]. Moreover, we require the non-positivity of the elements which should be equal to zero, in order to fulfill the condition described by the gangster operator. This amounts to the following stronger SDP relaxation for the QAP:

$$QAP_{R3} \qquad \begin{array}{ll} \min & tr((\hat{V}^t K \hat{V})R) \\ \text{subject to} & \\ & b^0 diag(\hat{V}R\hat{V}^t) = I \\ & o^0 diag(\hat{V}R\hat{V}^t) = I \\ & arrow(\hat{V}R\hat{V}^t) = E_0 \\ & \hat{V}R\hat{V}^t \geq 0 \\ & \mathcal{G}(\hat{V}R\hat{V}^t) \leq 0 \\ & R \succeq 0 \end{array}$$

In [232] it is shown that the optimal solution of the SDP relaxation QAP_{R3} produces a bound which is larger, and thus better, than that obtained by solving to optimality the continuous relaxation CAJ of the MILP formulation for the QAP proposed by Adams and Johnson [2]. As mentioned above, the SDP relaxations are basically solved by interior point methods. Unfortunately, it is extremely time-consuming to solve QAP_{R3} to optimality by means of such methods. Therefore a combined approach is applied, where QAP_{R1} is solved to optimality and then violated inequalities from QAP_{R3} are added, in the vain of a cutting plane approach. In order to maintain the computation time within certain limits, a maximum for the number of the added inequalities is defined previously. This maximum turns out to be too low for problems of size larger than 20. For such instances the bound obtained by solving QAP_{R2} outperforms the bound resulting from the approximate solution of QAP_{R3}, as described above.

However, the bounds obtained by semidefinite relaxations are competitive with the best existing bounds for QAPs, in terms of quality. As shown in [232], the SDP based bounds outperform the IPLP bounds in 12 out of 25 instances (from QAPLIB) for which both types of bounds are available. The experiments with these bounding techniques are performed on different machines and, to the best of our knowledge, no systematic investigations concerning the comparison of computation times have been made. However, the SDP based bounds, similarly as those obtained by Ramakrishnan et al., are very time-consuming. As an illustrative example consider that solving the QAP_{R3} relaxation for the Nugent et al. instance of size 20 takes 316.17 minutes of CPU time on a DEC 3000-900 Alpha AXP computer (275 MHz with 256 Mbytes and 643 Mbytes of RAM).

The quality of the SDP bounds and the related computation times could be improved if "stronger" cuts would be performed by the respective cutting plane routines. Most probably this requires, however, the derivation of "tighter" valid or facet defining inequalities for the QAP polytope. Clearly, LP-based lower bounds would take advantage of such results, too. Hence, it seems that, currently, the SDP techniques can not bring an essential improvement in the state of the art of the lower bound computation for the QAP.

2.1.6 Improving bounds by means of decompositions

The idea of applying the so-called *decompositions* to improve lower bounds for specially structured QAPs was initially proposed by Chakrapani and Skorin-Kapov [49], and then further elaborated by Karisch and Rendl [133]. Although the applicability of this approach seems to be restricted to QAPs with a very special structure, the so-called *grid QAPs* (or *rectilinear QAPs*) to be described below, we devote to this approach a section on its own. We do so for two reasons. First, this approach combines in a powerful bounding procedure different mathematical ideas, such as combinatorial reduction of problems to simple instances, and eigenvalue and convex optimization. This procedure yields the best existing bounds for almost all grid QAP instances from QAPLIB. Secondly, we hope that similar ideas can be applied to other specially structured QAPs, this approach being the only first step towards using easily and specially solvable QAPs for the solution of more general problems.

A grid QAP is a Koopmans-Beckmann QAP(A,B), where $A = (a_{ij})$ is the distance matrix of a uniform rectangular grid. That is, the indices $1 \le i, j \le n$ represent the vertices of the grid, numbered from the left to the right and from the bottom to the top, and a_{ij} is the number of edges in the shortest path connecting i and j on the grid. If $a_{ij} = a_{ik} + a_{kj}$, we say that k is on the shortest path connecting i and j. The triple $u = (i, j, k)$ is then called a *shortest path triple*. The shortest path triple $v = (i, j, k)$ for which $a_{ik} = a_{kj} = 1$ is called a *shortest triangle.*

We associate a matric $R_u = (p_{ij}^{(u)})$ to each shortest path triple $u = (k, l, m)$, and a matrix $T_v = (t_{ij}^{(v)})$ to each shortest triangle $v = (k', l', m')$, where R_u and T_v are defined by

$$R_{kl}^{(u)} = r_{lk}^{(u)} = r_{ml}^{(u)} = r_{lm}^{(u)} = 1,\ r_{km}^{(u)} = r_{mk}^{(u)} = -1, \text{ and } r_{ij}^{(u)} = 0 \text{ if } \{i,j\} \not\subset \{k,l,m\}$$

$$t_{k'm'}^{(v)} = t_{l'm'}^{(v)} = t_{m'l'}^{(v)} = t_{l'k'}^{(v)} = t_{k'm'}^{(v)} = t_{m'k'}^{(v)} = 1 \text{ and } t_{ij}^{(v)} = 0 \text{ if } \{i,j\} \not\subseteq \{k,l,m\}$$

The set of all shortest path triples is denoted by $\mathcal{R}$ and the set of all shortest triangles is denoted by $\mathcal{T}$.

The key observation is that, for each $u \in \mathcal{R}$ and for each $v \in \mathcal{T}$, $QAP(A, R_u)$ and $QAP(A, T_v)$ are solved to optimality by the identity permutation *id*, $id(i) = i$, $1 \le i \le n$. The optimal values for these QAPs are 0 and 8, respectively. These simple special QAPs can be used for improving the quality of lower bounds for an arbitrary grid QAP(A,B). Namely, the matrix B is decomposed as

$$B = \sum_{u \in \mathcal{R}} \alpha_u R_u + \sum_{v \in \mathcal{T}} \beta_v T_v + B_r \tag{2.14}$$

where B_r is the *residual* matrix given as

$$B_r := B - \sum_{u \in \mathcal{R}} \alpha_u R_u + \sum_{v \in \mathcal{T}} \beta_v T_v$$

For every choice of the parameters $\alpha_u \ge 0$, $u \in \mathcal{R}$, and $\beta_v \ge 0$, $v \in \mathcal{T}$, and for any permutation ϕ we have

$$Z(A, B, \phi) = \sum_{u \in \mathcal{R}} \alpha_u Z(A, R_u, \phi) + \sum_{v \in \mathcal{T}} \beta_v Z(A, T_v, \phi) + Z(A, B_r, \phi) \tag{2.15}$$

and therefore,

$$\min_{\phi} Z(A, B, \phi) \ge 8 \sum_{v \in \mathcal{T}} \beta_v + \min_{\phi} Z(A, B_r, \phi) \ge$$

$$8\sum_{v\in\mathcal{T}}\beta_v + LB(QAP(A,B_r)),$$

where $LB(QAP(A,B_r))$ is any lower bound for $QAP(A,B_r)$. Clearly, the expression on the right hand side of (2.15) is a lower bound for $QAP(A,B)$. This lower bound, which depends on the vectors $\alpha = (\alpha_u)$, $\beta = (\beta_v)$, is denoted by $h(\alpha,\beta)$. Then, $h(0,0)$ equals $LB(QAP(A,B))$, and therefore,

$$\max_{\alpha\geq 0,\beta\geq 0} h(\alpha,\beta) \geq LB\left(QAP(A,B)\right),$$

where a vector is said to be nonnegative if all its components are nonnegative. Hence, $\max_{\alpha\geq 0,\beta\geq 0} h(\alpha,\beta)$ is an improvement upon the bound $LB(QAP(A,B))$.

So far everything works analogously even if only one of the matrix classes $\{R_u\}_{u\in\mathcal{R}}$, $\{T_v\}_{v\in\mathcal{T}}$ is used for the decomposition, and an arbitrary lower bound, say LB, is involved. Chakrapani et al. [49] improve the Gilmore-Lawler bound (GL), and the elimination bound (ELI), by using only the matrices R_u, $u\in\mathcal{R}$, for the decomposition. Karisch et al. [133] use the decomposition scheme (2.14) to improve the elimination bound (ELI) (introduced in [113]). The authors show that the resulting bound $h(\alpha,\beta)$ is a concave function which, furthermore, attains the maximum for $\alpha\geq 0$, $\beta\geq 0$. So, the maximization of $h(\alpha,\beta)$ turns out to be a convex optimization problem over a convex set. It is worth noticing here that $h(\alpha,\beta)$ is not everywhere differentiable. (Consider that only *simple* eigenvalues of a symmetric matrix $A(x)$ are guaranteed to be differentiable on x.)
This optimization problem is solved approximately by a supergradient approach. The solution procedure involves one supergradient computation per iteration, which is performed in $O(n^3)$ time (cf. [133]), where n is the size of the considered QAP. The resulting bounds are often termed *triangle decomposition bounds (TDB)*.

The numerical results reported in [133] show that a very good trade-off between computation time and bound quality can be found for almost all instances of grid QAPs from QAPLIB. According to these results, TDB beats all existing bounds for 22 out of 23 grid QAP instances from QAPLIB. In general, the running times are surprisingly low in comparison with those of other bounding procedures which yield comparably good bounds. For example, for the Nugent et al. problem of size 30, the TDB bound, which is in the same time the best existing bound, equals 5772 and is computed in 200 iterations with 0.5 seconds per iteration, as reported in [133]. For the same problem, the LP relaxation bound of Ramakrishnan et al. [196] mentioned in Subsection 2.1.4, amounts to 4805 with a total computation time of 25589.5 seconds. Concluding, the TDB bounds, which cleverly exploit the combinatorial structure of the problem at hand, seems to be invincible for almost all grid QAPs from QAPLIB.

2.1.7 Bounding techniques: Conclusions

At this point, after having described more or less in detail the most frequently used bounding procedures for the QAP, the reader is probably confused by the most natural questions in this context: Which is the winner among all bounds for the QAP? Which procedure should be used for solving a given QAP instance, say from QAPLIB, by a branch and bound method?

Our current knowledge and understanding of the QAP, in general, and of the bounding approaches, in particular, do not enable us to give a mathematically rigorous answer to these questions. However, we can comment upon the obtained results and determine open questions to be handled in the near future.

Among the bounding techniques proposed in the literature, the best quality bounds seem to be produced by the most time-consuming approaches. The bounds of Ramakrishnan et al. [196], based on an LP relaxation, and those of Zhao et al. [232], based on SDP relaxations, beat all other bounds in terms of quality, for almost all (non-specially structured) test instances from QAPLIB. However, the time needed for computing these bounds is certainly prohibitive, when willing to incorporate them in branch and bound schemes as basic bounding procedures. Clearly, the LP-based bounds and the SDP-based bounds would benefit from a better understanding of the QAP polytope, in general, and the new results concerning valid and (possibly) facet defining inequalities, in particular. Considering the progress made recently in this direction, there is room and hope for improvement in terms of cost and quality.

For almost all grid QAPs from QAPLIB the triangle decomposition bounds (TDB) [49, 133] are by far the best ones in terms of solution quality and computation time-solution quality trade-off. The success of such decomposition approaches suggests that the combinatorial properties of specially structured QAPs could be well useful for deriving strong bounds for special classes of the problem. However, a recent study of Clausen, Karisch, Perregaard and Rendl [57] shows that the TDBs are not very well suited for use within branch and bound algorithms. The quality of these bounds deteriorates relatively fast when moving towards higher levels of the branch and bound tree. Clausen et al. investigate the applicability of 5 different bounding techniques in branch and bound algorithms: The Gilmore-Lawler bound GL, the elimination bound ELI, the GL bound improved by decomposition techniques (TGL), the ELI improved by decomposition techniques (TELI), and the bound of Carraresi and Malucelli CM2. The numerical experiments are performed on the Nugent et al. test instances from QAPLIB [37]. It turns out that the CM2 is the best suited for solving "large problems", i.e. problems of size larger than 20.

This is due to the high quality of this bound, which is *preserved* also in higher levels of the branch and bound tree. For "large" instances, the branch and bound approaches involving CM2 show the lowest running times, although CM2 is the most expensive among the investigated bounds, and moreover, it is clearly beaten by TELI in terms of quality, when computed at the root of the branch and bound tree[10].

The somewhat surprising results presented in [57] show that the applicability of different bounding techniques within branch and bound schemes is a well worth studying and by far non-trivial topic. Moreover, these results testimony that the behavior of the bounding techniques used in a branch and bound scheme may significantly depend on the size of the problem we are dealing with. This fact, which is not surprising, suggests the investigation of the performance of bounding techniques depending on the size of the problems we are willing to solve. To the best of our knowledge, Clausen et al. are the first authors addressing such topics, and much more remains to be done concerning the investigation of the applicability of other bounding techniques on larger sets of test instances.

For general QAPs, the dual procedure (DP) proposed recently by Hahn et al. [108] seems to offer the best trade-off between bound quality and computation time. The DP bounds are almost as good as the LP-based bounds of Ramakrishnan et al., and require about 70 times less running time, as reported by Hahn et al. Moreover, preliminary results show that the DP bounds are well suited for use in branch and bound algorithms (cf. [109]). However, more experimental work is needed in order to investigate their applicability.
The interesting thing about the DP bounds, is that they seem to combine in a single approach many ingredients from other bounding procedures such as reduction or reformulation ideas, GL bound computation, and LP-based techniques. Recall that Adams and Johnson [2] unify and generalize many bounding procedures in a dual ascent approach for the Lagrangean dual of an MILP formulation for the QAP. It remains an interesting question to investigate whether the DP bounds fit somehow in the scheme of Adams et al. Such investigations would also help for a better understanding of the DP procedure, in order to exploit the potential of this method as well as possible.

[10]Nowadays, in general, QAPs of size larger than 22 cannot be solved to optimality. The numerical results related to QAPs of size larger than 22 are based on estimations obtained by partially searching the branch and bound tree.

2.2 BRANCH AND BOUND APPROACHES

Many authors have proposed sequential and parallel implementations of branch and bound approaches for the QAP. These approaches can be classified in three main groups with respect to the branching rule:

1. Single assignment methods
2. Pair assignment methods
3. Relative positioning methods.

All these algorithms start with an *empty partial permutation* (i.e. in the facility location context, no facility is assigned to any location) and step by step extend it to a full permutation (a permutation which assigns all facilities to locations).

The single assignment methods were the first approaches applied to QAPs. Algorithms belonging to this group assign a single (not yet located) facility, say i, to a not yet occupied location, say j, at each branching step. From a historical point of view, single assignment methods date back to the algorithm proposed by Gilmore [92] for the QAP(A,B). This algorithm was then generalized by Lawler [152] for the more general Lawler QAP defined in (1.2). Other single assignment branch and bound algorithms for QAPs have been proposed by several authors [35, 71, 131, 183, 164].
The choice of the above mentioned pair of indices (i, j) is done according to the *selection rule* and usually depends on the used bounding technique. As almost all bounding techniques end up by solving a linear assignment problem, the choice of the pair (i, j) is often based on the so-called *alternative costs* p_{ij} given as

$$p_{ij} = \min\{\bar{c}_{ik}: k \neq j,\ 1 \leq k \leq n\} + \min\{\bar{c}_{kj}: k \neq i,\ 1 \leq k \leq n\},$$

where the coefficients $\bar{c}_{ij}$ are the reduced cost of the last linear assignment problem solved when calculating the current lower bound. The alternative cost p_{ij} is a lower bound for the increment of the current lower bound, in the case that the current permutation is extended by assigning facility i to location j. In [28] Burkard proposed the maximization of the alternative costs as a natural criterion for choosing the pair of indices (i, j). Another rule to choose the pair (i, j) has been considered by Bazaraa and Kirca [17]. They use a combination of the following two criteria:

- Try to minimize the sum of all lower bounds at the next branch and bound tree level.

- Try to minimize the number of branches at the next branch and bound tree level.

Another interesting criterion for the choice of the pair (i, j) is proposed by Mautor and Roucairol in [164]. They also calculate the alternative costs p_{ij} as above. Assume that $\underline{z}$ and $\bar{z}$ are the lower and the upper bound at the current node of the branch and bound tree, respectively. If $p_{ij} + \underline{z} \geq \bar{z}$ then the pair (i, j) is forbidden. i is chosen to be the index with the largest number of forbidden pairs having i as first element. The chosen index j is an allowed index, i.e. the pair (i, j) is not a forbidden one, for which the sum of the allowed entries of the corresponding column of the matrix (p_{ij}) is maximized. Ties are broken by maximizing the alternative cost. This selection rule seems to work pretty well and seems to produce a much smaller branch and bound tree than other rules.

The pair (i, j) could also be selected with respect to a previously fixed order, as proposed for example in [17]. The advantageous consequence of using this rule is that the partial permutation is completely determined by the position of the corresponding node in the branch and bound tree.

The pair assignment methods allocate a pair of facilities to a pair of locations at each branching step. Computational experiments developed by a number of authors [89, 148, 174] have shown that these algorithms do not produce good results.

Relative positioning methods. This type of branch and bound method was proposed by Mirchandani and Obata [168]. Here, the levels of the branch and bound tree do not correspond to the assignment of facilities to locations. The partial permutations at each level are determined in terms of distances between facilities, i.e. their relative positions. This approach is claimed to be appropriate for QAPs with sparse matrices, henceforth called *sparse QAPs*.

Another interesting branching rule, which does not belong to any of the above groups, was developed by Roucairol [200]. It is called the *polytomic* or *k-partite branching rule* and it is well suited for use in parallel implementation of branch and bound algorithms for the QAP. The branch and bound produced by this algorithm is not a binary tree as in most of the other approaches. In this case, the solution of the last linear assignment problem solved for calculating the lower bound at the current note of the branch and bound tree is considered. Assume that this solution is the permutation $\rho \in \mathcal{S}_n$ (for a problem of size n). Let $\mathcal{S}_n^{(i)}$ be the subset of $\mathcal{S}_n$ consisting of those permutations $\pi \in \mathcal{S}_n$, such that $\pi(i) = \rho(i)$. Analogously, $\bar{\mathcal{S}}_n^{(i)}$ is the set of permutations $\pi \in \mathcal{S}_n$, such that $\pi(i) \neq \rho(i)$. Then, the current node is branched into $n + 1$ new nodes, whose corresponding sets of feasible solutions are given as follows $\mathcal{S}_n^{(1)}, \mathcal{S}_n^{(1)} \cap \bar{\mathcal{S}}_n^{(2)}, \ldots, \mathcal{S}_n^{(1)} \cap \mathcal{S}_n^{(2)} \cap \ldots \cap \mathcal{S}_n^{(n-1)} \cap \bar{\mathcal{S}}_n^{(n)}, \mathcal{S}_n^{(1)} \cap \mathcal{S}_n^{(2)} \cap \ldots \cap \mathcal{S}_n^{(n)}$. (Clearly the nodes corresponding to the two last feasible sets are trivial in the sense that the first of these sets is empty and the second one consists only of permutation ρ.)

Concluding this paragraph, let us remark that better results on solving large size problems have been achieved lately by using parallel implementations [26, 58, 150, 183]. The above mentioned papers show that a linear speed-up can typically be realized as the number of the involved processors increases. This holds for a relatively low number of processors. The author is not aware of any experiments on solving QAPs with massively parallel computers. However, a less than linear speed-up can be expected due to the communication load between the processors.

Some of the most celebrated branch and bound algorithms for QAPs are listed in Table 2.1 together with the size of the Nugent et al. test instance which can be solved by them[11]. Recently the Nugent et al. test instance of size 22 (see QAPLIB) was solved to optimality by a parallel branch and bound algorithm [26] which uses basically the GL bound. To the best of our knowledge, this is the largest Nugent et al. test instance which has been ever solved to optimality. The progress in solving

Table 2.1 The evolution of branch and bound approaches for the QAP

Authors	Year	Size
Gilmore	1962	8
Lawler	1963	8
Gavett and Plyter	1965	8
Burkard	1973	8
Bazaraa and Sherali	1980	6
Burkard and Derigs	1980	15
Bazaraa and Kirca	1983	15
Roucairol	1987	12
Pardalos and Crouse	1988	15
Mautor and Roucairol	1992	19
Clausen and Perregaard	1994	20
Bruenegger et al.	1996	22

[11] The Nugent et al. test instances which are considered as very stubborn QAPs, are the obvious challenge for every new algorithm designed for solving QAP to optimality.

always larger QAP instances with state of the art algorithms is probably mainly due to hardware improvements. However, recent results confirm the effectiveness of the appropriate combination of the available algorithmic ideas with the most suitable hardware.

2.3 CUTTING PLANE METHODS

Among cutting plane methods one has to distinguish between traditional approaches and the so-called polyhedral cutting planes which are used in complex branch and cut algorithms. Several traditional cutting plane type algorithms have been developed for solving QAPs, see for example Kaufman and Broeckx [138], Bazaraa and Sherali [18, 19] and Balas and Mazzola [13]. As for polyhedral cutting planes, the experience of applying such methods to the QAP is still very limited. This is due to a lack of deep understanding of the QAP polytope and its structural combinatorial properties. Only recently, the first nice results in this direction have been obtained.

2.3.1 Traditional cutting planes

This type of algorithms makes use of mixed integer linear programming (MILP) formulations for the QAP which lend themselves very well to Benders' decomposition, and either solve the QAP to optimality or compute a lower bound. In the vein of Benders, the MILP formulation is divided into a master problem and a subproblem, often called also *slave problem*. The master problem contains the original assignment variables (these are the x variables in our MILP formulations in Chapter 1), and the slave problem contains the additional variables introduced to linearize the objective function. For a (fixed) solution of the initial problem, the slave problem is a linear program and hence, solvable in polynomial time. This problem is solved and the optimal values for its dual variables are determined. The master problem is a linear program formulated in terms of the original assignment variables and of the dual variables of the slave problem. Hence, the master problem is solvable in polynomial time for fixed values of the dual variables. The algorithms work typically as follows. First, a heuristic is used to make a feasible choice for the assignment variables. (The QAP has always a feasible solution and it is trivial to find one.) Then the slave problem is solved for this fixed choice of the assignment variables in order to compute the corresponding optimal primal and dual variables. If the dual solution of the slave problem satisfies all constraints of the master problem, we have an optimal solution for the original MILP formulation of the QAP.

Otherwise, at least one of the constraints of the master problem is violated. In this case, the master problem is solved with fixed values for the dual variables of the slave problem, the obtained solution (in terms of the assignment variables) is given to the slave problem, and the procedure is repeated until the solution of the slave problem fulfills all constraints of the master problem.

Any solution of the master problem obtained by fixing the dual variables of the slave problem to some feasible values, is a lower bound for the objective function of the considered QAP. On the other side, the objective function value of the QAP corresponding to any setting of the assignment variables is un upper bound. The algorithm terminates when the lower and the upper bounds coincide. From a computational point of view these methods perform badly even for problems of moderate size, since the time needed for the upper and the lower bounds to converge to a common value is too large. However, heuristics derived from cutting plane approaches produce good suboptimal solutions in early stages of the search. Consider, for example, the cutting plane type heuristics proposed in [29, 19]. Such approaches can be also adapted so as to compute a lower bound for the QAP, by combining them with a heuristic method which produce a relatively good solution, i.e. upper bound. The algorithm terminates when the gap between the lower bound produced by cutting plane algorithm, and the upper bound produced by the heuristic, is less than some prespecified limit.

Such an approach was applied by Kaufman and Broeckx to the MILP formulation (1.12) presented in Section 1.4. For a set of variables (x_{ik}), $1 \le i, k \le n$, which fulfill the assignment constraints (1.5), the following slave problem is solved:

$$\begin{array}{ll} \text{minimize} & \sum_{i=1}^{n} \sum_{k=1}^{n} y_{ik} \\ \text{subject to} & \\ & y_{ik} \geq d_{ik}(x_{ik} - 1) + \sum_{j=1}^{n} \sum_{l=1}^{n} a_{ij} b_{kl} x_{jl} \quad 1 \leq i, k \leq n \\ & y_{ik} \geq 0 \quad 1 \leq i, k \leq n \end{array}$$

Consider the dual of the slave problem, and denote by P the set of indices for the vertices of its feasible region. Let these vertices be given by the vectors $(u_{ik}^{(p)})$, $p \in P$. In terms of the extremal values of the dual variables $u_{ik}^{(p)}$, $p \in P$, of the slave problem, the master problem is formulated as follows:

$$\begin{array}{ll} \text{minimize} & z \\ \text{subject to} & \end{array}$$

$$z \geq \sum_{i=1}^{n} \sum_{k=1}^{n} u_{ik}^{(p)} \left(a_{ij} b_{kl} x_{ik} - d_{ik}\right) \quad p \in P$$

$$\sum_{i=1}^{n} x_{ik} = 1 \quad 1 \leq k \leq n$$

$$\sum_{k=1}^{n} x_{ik} = 1 \quad 1 \leq i \leq n$$

$$x_{ik} \in \{0,1\} \quad 1 \leq i,k \leq n$$

$$z \geq 0$$

Bazaraa and Sherali [18] use the following MILP formulation for the QAP, involving n^2 integer and $n^2(n-1)^2/2$ continuous variables and $2n^2$ linear constraints.

$$\text{minimize} \quad \sum_{i=1}^{n-1} \sum_{k=1}^{n} \sum_{j=i+1}^{n} \sum_{\substack{l=1 \\ l \neq k}}^{n} a_{ij} b_{kl} y_{ikjl}$$

subject to

$$\sum_{j=i+1}^{n} \sum_{\substack{l=1 \\ l \neq k}}^{n} y_{ikjl} - (n-i) x_{ik} = 0 \qquad \begin{array}{c} 1 \leq i \leq n-1 \\ 1 \leq k \leq n \end{array} \tag{2.16}$$

$$\sum_{i=1}^{j-1} \sum_{\substack{k=1 \\ k \neq l}}^{n} y_{ikjl} - (j-1) x_{jl} = 0 \qquad \begin{array}{c} 2 \leq j \leq n \\ 1 \leq l \leq n \end{array} \tag{2.17}$$

$$\sum_{i=1}^{n} x_{ik} = 1 \quad quad \; 1 \leq k \leq n$$

$$\sum_{k=1}^{n} x_{ik} = 1 \qquad 1 \leq i \leq n$$

$$x_{ij} \in \{0,1\} \qquad 1 \leq i,j \leq n$$

$$0 \leq y_{ikjl} \leq 1 \qquad \begin{array}{ll} 1 \leq i \leq n-1, & i+1 \leq j \leq n \\ 1 \leq k,l \leq n, & k \neq l \end{array} \tag{2.18}$$

where $y_{ikjl} = x_{ik} x_{jl}$, for $1 \leq i \leq n-1$, $i+1 \leq j \leq n$ and $1 \leq k,l \leq n$, $k \neq l$. For a fixed set of variables (x_{ij}) which fulfill the assignment constraints (1.5), the slave problem is given as follows:

$$\min \left\{ \sum_{i=1}^{n-1} \sum_{j=i+1}^{n} \sum_{k=1}^{n} \sum_{\substack{l=1 \\ l \neq k}}^{n} a_{ij} b_{kl} y_{ikjl} \colon y \text{ fulfills (2.16), (2.17), (2.18)} \right\}$$

The master problem, formulated in terms of the assignment variables (x_{ij}) and the vertices $\left(u_{ik}^{(p)}, v_{jl}^{(p)}, w_{ikjl}^{(p)}\right)$, $p \in P$, of the dual feasible region for the slave problem, where P is defined as before, is given as follows:

$$
\begin{array}{lrcl}
\text{minimize} & z & & \\
\text{subject to} & & & \\
& z & \geq & \sum\limits_{i=1}^{n-1} \sum\limits_{k=1}^{n} u_{ik}^{(p)}(n-i)x_{ik} + \\
& & & \sum\limits_{j=2}^{n} \sum\limits_{l=1}^{n} v_{jl}^{(p)}(j-1)x_{jl} - w^{(p)} \quad p \in P \\
& (x_{ij}) & \text{fulfill} & (1.5)
\end{array}
$$

where $w^{(p)} = \sum\limits_{i=1}^{n-1} \sum\limits_{j=i+1}^{n} \sum\limits_{k=1}^{n} \sum\limits_{\substack{l=1 \\ l \neq k}}^{n} w_{ikjl}^{(p)}$. According to the numerical results reported in [18], the developed cutting plane method produces relatively good bounds for several QAP instances, whereas it fails on proving optimality even for the Nugent QAP instance of size 7 (see [37], for a detailed information on QAP test instances).

A quite different approach is proposed by Balas and Mazzola [9, 10, 11]. In [9] the authors give the following formulation for the QAP with positive coefficient products $a_{ij}b_{kl} \geq 0$.

$$
\begin{array}{lrcl}
\text{minimize} & z & & \\
\text{subject to} & & & \\
& z & \geq & \sum\limits_{j,l=1}^{n} \left(f_{jl}y_{jl} + \sum\limits_{i,k=1}^{n} a_{ij}b_{kl}y_{ik} \right) x_{jl} - \\
& & & \sum\limits_{j,k=1}^{n} f_{jl}y_{jl} \\
& (x_{ik}), \; (y_{ik}) & & \text{fulfill } (1.5)
\end{array}
\qquad (2.19)
$$

where $f_{jl} = \max\{\sum_{j,l=1}^{n} a_{ij}b_{kl}x_{ik} \colon (x_{ij}) \text{ fulfill } (1.5)\}$. In the later papers [10, 11] the authors presented a cutting plane algorithm for general nonlinear 0-1 programming problems with linear objective function. Clearly, this algorithm can also be applied to the above formulation of the QAP (2.19). In this algorithm, for every nonlinear constraint an equivalent family of linear constraints is defined. This family contains among others the so-called *cover inequalities*. A linear programming relaxation of the given nonlinear program is build up by considering a number of these linear constraints. This relaxation is solved to optimality. If the optimal solution of the relaxation is feasible for the original problem, this is also an optimal solution of it. Otherwise, the solution of the linear relaxation violates some constraint of the original problem. Some of the corresponding linear inequalities are added then to build a "stronger" linear relaxation. The latter is solved, and the procedure is repeated until the optimal solution of the linear relaxation is also a feasible solution for the nonlinear problem.

Based on the linearization approach of Balas et al., Burkard and Bönniger [29] derive a heuristic for general 0-1 quadratic programming problems. The heuristic is an iterative approach which solves a linear program with only one constraint in each iteration. This constraint is associated with a feasible solution of the original problem, and gives a certain measure of its quality. The feasible solution, and with it the corresponding constraint of the linear program, is updated in each iteration. The update is based on an empirical rule which aims to improving the objective function value corresponding to the current feasible solution. The procedures stops after a prespecified number of iterations, and gives as output the best feasible solution found so far. This heuristic was applied to the MILP formulation (2.19) for the QAP, and turned out to be competitive with other heuristic approaches known at that time.

2.3.2 Polyhedral cutting planes

As we saw in the first chapter, only a few properties in general, and few facet defining inequalities of the QAP polytope in particular, are known yet. Hence, polyhedral cutting plane methods for the QAP are not yet backed by a strong theory. As a matter of fact, the author is aware of a single effort to design such an algorithm for the QAP, described in [177] and tested on small test instances from QAPLIB [37]. However, since a lot of promising results on the QAP polytope have been obtained recently [128, 129, 177] and a lot of research work is going on in this direction, we find it useful to shortly describe here generic polyhedral cutting plane approaches, or *branch and cut algorithms* [12], in contrast with traditional cutting plane algorithms.

Both traditional cutting plane approaches and branch and cut algorithms are general methods which make use of an LP or MILP relaxation of the combinatorial optimization problem to be solved, in our case the QAP. Additionally, polyhedral cutting plane methods make use of a class of (nontrivial) valid inequalities known to be fulfilled by all feasible solutions of the original problem and which, in the most favorable case, are facet defining for the polytope of the original problem.

[12]This term was originally used by Padberg and Rinaldi [179].

If the solution of the relaxation is feasible for the original problem, we are done. Otherwise, some of the above mentioned *valid* inequalities are probably violated. In this case a "cut" is performed, that is, one or more of the violated inequalities are added to the LP or MILP relaxation of our problem. The latter is solved again and the whole process is repeated. In the case that none of the valid inequalities is violated, but some integrality constraint is violated, the algorithm performs a branching step by fixing (feasible) integer values for the corresponding variable. The branching steps produce the so-called *search tree* like in branch and bound algorithms. Each node of this tree is processed as described above by performing cuts and then by branching it, if necessary. Other elements of a branch and bound algorithm such as fathoming and upper bounds (in the case of minimization problems) are also maintained, as well as questions like "Which node is to be processed next?" or "Which variable should be fixed next"? Hence, such an approach really combines elements of cutting plane and branch and bound methods and the term "branch and cut" is fully justified.

The main difference between polyhedral and traditional cutting planes is that whereas polyhedral cutting planes use cuts generated by inequalities which are valid for the whole polytope of the feasible solutions and possibly facet defining, traditional cutting planes frequently rely on cuts generated by inequalities which are not valid for the whole polytope of the feasible solutions, let alone being facet defining. For example, if Gomory cuts are used, the cuts are not valid for the whole polytope after some variable fixing. This feature causes one of the main drawbacks of traditional cutting plane algorithms versus polyhedral methods. If the cuts are not valid for the whole polytope, the whole computation is to be done from scratch for different variable fixings. This means that the cuts should be stored and treated separately for different nodes of the search tree. This requires additional running time and additional amounts of memory. The second and not less important drawback of traditional cutting plane algorithms is their slow convergence. Mathematically, this is due to the "weakness" of the cuts generated by these algorithms. In contrast with cuts defined by the facet defining inequalities, the weak cuts generated by traditional algorithms are not able to *quickly* direct the search towards interesting regions of the polytope of the feasible solutions.

It is beyond the scope of this monograph to give a detailed description of cutting plane methods, either traditional or polyhedral, and to go through the numerous questions which arise while designing such an algorithm for a given problem. For more information on these topics the reader is referred among others to [120, 173, 178, 179]. We would like to conclude this section by mentioning the four main ingredients of a polyhedral cutting plane algorithm, in the vein of Padberg and Rinaldi [178]: a heuristic procedure to produce a good initial solution, and hence good upper bounds, a separation algorithm which identifies violated valid (facet defining) inequalities and adds them to the relaxation of the problem, a good LP solver interface, and a branch and bound procedure which will appropriately answer questions like "Which node to process next?", "Which variable to fix next?". A first effort to combine these ingredients in an algorithm for the QAP has been made by Padberg and Rijal in [177]. The algorithm was tested on sparse QAPg instances from QAPLIB and the numerical results are encouraging, although the developed software is of preliminary nature, as claimed by the authors. The size of the linear programs to be solved within the branch and cut solver seem to be reasonable in comparison with the overall number of variables and constraints. Concluding, it is certainly worthwhile spending more efforts in developing the appropriate theoretical background for a better suited branch and cut algorithm for the QAP.

3

HEURISTICS AND ASYMPTOTIC BEHAVIOR

We saw in the previous section that only small QAP instances (instances of size up to 22) can be solved to optimality. On the other side, the large spectrum of applications of the QAP often leads to instances of much larger size. Under these conditions, polynomial time heuristics providing suboptimal solutions to the QAP abound in the literature. Without pretending to mention all kinds of heuristic approaches which have been applied to QAPs, we try to review the main and most fruitful ideas in this research direction. There are five main streams of heuristic approaches to QAP, listed in chronological order below.

1. Construction methods
2. Limited enumeration methods
3. Improvement methods
4. Tabu search algorithms
5. Simulated annealing approaches
6. Genetic algorithms
7. Greedy randomized search

In the second part of this chapter we present some basic results on the asymptotic behavior of the QAP. More specifically it is shown that, under natural probabilistic conditions on the entries of the coefficient matrices, the ratio between the best and the worst value of the objective function of the QAP approaches almost surely 1, as the size of the problem tends to infinity. Hence, the QAP - which is widely considered as a very difficult problem from a theoretical and from a practical point of view - becomes somehow trivial when its size approaches infinity. There is a whole class of combinatorial optimization problems which show such an intriguing asymptotic behavior. All these problems share a combinatorial property identified by Burkard and Fincke [36]. The QAP is a notoriously hard problem which belongs to this class, whereas the TSP is a famous combinatorial optimization problem which does not belong to it.

3.1 CONSTRUCTION AND LIMITED ENUMERATION METHODS

Construction methods. Construction methods are considered to be the simplest heuristic approaches to the QAP, from a conceptual and an implementation point of view. This simplicity is often associated with a poor quality of the resulting solutions. Basically, all construction methods start with an empty partial solution (permutation) and recursively assign locations to facilities according to certain criteria until all facilities have been assigned. These methods are probably the oldest ones, dating back to the early 60s with a heuristic proposed by Gilmore [92]. A refined version of the simple construction method introduced by Gilmore is suggested by Burkard [28]. Another construction method which yields relatively good results in comparison with other construction methods is proposed by Müller-Merbach [172]. This is the so-called method of *the increasing degree of freedom*. It works with an iteratively updated *partial permutation* and completes it into a permutation of $\{1, 2, \ldots, n\}$. A *partial permutation* of $\{1, 2, \ldots, n\}$ is an injective mapping π of a subset $X \subset \{1, 2 \ldots, n\}$ into $\{1, 2, \ldots, n\}$, $\pi: X \to \{1, 2, \ldots, n\}$, where $X \neq \{1, 2, \ldots, n\}$. The approach of Müller-Merbach starts with the *empty* partial permutation, i.e. set X is the empty set, and with a fixed order of the indices $1, 2, \ldots, n$, say $r_1, r_2, \ldots, r_n$, where n is the size of the considered QAP. For some $M \subseteq \{1, 2, \ldots, n\}$ let $\pi(M) = \{\pi(i): i \in M\}$. Let $\pi: M_\pi \to \{1, 2, \ldots, n\}$ be the current partial permutation, where $M_\pi = \{r_1, r_2, \ldots, r_{k-1}\}$. Construct a new permutation $\pi': M_{\pi'} \to \{1, 2, \ldots, n\}$ with $M_{\pi'} = M_\pi \cup \{r_k\}$, where r_k is the first unassigned index in $\{r_1, r_2, \ldots, r_n\}$. Permutation π' is constructed as follows. First, assign r_k to some $j \notin \pi(M_\pi)$, and compute the corresponding increase of the value of the objective function ΔZ_j. Then consider the assignment of r_k to an index $j \in \pi(M_\pi)$. Let $r_i \in \{r_1, r_2, \ldots, r_{k-1}\}$ such that $\pi(r_i) = j$. Denote by ΔZ_{jl} the change in the objective function which results from assigning r_k to j and r_i to l, for some $l \notin \pi(M_\pi)$. Among these $k(n - k + 1)$ possibilities for constructing π', choose the one which yields the smallest increase or change in the value of the objective function. Repeat this procedure until each facilities r_i, $1 \leq i \leq n$, is assigned to some location in $\{1, 2, \ldots, n\}$.

As mentioned above, the quality of the solutions produced by construction methods is generally not satisfactory. However, these approaches are generally simple to implement and exhibit modest requirements on computing resources. Due to these properties, construction methods may well be used as parts of more clever algorithms for the QAP.

Limited enumeration methods. Limited enumeration methods are strongly related to exact methods such as branch and bound approaches or cutting plane algorithms. Limited enumeration methods based on branch and bound algorithms basically rely on the fact that a good solution, i.e. a solution which is either optimal or close to the optimum, is often found at an early stage of the search. Then, a relatively large amount of time is spent to prove the optimality or to (slightly) improve the quality of this early found solution. Limit enumeration methods try to take advantage from this behavior in order to produce good solutions for the QAP (solutions which are either optimal or close to the optimum) in a reasonable time.

A first approach is to impose so-called *time limits*. We can stop the enumeration process either after a prespecified time limit has been reached or in the case that no improvement has been made during a time interval which is longer than a prespecified one. Obviously, the prespecified parameters need to be chosen according to the problem size.

A second approach is to increase the lower bounds in those nodes of the branch and bound tree which have not been branched yet, once certain criteria are fulfilled. For example, if no improvement has been found after a certain prespecified time interval, then the lower bounds are increased by a certain percentage. Obviously, such an approach may cut off the optimal solution and the optimal value *opt* of the objective function may be never reached, but in any case it speeds up the search. Moreover, the suboptimal solution produced by such an algorithm yields a value of the objective function which is smaller than or equal to all values Z of the objective function with $Z \geq opt + \Delta$. Here Δ is the difference between the largest among the increased bounds and the smallest lower bound before the increase, where the maximum and the minimum are taken over all nodes of the branch and bound tree which have not been branched yet.

Methods based on statistical evaluations of the value of the objective function can be also classified as limited enumeration methods. For instance, Graves and Whinston [102] calculate the expectation and the variance of the objective function value of solutions obtained when completing a given partial permutation. West [228] uses these ideas for constructing the starting solution of an improvement method for the QAP (see below).

Limited enumeration has been also applied to speed up cutting plane algorithms. Heuristics of this type have been developed, for example, by Bazaraa and Sherali [18] and Burkard and Bönniger [29]. As described in Section 2.3 Bazaraa and Sherali [18] formulate the QAP as a mixed integer linear program (MILP) and apply Benders' decomposition [20] for solving it. One of the resulting problems (the *slave* problem) is a transportation problem and can be solved to optimality in polynomial time. The other problem (the *master* problem) is a 0-1 MILP and is solved by a cutting plane method. A heuristic approach is derived by applying a limited number of cuts for solving the master problem. Although this approach is quite sensitive as regards the initial solution, it yields suboptimal solutions of good quality. Rather than computation time requirements, large memory requirements are the most significant drawback of this approach. Such large memory requirements cause sometimes the failure of the exact algorithm, i.e. the approach without time or cut limits. Burkard and Bönniger [29] make use of a 0-1 linear programming formulation of the QAP proposed by Balas and Mazzola [9]. The corresponding linear program (LP) has an exponential number of constraints, where each feasible solution of the QAP leads to one constraint of this LP. Burkard and Bönniger propose a heuristic to solve this LP. The proposed heuristic is an iterative approach which starts with an initial feasible solution and solves a relaxation of the LP formulation of the QAP in each iteration. In each iteration, the relaxation to be solved is obtained from the LP formulation by considering only one constraint, namely, the constraint which correspond to the current feasible solution. The other constraints are currently discarded. The optimal value of the current relaxation is used to update the current feasible solution according to a prespecified rule. (The goal is to determine an update rule which leads to a small optimal value of the LP relaxation to be solved in the subsequent iteration.) The computation of the new feasible solution involves the solution of a linear assignment problem (LAP). In the subsequent iteration a new relaxation of the LP formulation - determined by the new feasible solution - is solved, and the whole procedure is iteratively repeated. Usually the stop criterion is a maximum number of iterations. In contrast to the approach of Bazaraa and Sherali described in [18] this method seems to be quite stable in the sense that the quality of the produced suboptimal solution only marginally depends on the starting solution. From a point of view of the solution quality both approaches, the one of Bazaraa and Sherali [18] and the one of Burkard and Bönniger [29], are comparable, whereas the latter is advantageous with respect to memory and running time requirements.

3.2 IMPROVEMENT METHODS

Improvement methods are already classical approaches to solve difficult combinatorial optimization problems. These methods belong to the larger class of *local search* algorithms.

As described in the first chapter, a local search procedure starts with an initial feasible solution. It is an iterative approach which tries to improve the current solution by substituting it with a (better) feasible solution from the neighborhood of the current one. This iterative step is repeated until no further improvement can be found. Clearly, the definition of the neighborhood is a crucial point for this type of heuristics. Frequently used neighborhoods for QAPs are the *pair-exchange* neighborhood and the *cyclic triple-exchange* neighborhood. In the case of pair-exchanges the neighborhood of a given solution (permutation) consists of all permutations which can be obtained from the given one by applying a transposition to it. In this case, scanning the whole neighborhood takes $O(n^3)$ time, as the size of the neighborhood itself is $\binom{n}{2}$ and the objective function value $Z(A, B, \pi')$ for a neighbor π' of π can be calculated in $O(n)$ time, once the value $Z(A, B, \pi)$ is known. If the neighborhood of π is already scanned and π' is a neighbor of π, then the neighborhood of π' can be scanned in $O(n^2)$. This is a simple but important result of Frieze et al. [84] which can be used to save computation time in the case that a complete neighborhood evaluation is required.

In the case of cyclic triple-exchanges, the neighborhood of a solution (permutation) π consists of all permutations obtained from π by a cyclic exchange of three indices. For example, if the triple of indices to be exchanged is (i, j, k), we obtain two new permutations π' and π'' defined as follows:

$$\pi'(x) = \begin{cases} \pi(x) & x \notin \{i, j, k\} \\ \pi(k) & x = i \\ \pi(i) & x = j \\ \pi(j) & x = k \end{cases} \qquad \pi''(x) = \begin{cases} \pi(x) & x \notin \{i, j, k\} \\ \pi(j) & x = i \\ \pi(k) & x = j \\ \pi(i) & x = k \end{cases}$$

The size of this neighborhood is $2\binom{n}{3}$. This increase of the neighborhood size slows down the corresponding local search approaches. Moreover, cyclic triple-exchanges do not really lead to considerably better results when compared with pair-exchanges.

Some efforts have been invested on combining pair-exchange and cyclic triple-exchange neighborhoods. As an example, consider the results obtained by Mirchandani and Obata [168], where, along with all pair-exchanges, some cyclic triple-exchanges and cyclic quadruple-exchanges are evaluated. The important point here is that the size of the neighborhood remains $O(n^2)$.

Another important ingredient of local search algorithms and improvement methods is the order in which the neighborhood of the current solution is scanned. The scanning can be done either in a previously fixed order or in a randomly chosen order. For a fixed neighborhood structure and a fixed scanning order, there still remains one degree of freedom in the determination of a local search procedure. Namely, different updating rules can be used to update the current feasible solution (from an iteration to the subsequent one). In the case of pair-exchange neighborhood the following updating rules are frequently used:

- Method of first improvement
- Method of best improvement
- Heider's method [119]

The method of *first improvement* updates the current solution as soon as the first improving neighbor solution is found. The method of *best improvement* scans the whole neighborhood and chooses the best improving neighbor solution if there exists an improving neighbor solution at all. In other words an improving neighbor solution is chosen which yields the smallest value of the objective function among all improving neighbor solutions. *Heider's method* starts by scanning the neighborhood of the initial solution in a prespecified *cyclic* order. Namely, the order of the transpositions to be applied to the current solution for generating its neighbors is previously fixed. The transpositions are ordered cyclically, i.e.
the first transposition is the successor of the last one. The current solution is updated as soon as an improving neighbor solution is found. The neighbors of the new solution are generated by applying to it the transpositions in the prespecified order, starting with the successor of the last transposition which is applied to the previously current solution.

In order to get better results, local search algorithms are performed several times starting with different initial solutions. They may be easily combined with construction algorithms. A construction algorithm is run first in order to generate a reasonably good initial solution. However, there are no widely accepted strategies concerning the choice or the construction of an initial solution. The computational experiments of Bruijs [25] suggest that, in general, there is no strong argument in favor of initial solutions of good quality.

3.3 TABU SEARCH ALGORITHMS

The computational experience of many researches and practitioners suggests that tabu search is a useful heuristic for solving hard combinatorial optimization problems in general and the QAP in particular. A detailed description of specific aspects of tabu search and its applications to various optimization problems is given in [95]. Tabu search was introduced by Glover [93, 94] as a technique to overcome local optimality in local search approaches applied to combinatorial optimization problems. The main ingredients of tabu search are the *neighborhood structure*, the *moves*, the *tabu list* and the *aspiration criterion*. A *move* is an operation which, when applied to a certain solution π, generates a neighbor π' of it. In the case of QAPs the moves are usually transpositions and the neighborhood is the pair-exchange neighborhood. A *tabu list* is a list of forbidden moves, i.e. moves which are not allowed to be applied to the current solution. Forbidden moves are also called *tabu moves*. Clearly, the tabu list is updated during the search, i.e. the tabu status of the moves changes along with the search. An *aspiration criterion* is a condition which, when fulfilled by a tabu move, cancels its tabu status.

A generic tabu search procedure works as follows. It starts with an initial feasible solution S and selects a *best-quality* solution, i.e. a solution which yields the smallest value of the objective function, among (a part of) the neighbors of S obtained by non-tabu moves. Note that this neighboring solution does not necessarily improve the value of the objective function corresponding to the current solution. Then the current solution is updated, i.e. it is substituted by a solution selected as above, and the search in the neighborhood is repeated. Obviously, this procedure can *cycle*, i.e. visit some solution more than once. In an effort to avoid this phenomenon some identification criteria are introduced for moves which are expected to lead to cycles. Such moves are then added to the tabu list. As, however, forbidding certain moves could prohibit visiting "interesting" solutions, an aspiration criterion is introduced. It serves to distinguish the potentially interesting moves among the forbidden ones. The tabu list has a prespecified length which may vary during the search. In order to maintain the length of the tabu list within the prespecified limits, some tabu moves have to be canceled from the list. Usually, a first-in first-out (FIFO) rule is used to keep the length of the tabu list within the prespecified limits. The search procedure stops when a stop criterion is fulfilled. The stop criterion is often limit on the running time or a limit on the number of iterations.

Below we give a more formal description of a generic tabu search algorithm with fixed size of the tabu list, for a general combinatorial optimization problem as described in Section 1.5.2. The input consists of an initial feasible solution S_0 and two *control parameters*: the prespecified maximum length of the tabu list denoted by *size*, and a parameter r which determines the number of the neighbors of the current solution which have to be scanned before updating the latter. The variables used by the algorithms are: the current length of the tabu list denoted by *length*, the best value of the objective function found so far denoted by Z_{best}, the best value of the objective function among values which are yielded by the neighbors of the current solution scanned so far denoted by *loc_best*, the neighbor solution yielding *loc_best* denoted by *neigh_best*, a counter for the neighbors of the current solution scanned so far denoted by *loc_count*, and finally, the counter of iterations denoted by i. The algorithm involves also a parameter M, which is an arbitrary positive constant. The routines used by the algorithm are described below. For a feasible solution S, $Z(S)$ returns the corresponding value of the objective function, that is $Z(S) = \sum_{x \in S} f(x)$. For two feasible solution S, S', $\Delta(S, S')$ denotes the change in the value of the objective function when moving from S to S', i.e. $\Delta(S, S') = Z(S') - Z(S)$. $move(S)$ is a function which returns a neighboring solution of S.

Tabu Search($\mathbf{S_0}$, size, r)

```
S := S_0; /* generate some initial solution in S_0 ∈ F */
S_best := S_0; /* Initialize the best known solution */
Z_best := Z(S_0); /* Initialize the best known obj. func. value */
length := 0; /* Initialize the length of the tabu list */
i := 0; /* Initialize the iterations counter */
repeat
    loc_best := -M; /* initialize the best neighbor */
    loc_count := 0; /* initialize the local counter */
    repeat
        S' := move(S); /* find a neighbor S' of S, e.g. randomly */
        loc_count := loc_count + 1;
        if the move S -> S' is non-tabu and loc_best > Δ(S, S') then
            loc_best := Δ(S, S');
            neigh_best := S' /* update the best neighbor */
        endif
        if S -> S' is tabu and fulfills the aspir. criterion then
            if Δ(S, S') < loc_best then
                loc_best := Δ(S, S');
                neigh_best := S' /* update the best neighbor */
            endif
        endif
```

```
        update the tabu list;
    until loc_count = r
    S := neigh_best; /* update the current solution */
    if Z(S) < Z_best then
        S_best := S; /* update the best known solution */
        Z_best := Z(S); /* update the best known obj. func. val. */
    endif
    i := i + 1; /* Update the iteration counter */
until a stop criterion is fulfilled
return S_best, Z_best;
```

One of the first tabu search approaches for QAPs was proposed by Skorin-Kapov [213]. She used a tabu list of fixed size, where the size is a *control parameter*, a parameter belonging to the input, whose value influences the performance of the algorithm. The appropriate value of this parameter depends strongly on the problem instance. The mathematical nature of this dependence is unknown, and consequently it is quite difficult to tune this parameter. In [213], a move is declared as tabu if the transposition defining it has been applied for updating the current solution during the last ℓ iterations, where ℓ is another control parameter. The procedure stops after a fixed number of iterations have been performed, this number being the third control parameter of the procedure. An important drawback of this algorithm is the difficulty of tuning its control parameters. The dependence of suitable values of the control parameters on the problem instance is strong, but not clear. This leads to a bad behavior of the algorithm in terms of robustness.

In an effort to overcome this difficulty Taillard [218] proposes a new version of tabu search for the QAP, the so-called *robust tabu search*. His version differs from the version of Skorin-Kapov in two main points. First, a move is declared tabu if it locates both interchanged facilities to locations they had occupied within the s most recent iterations, where s is the length of the tabu list. Secondly, the length of the tabu list is frequently changed by choosing its value at random between a minimum and a maximum value. The minimum and the maximum value for the length of the tabu list are prespecified as control parameters. The following choice of the control parameters produces quite good solutions: $O(n^2)$ iterations and length of the tabu list between $\lfloor 0.9n \rfloor$ and $\lceil 1.1n \rceil$, where n is the size of the problem. This algorithm is more robust than the previous one, in the sense that its performance is less sensitive as regards the values of the control parameters.

Recently, another version of tabu search, the so-called *reactive tabu search (RTS)*, was proposed by Battiti and Tecchiolli [15]. RTS produces quite good results when applied to QAPs. Reactive tabu search aims at weakening the dependence of the performance of tabu search on the prespecified values of the control parameters. It involves a simple mechanism for adapting the length of the tabu list according to the properties of the considered instance of the problem. RTS notices when a cycle occurs, i.e. when a certain solution is revisited, and increases the tabu list size according to the length of the detected cycle. Once in a while the size of the tabu list is reduced in order to keep it within reasonable limits. If, moreover, the number of the solutions which are revisited a "large" number of times (this large number of times is a control parameter) exceeds a certain parameter called CHAOS (another control parameter), a random diversification of the search towards new feasible solutions is forced. Numerical results on test problems from the literature show that in most of the cases the reactive tabu search converges to the best known solution faster than any other known tabu search scheme.

The reader is referred to Taillard [219] for a comparison of different tabu search schemes for the QAP. The author proposes a classification of the most studied QAP instances. For each class of problems he points out the algorithm which shows the best performance when applied to problems of this class. More recently, also parallel implementations of tabu search have been proposed, see e.g. Chakrapani and Skorin-Kapov [48]. Tabu search algorithms allow a natural parallel implementation by dividing the burden of the search in the neighborhood among several processors.

3.4 SIMULATES ANNEALING APPROACHES

Simulated annealing approaches are another group of heuristic methods which try to overcome local optimality in solving hard combinatorial optimization problems. From a technical point of view, simulated annealing approaches inherit the structure and some properties of local search algorithms. From a theoretical point of view, simulated annealing approaches are based on the interesting analogy between combinatorial optimization problems and problems from statistical mechanics . Kirkpatrick, Gelatt and Vecchi [140] were among the first authors who recognized and exploited the similarities between these two fields. They showed how the Metropolis algorithm [167] used to simulate the behavior of a physical many-particle system can be naturally applied as a heuristic method in optimization. In an attempt to apply these ideas to the traveling salesman problem (TSP), they introduced the so-called *simulated annealing*. The same method for the TSP was developed independently by Černý [47]. Burkard and Rendl [40] showed that simulated annealing is a general approach which can be applied to each combinatorial optimization problem as soon as a neighborhood structure has been introduced on the set of the feasible solutions. The analogy between a combinatorial optimization problem and a many-particle physical system basically relies on the following two facts:

- Feasible solutions of the combinatorial optimization problem correspond to states of the physical system.
- The value of the objective function value for a feasible solution of the combinatorial optimization problem corresponds to the energy of the a state of the physical system.

In *condensed matter physics annealing* is known as a thermal process for obtaining lower energy states of a solid in a heat bath. The process runs through two phases. First, one increases the temperature of the heat bath to a maximum value at which the solid melts. Secondly, one decreases *carefully (slowly)* the temperature of the heat bath until the particles arrange themselves in the *ground state* of the solid. Notice that the ground state is characterized by a minimum of energy. The Metropolis algorithm which simulates the evolution of a solid in a heat bath in *thermal equilibrium* is based on Monte Carlo techniques for generating a sequence of states of the solid. Let i be the current state with energy E_i. A subsequent state j, with energy E_j, is generated by applying a small perturbation to the current state, say displacing one of the particles. If the energy difference $E_j - E_i$ is negative, then state j is accepted as the next current state. Otherwise, j is accepted with a certain probability given by $\exp(\frac{E_i - E_j}{k_B t})$, where t denotes the temperature and k_B is the so-called *Boltzmann constant*. If the temperature is decreased sufficiently slowly, the solid can reach the thermal equilibrium at each temperature. In the Metropolis algorithm this is achieved by generating a large number of states at a given temperature value. The thermal equilibrium is characterized by the so-called *Boltzmann distribution* which gives the probability of the solid being at a state i with energy E_i at temperature t:

$$P(\{X = i\}) = \frac{1}{Q(t)} \exp\left(\frac{-E_i}{k_B t}\right) ,$$

where X is a random variable denoting the current state of the solid. $Q(t)$ is the so-called *partition function* defined by

$$Q(t) = \sum_i exp(\frac{-E_i}{k_B t}) ,$$

where the summation extends over all possible states. The Boltzmann distribution plays an essential role in the theoretical analysis of the convergence of simulated annealing-based algorithms.

Like tabu search, simulated annealing belongs to the group of the so-called meta-heuristics, as it can be applied to any combinatorial optimization problem with a neighborhood structure specified on its set of feasible solutions. Below we give a more formal description of a generic simulated annealing algorithm for a general combinatorial optimization problem described in Section 1.5.2. The input consists of an initial solution S_0, the initial value t_0 of a control parameter t which is called temperature (for the sake of the analogy with the temperature of a physical system), and a function g which formalizes the rule for the update of the temperature during the annealing process. The variables involved in this algorithm are the following: the current feasible solution denoted by S, a feasible solution S' which is a neighbor of S, the best feasible solution known so far denoted S_{best}, the value of the objective function yielded by S_{best} denoted by Z_{best}, an iteration counter i, the value of the temperature at the i-th iteration denoted by t_i, and a decision variable *accept* which takes the values "yes" or "no" in the case that the current solution should be updated or not. The routines used by the algorithm are the following: $move(S)$, $\Delta(S, S')$, $Z(S)$ which are already introduced as routines of the generic tabu search algorithm, the function g which returns the value t_i of the temperature in the current iteration i and takes as argument the value t_{i-1} of the temperature in the previous iteration, and $Rand(0,1)$ which generates a random number between 0 and 1. The number generated by $Rand(0,1)$ serves to decide whether the current feasible solution will be updated, i.e. substituted by its most recently generated neighbor or not, in the case that this neighbor solution does not lead to an improvement of the value of the objective function. The sequence of values t_i, $0 \leq i$ is often referred to as a *temperature schedule*. Usual theoretical conditions to be fulfilled by the temperature schedule are the inequality $t_i < t_{i-1}$, for all $i \geq i$, and $\lim_{i \to \infty} t_i = 0$.

Simulated Annealing($\mathbf{S_0}$, $\mathbf{t_0}$, g)

> $S := S_0$; /* generate some initial solution in $S_0 \in \mathcal{F}$ */
> $S_{\text{best}} := S_0$; /* Initialize the best known solution */
> $Z_{\text{best}} = Z(S_0)$; /* Initialize the best known obj. func. value */
> $i := 0$; /* Initialize the iterations counter */
> $t_i := t_0$; /* Initialize the initial temperature t_0 */
> *repeat*
>   *repeat*
> $S' := move(S)$;
>   *if* $\Delta(S, S') < 0$ *then*
>   *accept* $:=$ *yes*;
>   *if* $Z(S') < Z_{\text{best}}$ *then*
> $S_{\text{best}} := S'$; $Z_{\text{best}} := Z(S')$;

```
            endif
        endif
        if Δ(S,S') ≥ 0 and exp(−Δ(S,S')/(t_i Q(t_i))) > Rand(0,1) then
            accept := yes;
            if accept = yes then
                S := S';
            endif
        endif
    until equilibrium is approached sufficiently closely;
    i := i + 1; /* Update the iteration counter */
    t_i := g(t_{i-1}); /* Update the current temperature */
until a stop criterion is fulfilled; /* "the system is frozen" */
return X_best, Z_best;
```

Different authors formalize in different ways the fact that "equilibrium is approached sufficiently closely". The simplest way is to run the corresponding "repeat - until" loop a prespecified (large) number of times, for each fixed value t_i of the temperature. A more complicated way is to run the corresponding "repeat-until" loop until the average of the values of the objective function yielded by the accepted solutions stabilizes in some sense, e.g. it changes only marginally during the last k iterations. Here k is a control parameter and the marginal changes are formalized by another control parameter.

The generic simulated annealing algorithm presented above can be modeled mathematically by an inhomogeneous ergodic Markov chain. (For an introduction to the theory of Markov chains see e.g. [125].) Its transition probabilities which are probabilities of moving from a feasible solution to some neighboring one depend on the change of the value of the objective function caused by the corresponding move and on the current value of the temperature. Clearly, they depend indirectly on the neighborhood structure as well. The theory of Markov chains has been used for the probabilistic analysis of the convergence of simulated annealing algorithms. Some authors have derived conditions on the neighborhood structure which guarantee the convergence of the generic algorithm to a (globally) optimal solution with probability equal to one, when the number of iterations approaches infinity, see for example [76]. Other authors essentially require a slow enough *cooling*, or more concretely $t_i = \frac{\Gamma}{\log i}$, where $\Gamma \geq d$ and d is a constant depending on the combinatorial structure of the considered problem. For a detailed discussion on different theoretical aspects of simulated annealing methods, the reader is referred to [1, 147]. The investigation of the speed of the above mentioned asymptotic probabilistic convergence and its practical impact on the performance of simulated annealing approaches for certain combinatorial optimization problems, remains an (apparently difficult) open problem.

Simulated annealing has been proven to be a useful approach to optimization problems arising in chip design, wiring, component placement in VLSI design, as well as to traditional combinatorial optimization problems as the traveling salesman problem and the QAP. Its main drawback is the relatively high number of control parameters and the absence of a widely accepted and well argued strategy for choosing their values. The first who derived a simulated annealing scheme for the QAP were Burkard and Rendl [40]. In [40] (and in all simulated annealing versions for the QAP known to the author) the pair-exchange neighborhood structure is used. The computational experiments in [40] confirm the expectations on the behavior of the algorithm: its performance strongly depends on the temperature schedule. However, a careful tuning of the control parameters often yields high quality solutions. A more sophisticated simulated annealing approach for the QAP was proposed by Wilhelm and Ward [229]. The algorithm proposed in [229] uses a so-called "equilibrium state", in an attempt to find a better mathematical model for the state of the "thermal equilibrium". The authors report solutions of quite good quality, but they do not motivate their choices for the control parameters. Connolly [59] tries to analyze the role of the optional components of simulated annealing approaches such as temperature schedule and random or sequential neighborhood search. He introduces a so-called *optimal value of the temperature* that is a temperature state where most of the search should be performed. This idea will be discussed in more details in Section 7.6, where three simulated annealing schemes for a generalization of the QAP, namely the biquadratic assignment problem (BiQAP), are briefly reviewed. Laursen [151] argues on the choice of several control parameters and investigates experimentally the benefit of different strategies for setting their values.

There is no general agreement concerning the comparison of the performance of simulated annealing approaches with that of tabu search approaches for the QAP. Battiti and Tecchiolli [16] have compared a standard simulated annealing scheme with a reactive tabu search algorithm on a small sample of instances from QAPLIB [37]. They conclude that in the long run both methods are competitive with respect to the quality of the produced solutions. When the running time limits increase (the stop criterion is formulated in terms of running time), the reactive tabu search provides in general better solutions, whereas simulated annealing gets more easily stucked in local optima.

3.5 GENETIC ALGORITHMS

The so-called *genetic algorithms*, henceforth abbreviated by GA, are another nature inspired approach to large scale combinatorial optimization problems. The underlying motivation of such algorithms is the attempt to borrow ideas from the selection process in nature which develops complex and well adapted species through relatively simple evolutionary mechanisms. The basic idea is to adapt these simple evolutionary mechanisms to combinatorial optimization problems. The first genetic algorithm for optimization problems was proposed by Holland [121] in 1975. Recently, a lot of research has been done in deriving good GAs for different combinatorial optimization problems. There are (at least) two evident reasons for the intensification of the research work in this direction: the generality of GAs and the simplicity of parallel implementations of GAs.

Basically, a genetic algorithm starts with a set of initial feasible solutions (generated randomly or by using some heuristic) called the *initial population*. The elements of a population are usually termed "individuals" or "members". The algorithm *selects* a number of *pairs of individuals* from the current population and uses so-called *cross-over rules* to produce a new feasible solution out of each pair of individuals. For each selected pair, the individuals are called *parents* and the new solution produced by applying cross-over rules is called a *child*. Further, a number of "bad" solutions is thrown out of the current population. This process is repeated until a *stop criterion* is fulfilled. The stop criterion may often be a time limit, a limit on the number of iterations, or a measure for the convergence of the algorithm, e.g. the current population consists of copies of a *small* number of pairwise different individual where "small" is formalized by the above mentioned measure. During the run of the algorithm, a *mutation* or an *immigration* is applied periodically to the current population, trying to improve its overall quality by modifying some of the individuals or replacing them by better ones, respectively. Very often local optimization tools are used periodically in order to improve the performance of the algorithm, in general, and to speed up its convergence, in particular. For the diversification of the search so-called *tournaments* are used. In principal, a tournament consists of applying several runs of a GA starting from different initial populations and stopping them before they converge. A "better" population is derived as a union of the final populations of these different runs. Then a new run of the GA is started with this union as an initial population. For a good coverage of theoretical and practical issues on genetic algorithms the reader is referred to [62, 97].

Below we present a general scheme of a genetic algorithm without tournaments for a minimization problem. The input of the algorithm consists of an initial population $\mathcal{J}_0$, generated e.g. randomly, and a parameter M which is the number of iterations the algorithm runs through. The variables involved in this algorithm are the following: the current population $\mathcal{J}$, the best individual S_{best} in the current population, i.e. the individual (solution) which yields the smallest value of the objective function among all individuals of the current population, Z_{best} which is the value of the objective function yielded by S_{best}, S_1, S_2, two individuals from the current population $\mathcal{J}$ selected as current parents, S_3 the child of S_1 and S_2, an iteration counter i, and finally, three decision variables *immig*, *mut* and *l_s* which are set to "true" or "false" in the case that immigration, mutation, local search should be performed or not, respectively. The routines used by this algorithm are described below. *Best* gets as input a set of feasible solutions of the considered optimization problem and outputs the best among them (with respect to the value of the objective function). Usually such a best solution is found by explicit or implicit enumeration. As in Sections 3.2, 3.4, $Z(S)$ returns the value of the objective function yielded by a feasible solution S. *Select* selects two parents out of the current population. The selection is made randomly or is biased towards high quality solutions according to some appropriate prespecified rules. *Cross* gets as input the parents S_1 and S_2 and outputs a child S_3 according to a prespecified cross-over operator. *Cleanse* gets as input the current population $\mathcal{J}$ and singles out a number of solutions which are to be deleted from the current population. The goal is that the cardinality of the population remains unchanged at the end of each iteration. The solutions to be deleted may be chosen randomly or in a biased way according to appropriate prespecified rules. The routine *Immigr* adds new solutions to the population, for the sake of diversification. The routine *Mut* changes the current population by applying some systematic perturbation which improves the quality of the affected solutions. The routine *L_S* is an attempt to intensify the search towards high quality solutions. It applies a local search procedure a number of times, starting at selected solutions from the current population. The there-by produced local minima are added to the current population. Finally, notice that because of the diversification of the search, e.g. by *immigration* and/or *mutation*, the quality of the best solution in the current population is not an increasing function. Therefore, it is necessary that the three last routines keep track of the best solution found so far by the algorithm and the corresponding value of the objective function. This is done by comparing each new solution with the best solution known so far.

Genetic Algorithm($\mathcal{J}_0$, M)

$\mathcal{J} := \mathcal{J}_0$; /* $\mathcal{J}_0$ initial population generated eg. randomly */
$S_{best} := Best(\mathcal{J})$; /* Initialize the best known solution */
$Z_{best} := Z(S_{best})$; /* Initialize the best known obj. func. value */
$i := 0$; /* Initialize the iterations counter */

```
repeat
    repeat
        i := i + 1;
        (S_1, S_2) := Select(J); /* select two individuals in J */
        S_3 := Cross_1, S_2); /* produce S_3 from S_1 and S_2 */
        J := J ∪ {S_3}; /* update population */
        if Z(S_3) < Z_best
            S_best := S_3; /* update the best known solution */
            Z_best := Z(S_3); /* update the best known value */
        endif
    until i = M; /* nr. of iterations reaches M */
    D := Cleanse(J); /* Generate a set of "bad" solutions */
    J := J \ D; /* Remove the "bad" solutions from J */
    if immig = true
        J := Immigr(J); /* Perform immigration */
    endif
    if mut = true then
        J := Mut(J); /* Perform mutation */
    endif
    if l_s = true then
        J := L_S(J); /* Perform local search */
    endif
until a stop criterion is fulfilled;
return S_best, Z_best;
```

Several genetic algorithm approaches have been applied to the QAP. The algorithm developed by Tate and Smith [222], a more or less standard one, reveals some of the drawbacks of such algorithms, despite encouraging numerical results. This algorithm does not find the best known solution for the Nugent problems of size 20 and 30. For larger problems of size up to 100 it cannot really compete with tabu search procedures. In an attempt to overcome these difficulties, Fleurent and Ferland [79] propose so-called hybrid approaches which are derived by combining genetic algorithms with other heuristic procedures for QAP, in particular local search and tabu search. They manage to improve the best solutions known at that time for most of the large scale test problems of Taillard and Skorin-Kapov (see QAPLIB), respectively. However, the computational times spent for these improvements are very large, sometimes reaching more than 20 hours of CPU time for large problems. Recently, Ahuja, Orlin and Tivari [4] obtained very promising results on large scale QAPs from QAPLIB by applying a version of GA called by them a *greedy genetic algorithm*. The greedy genetic algorithm is tested on all problems from QAPLIB, obtaining the best known solutions for 103 out of 132 test problems. The solutions obtained for 28 out of the 29 remaining problems deviate less than 1% from the best known ones. These results are obtained by applying the algorithm only once and within reasonable limits on the computation time. The greedy elements incorporated in this algorithm seem to help in maintaining the balance between biased search and diversity of the population.

3.6 GREEDY RANDOMIZED ADAPTIVE SEARCH

The greedy randomized adaptive search procedure, so-called GRASP, was introduced by Feo and Resende [77] as a heuristic approach to hard combinatorial optimization problems. GRASP was applied to the QAP by Li, Pardalos and Resende [157] in 1994. GRASP is a combination of greedy elements with random search elements in a two phase heuristic. It consists of a construction phase and a local improvement phase. In the construction phase GRASP starts by assigning two facilities i_0, j_0 to two locations k_0, l_0, respectively. According to the greedy component, GRASP selects this pair of assignments among those with minimal cost. That is, for a $QAP(A,B)$ of size n, the greedy component would select i_0, j_0, k_0, l_0 such that

$$a_{i_0j_0}b_{k_0l_0} = \min\left\{a_{ij}b_{kl}: i,j,k,l \in \{1,2,\ldots,n\},\ \ i \neq j,\ \ k \neq l\right\}$$

The random element gives some freedom to the search procedure and modifies this selection with the hope of avoiding to get trapped in locally optimal solutions of poor quality. To this end a control parameter $0 < \beta < 1$ is introduced. The non-diagonal entries of the matrix $A = (a_{ij})$ are sorted non-decreasingly and the r smallest entries are denoted by $a_{i_1j_1}, a_{i_2j_2}, \ldots, a_{i_rj_r}$, where $r = \lfloor\beta(n^2-n)\rfloor$. Analogously, the non-diagonal entries of the matrix $B = (b_{ij})$ are sorted non-increasingly and the r largest entries are denoted by $b_{k_1l_1}, b_{k_2l_2}, \ldots, b_{k_rl_r}$. Then we get:

$$a_{i_1j_1} \leq a_{i_2j_2} \leq \cdots \leq a_{i_rj_r}$$

$$b_{k_1l_1} \geq b_{k_2l_2} \geq \cdots \geq b_{k_rl_r}$$

The costs of the possible pairs of assignments

$$a_{i_1j_1}b_{k_1l_1}, a_{i_2j_2}b_{k_2l_2}, \ldots, a_{i_rj_r}b_{k_rj_r}\ ,$$

are then sorted non-decreasingly and the smallest $r' = \lfloor \alpha\beta(n^2 - n) \rfloor$ are taken further into consideration. Here $0 < \alpha < 1$ is a second control parameter. Then the locations i_0, j_0 and the facilities k_0, l_0, are chosen randomly among (i_t, j_t) and (k_t, l_t), respectively, for $1 \le t \le r'$, and the initial assignments $i_0 \mapsto k_0$, $j_0 \mapsto l_0$ are made. In the next steps of the construction phase the remaining facilities are assigned to the remaining locations, one facility to one location at a time. Assume that we are performing the (t + 1)-th iteration of the construction phase and $\Gamma = \{(i_1, k_1), (i_2, k_2), \ldots, (i_t, k_t)\}$ is the set of the already made assignments. The cost c_{jl} of locating some facility j, $j \notin \{i_1, i_2, \ldots, i_t\}$, to some location l, $l \notin \{k_1, k_2, \ldots, k_t\}$, with respect to the already made assignments is given by

$$c_{jl} = \sum_{(i,k)\in\Gamma} (a_{ij}b_{kl} + a_{ji}b_{lk}).$$

c_{jl} is referred to as an *intermediate* cost. If there are m pairs (j, l) of unassigned facilities and locations, select one pair at random among pairs which yield the $\lfloor \alpha m \rfloor$ smallest intermediate costs c_{jl}. Add the selected pair to Γ and repeat this step until a permutation of $\{1, 2, \ldots, n\}$ has been constructed. Here terminates the construction phase of GRASP.

The local improvement phase consists of a standard local search algorithm starting with the solution constructed in the first phase. The whole procedure is repeated a certain number of times, where this number is the third and last control parameter. In [157] GRASP is tested on almost all of the test problems from QAPLIB as well as on two classes of QAP instances with known optimal solution generated by Li and Pardalos in [154]. (The generation of QAP instances with known optimal solution is discussed in the next section.) The version of GRASP tested in [157] involves a local search approach with the pair-exchange neighborhood. The control parameters α and β are equal to 0.5 and 0.1, respectively, and the maximum number of iterations is equal to 100000. This implementation of GRASP yields best known solutions for most of the tested instances and even improves the best known solutions in a few cases. Moreover, its running times are reasonable compared to other heuristic approaches. Some more recent experimental results presented in [4] show that the greedy genetic algorithm of Ahuja et al. mentioned in the previous section outperforms GRASP in terms of the quality of the solutions produced within a reasonably long, common running time limit for both algorithms. However, the same experiments show that GRASP produces better solutions in early stages of the search, i.e. GRASP seems to be advantageous in the short run.

3.7 QAP INSTANCES WITH KNOWN OPTIMAL SOLUTIONS

In this section we discuss two *generators of QAP instances with known optimal solution*. Such generators are important since QAP instances with known optimal solution can be used as test problems for the evaluation and the comparison of heuristics.

Palubeckis' generator

The first method for generating QAP instances with a known optimal solution was proposed by Palubeckis [180] in 1988. The input of the Palubeckis' algorithm consists of the size n of the instance to be generated, the optimal solution (permutation) π of the output instance, two control parameters w and z, where $z < w$, and the distance matrix A of an $r \times s$ grid with $rs = n$. Throughout the rest of this section, such a grid has r knots per row and s knots per column, the knots being labeled from the left to right and then from the top to the bottom. We use *rectilinear distances*, also called *Manhattan distances*, i.e. the distance a_{ij} between two given knots i, j lying in rows r_i, r_j and in columns c_i, c_j, respectively, is given by $a_{ij} = |r_i - r_j| + |c_i - c_j|$. The algorithm starts with a trivial QAP instance with known optimal solution $QAP(A, B)$, where $B = (b_{ij})$ is a *constant matrix* with $b_{ij} = w$. Notice that such a $QAP(A, B)$ is a trivial problem, in the sense that all permutations yield the same value of the objective function and thus, are optimal solutions. Hence, also the identity permutation id is an optimal solution of $QAP(A, B)$. The algorithm leaves matrix A unchanged and transforms step-by-step matrix B so that it is not a constant matrix any more. This transformation is done so that the identity permutation id remains an optimal solution to all intermediate QAP instances. In the last step, the algorithm uses a simple standard trick for transforming the QAP instance with optimal solution id to another QAP instance with optimal solution π. A more formal description of Palubeckis' algorithm is given below. The input of the algorithm consists of the dimensions r, s of the grid, two parameters z, w with $z < w$, and a permutation π of $\{1, 2, \ldots, n\}$ which will be the optimal solution of the QAP instance to be generated by the algorithm, where n is the size of the QAP to be generated given by $n = rs$. The variables involved in this algorithm are the following: the entries a_{ij}, b_{ij} of the coefficient matrices A and B of the QAP instance to be generated, the optimal value of this QAP instance denoted by $Z_{\rm opt}$, the auxiliary variables i, j, k, l, m, Δ, $\bar{i}$, $\bar{j}$, l_1, l_2, u_1, u_2, the auxiliary matrices $G = (g_{ij})$, $T = (t_{ij})$, and a decision variable *decide* which serves to decide whether to perform further changes on matrix A or not. The routines called by the algorithm are: *Choose_max* with arguments A, G

which returns a pair of indices (l, m), $1 \le l, m \le n$, such that $a_{lm} = \max_{(i,j):g_{ij} \le 0} \{a_{ij}\}$, *Choose_rand* with arguments n, l_1, l_2, u_1, u_2 which chooses uniformly at random an index k, $1 \le k \le n$, such that the following inequalities are fulfilled

$$l_1 \le \lceil k/r \rceil \le u_1 \qquad l_2 \le k - r(\lceil k/r \rceil - 1) \le u_2 \tag{3.1}$$

Finally, *Rand* with argument z which generates a number between 0 and z uniformly at random.

```
Palubeckis' Generator(r, s, z, w, π)
    n := rs; /* fix the QAP size n */
    for i = 1 to n do
        ī := ⌈i/r⌉; /* fix the auxiliary variable ī */
        for j = 1 to n do
            j̄ := ⌈j/r⌉; /* fix the auxiliary variable j̄ */
            a_ij := |ī−j̄|+|(i−j)+(j̄−ī)r|; /* set a_ij as rectilinear distance
                    between knot i and j on the grid */
            b_ij := w; /* initialize entry b_ij of matrix B */
            g_ij := 2 − a_ij; /* compute entry g_ij of the auxiliary matrix G */
        endfor
    endfor
    decide := false; /* initialize the decision variable */
    for i = 1 to n do
        for j = 1 to n do
            if g_ij ≤ 0 then
                decide := true; /* update the decision variable */
            endif
        endfor
    endfor
    while decide do
        (l, m) := Choose_max(A, G); /* choose two indices 1 ≤ l, m ≤ n */
        l_1 := min{⌈l/r⌉, ⌈m/r⌉}; /* first aux. lower bound */
        u_1 := max{⌈l/r⌉, ⌈m/r⌉}; /* first aux. upper bound */
        l_2 := min{l − r(⌈l/r⌉ − 1), m − r(⌈m/r⌉ − 1)}; /* second aux.
                lower bound */
        u_2 := max{l − r(⌈l/r⌉ − 1), m − r(⌈m/r⌉ − 1)}; /* second aux.
                upper bound */
        k := Choose_rand(n, l_1, l_2, u_1, u_2); /* choose 1 ≤ k ≤ n randomly */
```

```
        Δ := Rand(z); /* choose 0 ≤ Δ ≤ z uniformly at random */
        b_lm := Δ; b_lk := b_lk + (w − Δ); /* update matrix B */
        b_km := b_km + (w − Δ); /* update matrix B */
        g_lm := 1; g_km := 1; g_lk := 1; /* update matrix G */
        b_ml := Δ; b_mk := b_mk + (w − Δ); /* update matrix B */
        b_kl := b_kl + (w − Δ); /* update matrix B */
        g_ml := 1; g_mk := 1; g_kl := 1; /* update matrix G */
        decide := false; /* reinitialize the decision variable */
        for i = 1 to n do
            for j = 1 to n do
                if g_ij ≤ 0 then
                    decide := true; /* update the decision variable */
                endif
            endfor
        endfor
    endwhile
    for i = 1 to n do
        for j = 1 to n do
            t_ij := b_π(i)π(j); /* permute matrix B according to permut. π */
        endfor
    endfor
    for i = 1 to n do
        for j = 1 to n do
            b_ij := t_ij; /* update matrix B */
        endfor
    endfor
    Z_opt := w Σ_{i=1}^n Σ_{j=1}^n a_ij;
    return A = (a_ij), B = (b_ij), Z_opt;
```

We show next the correctness of the Palubeckis' generator.

Theorem 3.1 (Palubeckis [180], 1988)
Let A and B be the $n \times n$ matrices generated by Palubeckis' generator. The input permutation π is an optimal solution to QAP(A,B) with optimal value equal to $w\sum_{i=1}^{n}\sum_{j=1}^{n} a_{ij}$.

Proof. By induction on the number of repetitions of the while loop, we prove that the identity permutation *id* is an optimal solution of $QAP(A,B)$, where B is the matrix obtained after the completion of the "while" loop. Denote by $B^{(t)}$, $t \geq 0$, matrix B after the t-th execution of the "while" loop. At the moment when the while loop is executed for the first time matrix $B = B^{(0)}$ is a constant matrix and therefore $QAP(A, B^{(0)})$ is a constant QAP, i.e. $Z(A, B^{(0)}, \phi) = w\sum_{i=1}^{n}\sum_{j=1}^{n} a_{ij}$, for each $\phi \in \mathcal{S}_n$. (Recall that $\mathcal{S}_n$ is the set of permutations of $\{1,2,\ldots,n\}$. Thus, permutation *id* is an optimal solution to $QAP(A, B^{(0)})$. Assume that the identity permutation *id* is an optimal solution to $QAP(A, B^{(t)})$ with optimal value $w\sum_{i=1}^{n}\sum_{j=1}^{n} a_{ij}$. We show that *id* is also an optimal permutation to $QAP(A, B^{(t+1)})$. According to the update of the entries b_{ij} within the "while" loop, we have $B^{(t+1)} = B^{(t)} + \widehat{B}$, where $\widehat{B} = (\widehat{b}_{ij})$ is given by:

$$\widehat{b}_{ij} = \begin{cases} w - \Delta & \text{if } i = l \text{ and } j = k \\ w - \Delta & \text{if } i = m \text{ and } j = k \\ \Delta - w & \text{if } i = l \text{ and } j = m \\ 0 & \text{otherwise} \end{cases}$$

An elementary calculation reveals that $Z(A, \widehat{B}, id) = 0$. The last equality implies that *id* is an optimal solution to $QAP(A, \widehat{B})$. By considering additionally the fact that equality

$$Z(A, B^{(t+1)}, \phi) = Z(A, B^{(t)}, \phi) + Z(A, \widehat{B}, \phi),$$

holds for each $\phi \in \mathcal{S}_n$, it is easily seen that the identity permutation *id* is an optimal solution to $QAP(A, B^{(t+1)})$ with optimal value $w\sum_{i=1}^{n}\sum_{j=1}^{n} a_{ij}$. Assume that the "while" loop has been applied ℓ times. Thus after "endwhile" we have $B = B^{(\ell)}$ and *id* is an optimal solution of $QAP(A,B)$, as shown by induction. Finally the algorithm updates matrix B by permuting its rows and columns according to permutation π and outputs matrix $B^{\pi} = (b_{\pi(i)\pi(j)})$. Now the claim follows immediately, since $Z(A, B^{\pi}, \phi) = Z(A, B, \phi \circ \pi)$ for each $\phi \in \mathcal{S}_n$. □

Once a generator of test instances is proposed, a natural question arises concerning the difficulty of the generated instances. If these instances are in some sense "easy", while the considered problem is "difficult", then the generated instances are not representative for the general problem and hence, they are not appropriate for use in evaluation of heuristics. Cyganski, Vaz and Virball [56] have observed that the QAP instances generated by Palubeckis' generator are "easy" in the sense that their optimal value can be computed in polynomial time, by solving an auxiliary linear program. In order to show this result we need some more definitions. Consider all ordered pairs of the form $((i,j),k)$ where the indices i,j,k, $1 \leq i,j,k \leq n$, are pairwise distinct. There are $\binom{n}{2}(n-2)$ such pairs, where n is as usually the size of the considered QAP. Assume that these pairs are numbered by the integers $1, 2, \ldots, \binom{n}{2}(n-2)$ and that each pair is identified with its corresponding number. If p is the pair $((i,j),k)$, let $T_p^{(+)} = \{(i,k),(k,i),(j,k),(k,j)\}$ and $T_p^{(-)} = \{(i,j),(j,i)\}$. Now, the above mentioned result can be formulated as follows.

Theorem 3.2 *Consider QAP(A,B) of size n whose coefficient matrices A and B have been generated by Palubeckis' generator. Assume that we do not know the values of the input parameters z and w involved in the corresponding run of the Palubeckis' generator. The optimal value of QAP(A,B) is equal to* $\tilde{w}\sum_{i=1}^{n}\sum_{j=1}^{n} a_{ij}$, *where* $\tilde{w}$ *is the optimal value of the following linear program*

$$
\begin{array}{lll}
\min & w & \\
\textit{subject to} & & \\
& w - \sum\limits_{(i,j)\in T_p^{(-)}} \alpha_p + \sum\limits_{(i,j)\in T_p^{(+)}} \alpha_p = b_{ij} & 1 \le i,j \le n \\
& w - z \le \alpha_p \le w & p = 1, 2, \ldots, \binom{n}{2}(n-2) \\
& z \le w & \\
& w > 0 & \\
& z \ge 0 &
\end{array}
\tag{3.2}
$$

Proof. Let us consider a modification of Palubeckis' generator the so-called *weak generator*. The weak generator result by making two modifications. First, the routine *Choose_rand* of the Palubeckis' generator selects any k between 1 and n neglecting the inequalities (3.1) to be fulfilled by the chosen k. Secondly, we do not make use of the auxiliary matrix $G = (g_{ij})$ to determine the number of the executions of the "while" loop, but we just execute this loop a prespecified number of times. In this case, *Choose_max* returns a pair (l,m) such that $a_{lm} = \max\{a_{ij}: 1 \le i,j \le n\}$ Clearly, for matrices A, B generated by the weak generator the input permutation π is not necessarily an optimal solution of QAP(A,B). However, the optimal value of this QAP(A,B) would be larger than or equal to $w\sum_{i,j=1}^{n} a_{ij}$.
Now assume that the given QAP(A,B) of size n is obtained by applying Palubeckis' generator with values of the input parameters w, z, equal to w^*, z^*, $(z^* < w^*)$, respectively. As the matrix A is the distance matrix of an $r \times s$ grid, $n = rs$, we can assume w.l.o.g. that r and s are known. Let $\alpha_p^* := 0$ for all p. Let p be the number which corresponds to the pair $((l,m),k_0)$, where (l,m) is the pair of indices chosen by *Choose_max*, and k_0 the index chosen by *Choose_rand* during the first execution of the "while" loop. Set $\alpha_p^* := \alpha_p^* + w^* - \Delta$, where Δ is the random

number chosen by *Rand* during the first execution of the "while" loop. Update α_p^* in an analogous way after each execution of the "while" loop. After having done so we have $\alpha_p^* := 0$ for the indices p, $1 \le p \le \binom{n}{2}(n-2)$, which occur in no execution of the "while" loop. With these settings w^*, z^* and α_p^*, for $1 \le p \le \binom{n}{2}(n-2)$, constitute a feasible solution of the linear program (3.2). Further, the optimal value of QAP(A,B) equals $w^* \sum_{i,j=1}^n a_{ij}$. Now, let $\tilde{w}$, $\tilde{z}$, $\tilde{\alpha}_p$, for $1 \le p \le \binom{n}{2}(n-2)$, be an optimal solution of (3.2). Then, $\tilde{w} \sum_{i,j=1}^n a_{ij} \le w^* \sum_{i,j=1}^n a_{ij}$. Moreover, there exists a run of the weak generator which produces QAP(A,B) (or more precisely, matrix B) under the input $w := \tilde{w}$, $z := \tilde{z}$ and $A = (A_{ij})$. This run executes the "while" loop $\binom{n}{2}(n-2)$ times, and chooses $\Delta := \tilde{\alpha}_p$ and a triple (l, m, k_0) at the p-th execution such that $((l,m),k_0)$ is the p-th pair (see the last paragraph before Theorem 3.2). According to the observation made at the beginning of the proof we have then

$$Z(A,B,\pi^*) \le \tilde{w} \sum_{i,j=1}^n a_{ij} \le w^* \sum_{i,j=1}^n a_{ij} = Z(A,B,\pi^*),$$

where π^* is an optimal solution of $QAP(A,B)$. From the last inequalities follows $\tilde{w} \sum_{i,j=1}^n a_{ij} = Z(A,B,\pi^*)$, and this completes the proof. $\square$

It is worthy to notice that nothing is known about the computational complexity of QAP instances generated by Palubeckis' generator. We believe that *finding an optimal solution* to these QAPs is NP-hard, although the corresponding decision problem is polynomially solvable.

Finally, notice that instead of using the distance matrix of a grid graph, one can start Palubeckis' generator with A being an arbitrary Euclidean matrix. The indices l, m, k_0 at the "Random" step should correspond then to three point which are collinear in the respective Euclidean space.

The generator of Li and Pardalos

Li and Pardalos [154] propose another generator of QAP instances with known optimal solution, called Li&Pardalos' generator. This generator shares the basic idea of Palubeckis' generator. It starts with a trivial QAP instance with known optimal solution and transforms its coefficient matrices so that the resulting QAP instance has the same optimal solution as the initial QAP, but is not trivial any more. The strategy for transforming the coefficient matrices is based on Proposition 2.1. A more formal description of Li&Pardalos' generator is given below. The input of the algorithm consists of the dimensions n of the QAP instance to be generated, two positive parameters Δ_A and Δ_B, and a permutation π of $\{1,2,\ldots,n\}$ which will be the optimal solution of the QAP instance to be generated by the algorithm. The variables involved in this algorithm are the following: the coefficients a_{ij}, b_{ij} of the QAP instance to be generated, the optimal value of this QAP instance denoted by Z_{opt}, the auxiliary variables i, j, an auxiliary $n(n-1)$-dimensional vector $X = (x_i)$, and the auxiliary matrices $R = (r_{ij})$ with zeros on the diagonal. Matrix R is used to save the output of routine *Rank* which is introduced below. The routines called by the algorithm are *Rank*, *Sort* and *Rand*, introduced below. The routine *Rank* takes as argument an $n \times n$ matrix (matrix B in our case), considers its non-diagonal entries as the elements of an $n(n-1)$-dimensional vector, sorts these elements non-decreasingly (without changing the input matrix), and for each element b_{ij} saves its rank r_{ij}, i.e. the number showing its position in the above mentioned non-decreasing ordering. *Rank* outputs then the matrix $R = (r_{ij})$. The routine *Sort* takes as input an $n(n-1)$-dimensional vector ($X = (x_i)$ in our case) and sorts it non-decreasingly. The output is the updated (sorted) vector X. Finally, *Rand* takes as argument a real number z (in our case Δ_A or Δ_B) and generates a number between 0 and z uniformly at random.

Li&Pardalos' Generator(n, $\mathbf{\Delta_A}$, $\mathbf{\Delta_B}$, π)

```
for i = 1 to n do
    for j = 1 to n do
        if i ≠ j then
            a_ij := Δ_A; /* initialize non-diagonal entries a_ij */
            b_ij := Rand(Δ_B); /* initialize non-diagonal entries b_ij */
        endif
        else
            a_ij := 0; /* set to 0 the diagonal entries a_ii */
            b_ij := 0; /* set to 0 the diagonal entries b_ii */
        endelse
    endfor
endfor
R := Rank(B); /* Sort b_ij non-decreasingly and output the ranks */
for i = 1 to n(n − 1) do
    x_i := Rand(Δ_A); /* Generate X = (x_i) at random */
endfor
X := Sort(X) /* sort x_i, 1 ≤ i ≤ n(n − 1), non-decreasingly */
for i = 1 to n do
    for j = 1 to n do
        if i ≠ j then
            a_ij := a_ij − x_{r_ij} /* update the entries a_ij */
```

```
            endif
        endfor
    endfor
    Z_opt := Σ_{i=1}^n Σ_{j=1}^n a_ij b_ij;
    for i = 1 to n do
        for j = 1 to n do
            t_ij := b_π(i)π(j); /* permute matrix B according to permut. π */
        endfor
    endfor
    for i = 1 to n do
        for j = 1 to n do
            b_ij := t_ij; /* update matrix B */
        endfor
    endfor
    return A = (a_ij), B = (b_ij), Z_opt;
```

The following theorem, whose simple proof is not given here, states the correctness of the Li&Pardalos' generator.

Theorem 3.3 (Li and Pardalos [154], 1992)
Assume that the input of the Li&Pardalos' generator consists of a natural number n, a permutation π and two constants Δ_A and Δ_B. Then the generator outputs two $n \times n$ matrices A, B and a number Z_{opt} such that π is an optimal solution to QAP(A,B) and Z_{opt} is its optimal value. This is done in $O(n^2 \log n)$ time, assuming constant time for the generation of a random number. □

A more general scheme for generating QAP instances with known optimal solution is described in [154]. This scheme makes use of the so-called *sign-subgraphs* of a given graph. The application of this scheme with different sign-subgraphs yields different QAP generators. In this context, the Li&Pardalos' generator is obtained when the sign-subgraphs are as simple as possible: each of them consists of a single edge. It is interesting and surprising that QAP instances generated by applying the general scheme with more complex sign-subgraphs such as triangles and spanning trees are "easier" to solve than those generated by the Li&Pardalos' generator. Here a QAP instance is considered to be "easy", if most heuristics applied to it find a solution near to the optimal one in a relatively short time. Another interesting but less surprising observation is the following. Instances of rectilinear QAPs seem to be easier to solve than instances of symmetric QAPs, and the latter seem to be easier to solve than instances of asymmetric QAPs, in terms of the quality of solutions produced by different heuristics. On the other hand, the above order of simplicity seems to be reversed when the computation times needed to find good solutions are considered. An intuitive explanation for this behavior would be the large number of equal entries in symmetric matrices and in distance matrices of grids. This implies the existence of large groups of feasible solutions with the same value of the objective function, a property which confuses matters for most heuristics. Notice that Theorem 3.2 does not apply to QAP instances generated by the generator of Li and Pardalos. The proof of this theorem, as well as Palubeckis' generator itself, essentially exploit the triangle inequality fulfilled by the matrix A, whereas none of the two matrices generated by Li&Pardalos' generator needs to be Euclidean.

In Chapter 7 we will see how to use the generator of Li and Pardalos for the generation of instances of the BiQAP with known optimal solution.

3.8 ASYMPTOTIC BEHAVIOR

This section presents some results on the asymptotic behavior of QAPs. Among others, it is shown that if certain probabilistic conditions on the coefficient matrices of QAP(A,B) are fulfilled, the ratio between its "best" and "worst" values of the objective function approaches 1 almost surely, as the size of the problem approaches infinity. As most of the known results on QAPs confirm the general belief on the extreme difficulty of this problem, this kind of asymptotic behavior is somewhat astonishing. Actually, it suggests that the relative error of every heuristic method vanishes almost surely, as the size of the problem tends to infinity, i.e. every heuristic finds almost always an almost optimal solution when applied to QAP instances which are large enough. In other words, under certain probabilistic assumptions concerning the problem data, the QAP becomes in some sense trivial as the size of the problem increases. Burkard and Fincke [36] identify a common combinatorial property of a number of problems which, under natural probabilistic conditions on the problem data, behave as described above. This property seems to be also the key for the specific asymptotic behavior of the QAP. Later on, this general result of Burkard and Fincke and a stronger result of Szpankowski [217] will be used for showing that also the *biquadratic assignment problem (BiQAP)* behaves asymptotically as described above.

Next some general results on the asymptotic behavior of combinatorial optimization problems are presented and then the asymptotic behavior of the QAP is discussed.

3.8.1 On the asymptotic behavior of some combinatorial optimization problems

Consider a sequence P_n, $n \in \mathbb{N}$, of combinatorial optimization (minimization) problems with sum objective function, as described in Section 1.5.2. Let $\mathcal{E}_n$ and $\mathcal{F}_n$ be the ground set and the set of the feasible solutions of problem P_n, respectively. Moreover, let $c_n: \mathcal{E}_n \to \mathbb{R}^+$ and $f: \mathcal{F} \to \mathbb{R}^+$ be the nonnegative cost function and the objective function for problem P_n, respectively, For $n \in \mathbb{N}$, a *worst solution* $X_{\text{wor}} \in \mathcal{F}_n$ is defined as follows:

$$f(X_{\text{wor}}) = \sum_{x \in X_{\text{wor}}} c_n(x) = \max_{X \in \mathcal{F}} f(X) = \max_{X \in \mathcal{F}} \sum_{x \in X} c_n(x)$$

We will consider the ratio between the objective function values corresponding to an optimal (or best) and a worst solution. In [36] Burkard and Fincke show that this ratio is strongly related to the ratio between the cardinality of the set of the feasible solutions and the cardinality of an arbitrary feasible solution, under the assumption that all feasible solutions have the same cardinality. For the proof of this result we make use of a lemma of Rényi [195].

Lemma 3.4 *Let $Y_1, Y_2, \ldots Y_n$ be independent random variables with $|Y_k - E(Y_k)| \le K$, $1 \le k \le n$, where $E(Y_k)$ is the expected value of Y_k and K is some positive number. Let*

$$D := \sqrt{\sum_{k=1}^{n} \sigma^2(Y_k)},$$

where $\sigma^2(Y_k)$ is the variance of Y_k. Let μ be a positive real number with $\mu \le \frac{D}{K}$. Then

$$P\left\{\left|\sum_{k=1}^{n} (Y_k - E(Y_k))\right| \ge \mu D\right\} \le 2\exp\left(-\frac{\mu^2}{2(1+\mu K/2D)^2}\right).$$

Theorem 3.5 (Burkard and Fincke [36], 1985)
Let P_n be a sequence of combinatorial minimization problems with sum objective function as described above. Assume that the following conditions are fulfilled:

(BF1) *For all $X \in \mathcal{F}_n$, $|X| = |X^{(n)}|$, where $X^{(n)}$ is some feasible solution in $\mathcal{F}_n$.*

(BF2) *The costs $c_n(x)$, $x \in X$, $X \in \mathcal{F}_n$, $n \in \mathbb{N}$, are random variables identically distributed on $[0,1]$. The expected value $E = E(c_n(x))$ and the variance $\sigma^2 =$*

$\sigma^2(c_n(x)) > 0$ of the common distribution are finite. Moreover, for all $X \in \mathcal{F}_n$, $n \in \mathbb{N}$, the variables $c_n(x)$, $x \in X$, are independently distributed.

(BF3) *$|\mathcal{F}_n|$ and $|X^{(n)}|$ tend to infinity as n tends to infinity and moreover,*

$$\lim_{n\to\infty} \lambda_0|X^{(n)}| - \ln|\mathcal{F}_n| \to +\infty$$

where λ_0 is defined by $\lambda_0 := (\epsilon_0\sigma/(\epsilon_0 + 2\sigma^2))^2$ and ϵ_0 fulfills

$$0 < \epsilon_0 < \sigma^2 \quad \textit{and} \quad 0 < \frac{E+\epsilon_0}{E-\epsilon_0} \le 1+\epsilon, \tag{3.3}$$

for a given $\epsilon > 0$.

Then

$$P\left\{\frac{\max\limits_{X\in\mathcal{F}_n}\sum\limits_{x\in X} c_n(x)}{\min\limits_{X\in\mathcal{F}_n}\sum\limits_{x\in X} c_n(x)} < 1+\epsilon\right\} \ge 1 - 2|\mathcal{F}_n|\exp(-|X^{(n)}|\lambda_0) \to 1 \;\; \textit{as} \quad n\to\infty.$$

Proof. The proof basically relies on showing that

$$P\left\{\exists X \in \mathcal{F}_n : \left|\sum_{x\in X}(c_n(x)-E)\right| \ge \epsilon_0|X|\right\} \le 2|\mathcal{F}_n|\exp(-|X^{(n)}|\lambda_0) \to 0 \tag{3.4}$$

as $n \to \infty$. Let us consider the following chain of inequalities:

$$\begin{aligned} P\left\{\exists X \in \mathcal{F}_n : \left|\sum_{x\in X}(c_n(x)-E)\right| \ge \epsilon_0|X|\right\} &\le \\ \sum_{X\in\mathcal{F}_n} P\left\{\left|\sum_{x\in X}(c_n(x)-E)\right| \ge \epsilon_0|X|\right\} &\le \\ |\mathcal{F}_n| P\left\{\left|\sum_{x\in \widehat{X}}(c_n(x)-E)\right| \ge \frac{\epsilon_0\sqrt{|\widehat{X}|}}{\sigma}\sqrt{|\widehat{X}|}\sigma\right\} &=: \; P_1, \end{aligned}$$

where $\widehat{X}$ is the feasible solution for which the above considered probability becomes maximum. Now the lemma of Rényi [195] can be applied. Set

$$D = \sqrt{\sum_{x\in\widehat{X}} \sigma^2(c_n(x))} = \sqrt{|\widehat{X}|}\sigma \text{ and } K = 1$$

and define

$$\mu := \frac{\epsilon_0\sqrt{|\widehat{X}|}}{\sigma}.$$

(Due to assumption (3.3) μ is defined correctly as required in Rényi's lemma.) By the definition of λ_0, we get

$$\begin{aligned} P_1 &\leq 2|\mathcal{F}_n|\exp\left(-\left(\frac{\epsilon_0\sqrt{|\widehat{X}|}}{\sigma}\right)^2 \Big/ 2\left(1+\frac{\epsilon_0\sqrt{|\widehat{X}|}}{\sigma}\cdot\frac{1}{2\sqrt{|\widehat{X}|\sigma}}\right)^2\right). \\ &= 2|\mathcal{F}_n|\exp(-\lambda_0|\widehat{X}|) = 2|\mathcal{F}_n|\exp(-\lambda_0|X^{(n)}|) =: P_2 \end{aligned}$$

Assumption (BF3) implies $P_2 \to 0$ and consequently, $P_1 \to 0$ which proves (3.4). Note that (3.4) is equivalent to

$$P\left\{\forall X \in \mathcal{F}_n : \left|\sum_{x\in X}(c_n(x) - E)\right| < \epsilon_0|X|\right\} \geq 1 - 2|\mathcal{F}_n|\exp(-|X^{(n)}|\lambda_0) \to 1$$

as $n \to \infty$. The last inequality and (3.3) imply

$$\frac{\max\limits_{X\in\mathcal{F}_n}\sum\limits_{x\in X} c_n(x)}{\min\limits_{X\in\mathcal{F}_n}\sum\limits_{x\in X} c_n(x)} < \frac{\epsilon_0|X^{(n)}| + E|X^{(n)}|}{-\epsilon_0|X^{(n)}| + E|X^{(n)}|} \leq 1+\epsilon$$

with probability 1 as n approaches $+\infty$. □

The combinatorial condition represented by the limit in (BF3) means basically that the cardinality of the feasible solutions is large enough with respect to the cardinality of the set of feasible solutions. Namely, the result of the theorem is true if the following equality holds:

$$\lim_{n\to\infty}\frac{\ln|\mathcal{F}_n|}{|X^{(n)}|} = 0 \qquad (*)$$

The other conditions of Theorem 3.5 are natural probabilistic requirements on the coefficients of the problem. Two problems which fulfill condition (∗) are the quadratic assignment problem and the minimum perfect matching problem for graphs where the number of the perfect matchings is not "too large" in the sense of condition (∗). The traveling salesman problem (TSP) and the linear assignment problem (LAP) are two examples of problems which do not fulfill (∗).

Theorem 3.5 states that for each $\epsilon > 0$, the ratio between the best and the worst value of the objective function lies on $(1-\epsilon, 1+\epsilon)$, *with probability tending to 1*, as the "size" of the problem approaches infinity. Thus, we have convergence *with probability.* Under one additional natural (combinatorial) assumption (condition (S3) of the theorem below), Szpankowski strengthens this result and improves the range of the convergence to *almost surely*. In the almost sure convergence the probability that the above mentioned ratio tends to 1 is equal to 1. The almost sure convergence implies the convergence with probability but not vice-versa. (Detailed explanations on the probabilistic notions used in this section can be found in every textbook on the fundamentals of probability.)

Theorem 3.6 (Szpankowski [217], 1995)
Let P_n be a sequence of combinatorial minimization problems with sum objective function as above. Assume that the following conditions are fulfilled:

(S1) *For all $X \in \mathcal{F}_n$, $|X| = |X^{(n)}|$, where $X^{(n)}$ is some feasible solution in $\mathcal{F}_n$.*

(S2) *The costs $c_n(x)$, $x \in X$, $X \in \mathcal{F}_n$, $n \in \mathbb{N}$, are random variables identically and independently distributed on $[0,1]$. The expected value $E = E(c_n(x))$, the variance, and the third moment of the common distribution are finite.*

(S3) *The worst values of the objective function, $\max\limits_{X\in\mathcal{F}_n} \sum\limits_{x\in X} c_n(x)$, form a nondecreasing sequence for increasing n.*

(S4) *$|\mathcal{F}_n|$ and $|X^{(n)}|$ tend to infinity as n tends to infinity and moreover, $\ln|\mathcal{F}_n| = o(|X^{(n)}|)$.*

Then, the following equalities hold almost surely:

$$\min_{X\in\mathcal{F}_n} \sum_{x\in X} c_n(x) = |X^{(n)}|E - o(|X^{(n)}|)$$

$$\max_{X\in\mathcal{F}_n} \sum_{x\in X} c_n(x) = |X^{(n)}|E + o(|X^{(n)}|)$$ □

3.8.2 On the asymptotic behavior of the QAP

Theorems 3.5 and 3.6 can be applied to QAP instances whose coefficients fulfill natural probabilistic conditions. The main point is that the QAP fulfills the combinatorial condition (S4) in Theorem 3.6 (and therefore, also condition (BF3) in

Theorem 3.5). Indeed, consider a QAP(A,B) of size n as a general combinatorial optimization problem, following the description given in Section 1.5.2. The ground set $\mathcal{E}_n$ consists of all quadruples of indices (i,j,k,l), $1 \le i,j,k,l \le n$, $\mathcal{E}_n = \{(i,j,k,l) : 1 \le i,j,k,l \le n\}$. The feasible solutions are permutations of $\{1,2,\ldots,n\}$. The $c_n : \mathcal{E}_n \to \mathbb{R}$ are given by $c_n(i,j,k,l) = a_{ij}b_{kl}$[1]. For each permutation π we get a feasible solution X_π as a subset of $\mathcal{E}_n$, $X_\pi = \{(i,j,\pi(i),\pi(j) : 1 \le i,j \le n\}$. The set $\mathcal{F}_n$ of the feasible solutions consists of all feasible solutions X_π for $\pi \in \mathcal{S}_n$. Thus, $|\mathcal{F}_n| = n!$ and $|X| = n^2$, for each $X \in \mathcal{F}_n$. By applying Stirling's formula it is easy to see that $\lim\limits_{n\to\infty} \frac{\ln(n!)}{n^2} = 0$. Moreover, notice that both in Theorem 3.5 and in Theorem 3.6, we can assume w.l.o.g. that the values of the cost functions c_n are random variables, identically and independently distributed on $[0,M]$, instead on $[0,1]$, where M is some positive constant. So, we immediately get the following corollary:

Corollary 3.7 *Consider a sequence of problems $QAP(A^{(n)},B^{(n)})$, $n \in \mathbb{N}$, with $n \times n$ coefficient matrices $A^{(n)} = \left(a_{ij}^{(n)}\right)$ and $B = \left(b_{ij}^{(n)}\right)$. Assume that $a_{ij}^{(n)}$ and $b_{ij}^{(n)}$, $n \in \mathbb{N}$, $1 \le i,j \le n$, are independently distributed random variables on $[0,M]$, where M is a positive constant. Moreover, assume that the entries $a_{ij}^{(n)}$, $n \in \mathbb{N}$, $1 \le i,j \le n$, have the same distribution and the entries $b_{ij}^{(n)}$, $n \in \mathbb{N}$, $1 \le i,j \le n$, have also the same distribution (which does not necessarily coincide with that of $a_{ij}^{(n)}$). Furthermore, let these variables have finite expected values, variances and third moments.*
Let $\pi_{\text{opt}}^{(n)}$ and $\pi_{\text{wor}}^{(n)}$ denote an optimal and a worst solution of $QAP(A^{(n)},B^{(n)})$, respectively, i.e.

$$Z\left(A^{(n)},B^{(n)},\pi_{opt}^{(n)}\right) = \min_{\pi\in\mathcal{S}_n} Z\left(A^{(n)},B^{(n)},\pi\right)$$

and

$$Z\left(A^{(n)},B^{(n)},\pi_{wor}^{(n)}\right) = \max_{\pi\in\mathcal{S}_n} Z\left(A^{(n)},B^{(n)},\pi\right)$$

Then the following equality holds almost surely:

$$\lim_{n\to\infty} \frac{Z\left(A^{(n)},B^{(n)},\pi_{opt}^{(n)}\right)}{Z\left(A^{(n)},B^{(n)},\pi_{wor}^{(n)}\right)} = 1$$

□

[1] The mathematically correct notation would be $c((i,j,k,l))$, but we remove the inner parenthesis for the sake of readability.

The above result suggests that the objective function value of $QAP(A^{(n)}, B^{(n)})$ (corresponding to an arbitrary feasible solution) gets somehow close to its expected value $n^2E(A)E(B)$, as the size of the problem increases, where $E(A)$ and $E(B)$ are the expected values of $a_{ij}^{(n)}$ and $b_{ij}^{(n)}$, $n \in \mathbb{N}$, $1 \le i,j \le n$, respectively. Several authors provide different analytical evaluations for this "getting close", by making use of different probabilistic conditions. The papers of Frenk, Houweninge and Rinnooy Kan [82] and the papers of Rhee [197, 198] can be mentioned here. The following theorem states two important results proved in [82] and [198], respectively.

Theorem 3.8 (Frenk et al. [82], 1986, Rhee [198], 1991)
Consider the sequence of $QAP(A^{(n)}, B^{(n)})$, $n \in \mathbb{N}$, as in Corollary 3.7. Assume that the following conditions are fulfilled:

(C1) $a_{ij}^{(n)}$, $b_{ij}^{(n)}$, $n \in \mathbb{N}$, $1 \le i,j \le n$, *are random variables independently distributed on* $[0, M]$.

(C2) $a_{ij}^{(n)}$, $n \in \mathbb{N}$, $1 \le i,j \le n$, *have the same distribution on* $[0, M]$. $b_{ij}^{(n)}$, $n \in \mathbb{N}$, $1 \le i,j \le n$, *have also the same distribution on* $[0, M]$.

Let $E(A)$, $E(B)$ be the expected values of the variables $a_{ij}^{(n)}$ and $b_{ij}^{(n)}$, respectively. Then, there exists a constant K_1 (which does not depend on n), such that the following inequality holds almost surely, for $\pi \in S_n$, $n \in \mathbb{N}$

$$\limsup_{n\to\infty} \frac{\sqrt{n}}{\sqrt{\log n}} \left| \frac{Z(A^{(n)}, B^{(n)}, \pi)}{n^2E(A)E(B)} - 1 \right| \le K_1$$

Moreover, let Y be a random variable defined by

$$Y = Z\left(A^{(n)}, B^{(n)}, \pi_{\text{opt}}^{(n)}\right) - n^2E(A)E(B),$$

where $\pi_{\text{opt}}^{(n)}$ is an optimal solution of $QAP(A^{(n)}, B^{(n)})$. Then there exists another constant K_2, also independent of the size of the problem, such that

$$\frac{1}{K_2} n^{3/2}(\log n)^{1/2} \le E(Y) \le K_2 n^{3/2}(\log n)^{1/2}$$

$$P\{|Y - E(Y)| \ge t\} \le 2\exp\left(\frac{-t^2}{4n^2\|A\|_\infty^2\|B\|_\infty^2}\right)$$

for each $t \ge 0$, where $E(Y)$ denotes the expected value of variable Y and $\|A\|_\infty$ ($\|B\|_\infty$) is the so-called row sum norm *of matrix A (B) defined by $\|A\|_\infty =$ $\max 1 \le i \le n \sum_{j=1}^n |a_{ij}|$.* □

4

QAPS ON SPECIALLY STRUCTURED MATRICES

In this chapter and in the two next chapters we consider polynomially solvable and provably difficult (NP-hard) cases of the QAP. As there is no hope to find a polynomial time algorithm for solving the general QAP, and as QAP instances arising in different practical applications may often have a special structure, it is interesting to derive polynomial time algorithms for solving special cases of the problem. On the other hand, any information on provably difficult (NP-hard) cases of the problem is of particular relevance for a better understanding of the problem and its complexity. Nowadays there exist only few, sporadic results concerning this challenging but difficult aspect of research on the QAP. We try to give a systematic presentation of the already existing results on complexity questions related to special cases of the QAP, focusing on methodology issues. Further, we formulate a number of open problems which we consider to be a promising object of further research in this direction.

A *special case* or a *restricted version of the QAP* consists of all QAP instances whose coefficient matrices fulfill certain prespecified properties. In this chapter we present results on versions of the QAP whose coefficient matrices have specific combinatorial properties, e.g. they belong to some of the following classes of matrices: Monge and Anti-Monge matrices, Toeplitz and circulant matrices, sum and product matrices, graded matrices. Among polynomially solvable restricted versions of QAPs two notable groups can be distinguished: *constant QAPs* and *constant permutation QAPs*. A *constant QAP* is a QAP whose objective function value does not depend on the feasible solution, i.e. all feasible solutions yield the same value of the objective function. More specifically, if we denote the objective function of $QAP(A,B)$ by $Z(A,B,\phi)$ (as in the previous chapters), then QAP(A,B) is called a constant QAP if and only if the function $Z(A,B,\phi)$ does not depend on the permutation ϕ. Clearly, in this case each feasible solution is an optimal solution and the problem is trivial.

A restricted version of the QAP is called a *constant permutation QAP* if it has the following property: For each problem instance there exists an optimal solution which depends only on the instance size, but not on the entries of the coefficient matrices. For such QAPs, there is usually a rule telling us how to construct the above mentioned optimal solution. The permutation constructed according to this rule is called *constant permutation* with respect to the considered QAP version. Obviously, the constant permutation is not necessarily the unique optimal solution of the constant permutation QAP. The terms *constant QAP* and *constant permutation QAP* maintain their meaning throughout this chapter and the next one.

This chapter is organized as follows. In the first section we introduce the matrix classes which we are dealing with and related notations. The second section summarizes some preliminary results and elementary observations. In the third section some simple polynomially solvable cases of the QAP are listed. Then, the QAP on *Monge* and *Monge-like* matrices is investigated in Section 4.4. In Section 4.5 some QAPs with *circulant* and *limited bandwidth* matrices are considered. Results on the so-called *taxonomy problem* are included in this section, too. This series of investigations on special cases of the QAP continues then in the next chapter with the Anti-Monge–Toeplitz QAP.

4.1 DEFINITIONS AND NOTATIONS

In this section we define the classes of matrices we are going to deal with and introduce the corresponding notations.

Monge and Monge-like matrices

It will turn out that special versions of the QAP with coefficient matrices having Monge or Monge-like properties are interesting problems, especially form a theoretical point of view. Let us first recall the definition of Monge and Anti-Monge[1] matrices as they usually occur in the literature.

[1]Different authors use different names for Anti-Monge matrices. Some frequent alternative names are "inverse Monge" and "contra Monge".

Definition 4.1 *A matrix $A = (a_{ij})$ is called a* Monge matrix *if its elements satisfy*

$$a_{ij} + a_{rs} \le a_{is} + a_{rj}, \quad \text{for } 1 \le i < r \le n \text{ and } 1 \le j < s \le n. \tag{4.1}$$

The class of Monge matrices is denoted by MONGE.

A matrix $A = (a_{ij})$ is called an Anti-Monge matrix *if its elements satisfy*

$$a_{ij} + a_{rs} \ge a_{is} + a_{rj}, \text{for } 1 \le i < r \le n \text{ and } 1 \le j < s \le n \tag{4.2}$$

The class of Anti-Monge matrices is denoted by A-MONGE.

A matrix $A = (a_{ij})$ is called a Kalmanson matrix if A is symmetric and its elements satisfy the following inequalities

$$a_{ij} + a_{kl} \le a_{ik} + a_{jl} \quad \text{for } 1 \le i < j < k < l \le n \tag{4.3}$$
$$a_{il} + a_{jk} \le a_{ik} + a_{jl} \quad \text{for } 1 \le i < j < k < l \le n \tag{4.4}$$

This class of matrices is denoted by KALMANSON.

The classes MONGE, A-MONGE and KALMANSON are closely related to each other. It is easy to see that if $A = (a_{ij})$ is a Monge matrix, $-A = (-a_{ij})$ is an Anti-Monge matrix, and vice-versa. Further, inequality (4.4) in the definition of Kalmanson matrices is the same as that in the definition of Anti-Monge matrices. (The latter inequality is sometimes called *the Anti-Monge inequality*). In a Kalmanson matrix the Anti-Monge inequality is fulfilled for quadruples of entries which lie strictly over the main diagonal (see (4.4)). Since a Kalmanson matrix is symmetric, the Anti-Monge inequality is also fulfilled for quadruples of entries which lie strictly under the main diagonal. In this sense, a Kalmanson matrix is "almost" an Anti-Monge matrix. The only quadruples of entries for which the Anti-Monge inequality may be violated are situated "around" the diagonal.

Other interesting classes of matrices with simple combinatorial structure, which are related to Monge-like matrices and lead to well solvable versions of the QAP, are defined below.

Definition 4.2 *A matrix $A = (a_{ij})$ is called a* sum matrix (product matrix) *if there exist real numbers α_i^r and α_i^c, $1 \le i \le n$, such that $a_{ij} = \alpha_i^r + \alpha_j^c$ ($a_{ij} = \alpha_i^r \alpha_j^c$), for $1 \le i, j \le n$. The classes of sum and product matrices are denoted by* SUM *and* PROD, *respectively.* (α_i^r) *and* (α_i^c) *are called* row *and* column generating vectors, *respectively.*

A matrix $A = (a_{ij})$ *is called a* small matrix (large matrix) *if there exist real numbers* α_i^r *and* α_i^c, $1 \le i \le n$, *such that* $a_{ij} = \min(\alpha_i^r, \alpha_j^c)$ ($a_{ij} = \max(\alpha_i^r, \alpha_j^c)$), *for* $1 \le i, j \le n$. *The classes of small and large matrices are denoted by* SMALL *and* LARGE, *respectively. Again,* (α_i^r) *and* (α_i^c) *are called* row *and* column generating vectors, *respectively.*

The matrix $A = (a_{ij})$ *with* $a_{ij} = (-1)^{i+j}$, $1 \le i, j \le n$, *is called a* chess-board matrix. *The class of chess-board matrices is denoted by* CHESS.

The matrix $A = (a_{ij})$ *with all entries being equal to some constant* κ, $a_{ij} = \kappa$, *for all* $1 \le i, j \le n$, *is called a* constant matrix.

Clearly, a constant matrix is a sum matrix with generating row and column vectors given by $\alpha_i^r = \alpha_j^c = \kappa/2$, for all i, where κ is the constant in the definition of constant matrices. It is easy to see that every sum matrix belongs MONGE and A-MONGE. Moreover, it is not difficult to see that sum matrices are the only matrices which belong simultaneously to MONGE and A-MONGE. Thus,

$$\text{SUM} = \text{MONGE} \cap \text{A-MONGE}$$

Moreover, it is easy to check that a SUM matrix A with generating vectors (α_i^r) and (α_i^c) is symmetric if and only if the generating vectors fulfill the equalities $\alpha_j^c - \alpha_i^c = \alpha_j^r - \alpha_i^r$, for all $1 \le i, j \le n$, or equivalently, if and only if there exists an n-dimensional vector (λ_i) such that $\alpha_i^c = \alpha_1^c + \lambda_i$ and $\alpha_i^r = \alpha_1^r + \lambda_i$ for $1 \le i \le n$. Consequently, every symmetric sum matrix is also a Kalmanson matrix. Similarly, a product matrix A with generating vectors (α_i^r) and (α_i^c) is symmetric, if and only if there exists a vector (γ_i), such that $\alpha_i^c = \gamma_i \alpha_1^c$, $\alpha_i^r = \gamma_i \alpha_1^r$. With this observation it can be easily checked that every symmetric product matrix is a *permuted* Anti-Monge matrix, i.e. there exists a permutation π such that $A^{(\pi)} = (a_{\pi(i),\pi(j)})$ is an Anti-Monge matrix. In our case a permutation which orders the vector (γ_i) non-increasingly, i.e. $\gamma_{\pi(1)} \ge \gamma_{\pi(2)} \ge \ldots \ge \gamma_{\pi(n)}$, does the job.

Graded matrices

For the so-called graded matrices we adopt the definition used in [227].

Definition 4.3 *A matrix A is* graded on its rows (graded on its columns) *if all its rows (columns) have the same monotonicity, i.e. either all of them are non-decreasing, or all of them are non-increasing.*

A matrix $A = (a_{ij})$ *is called* left-higher graded *if all its rows and columns are non-decreasing, i.e.* $a_{ij} \leq a_{i,j+1}$, $1 \leq j \leq n-1$, $1 \leq i \leq n$, *and* $a_{ij} \leq a_{i+1,j}$, $1 \leq i \leq n-1$, $1 \leq j \leq n$. *The matrix* A *is called* right-lower graded *if both its rows and its columns are non-increasing, i.e.* $a_{ij} \geq a_{i,j+1}$, $1 \leq j \leq n-1$, $1 \leq i \leq n$, *and* $a_{ij} \geq a_{i+1,j}$, $1 \leq i \leq n-1$, $1 \leq j \leq n$. *The terms* left-lower graded *and* right-higher graded *are defined analogously. The above defined properties are abbreviated as* LHG, RLG, LLG *and* RHG, *respectively. In a pictorial setting, the two first letters of these abbreviations describe the position of the smallest entry in the matrix.*

Diagonally structured matrices

It turns out that some versions of the QAP with diagonally structured matrices can be solved in polynomial time. The following definition introduces some interesting classes of diagonally structured matrices.

Definition 4.4 *A matrix* $A = (a_{ij})$ *is a* Toeplitz matrix, *if there exist* $2n-1$ *real numbers* $c_{1-n}, \ldots, c_0, \ldots, c_{n-1}$ *such that* $a_{ij} = c_{i-j}$, *for all* $1 \leq i, j \leq n$. *The class of Toeplitz matrices is denoted by* TOEPLITZ.

If the numbers c_i *in the definition of a Toeplitz matrix* A *fulfill the equalities* $c_k = c_{k-n}$, *for all* $1 \leq k \leq n-1$, *then* A *is called a* circulant matrix. *This class of matrices is denoted by* CIRC.

Consider a matrix $A = (a_{ij})$. *In the case that* d *is the smallest integer having the property* $a_{ij} = 0$ *for all* $1 \leq i, j \leq n$ *with* $|i-j| \geq d$, *then matrix* A *is called a* bandwidth-d matrix. *The class of bandwidth-d matrices is denoted* BAND-d.

More notations

We use the following definition which does not occur frequently in the literature: An $n \times n$ matrix A is called an *odd matrix* if n is an odd integer, and is called an *even matrix* if n is even. Further, a matrix will all entries being positive (nonnegative) numbers is called a *positive matrix* (*nonnegative matrix*). The abbreviations SYM, SKEW, ODD, EVEN, POS, NNEG, are used to denote the properties symmetric, skew symmetric, odd, even, positive and nonnegative respectively. Symmetric and skew symmetric matrices are defined as usually: matrix $A = (a_{ij})$ is symmetric (skew symmetric) if $a_{ij} = a_{ji}$ ($a_{ij} = -a_{ji}$), for all i, j.

Throughout the rest of this chapter we use the notation CLASS(PROP1, PROP2, PROP3) for the subclass of matrices in CLASS with properties PROP1, PROP2 and PROP3. For example, the notation MONGE(SYM,LHG,ODD) will be used to denote the class of symmetric Monge matrices of odd size with non-decreasing rows and columns.

For a class CLASS of matrices, we denote by PERMCLASS the class consisting of matrices $A^{\pi} = (a_{\pi(i)\pi(j)})$, for all $A \in$ CLASS and for all $\pi \in \mathcal{S}_n$. Obviously, for two matrix classes CLASS1 and CLASS2 the inclusions PERMCLASS1$\subseteq$PERMCLASS2 and CLASS1$\subseteq$PERMCLASS2 are equivalent. If these inclusions hold, we write CLASS1$\subseteq_{\pi}$CLASS2. For example, it is easily seen that PROD(SYM) $\subseteq_{\pi}$A-MONGE. The case where both inclusions CLASS1$\subseteq_{\pi}$CLASS2 and CLASS2$\subseteq_{\pi}$CLASS1 hold simultaneously will be abbreviated by CLASS1$=_{\pi}$CLASS2. As an example note that LHG$=_{\pi}$RLG and RHG$=_{\pi}$LLG. The significance of these notations becomes clear when considering that two problems QAP(A, B) and QAP(A_1,B_1) with matrices $A \in$CLASS1, $B \in$CLASS2 and $A_1 \subseteq_{\pi}$CLASS1, $B_1 \subseteq_{\pi}$CLASS2 are equivalent. From now on the sentence "the problems *QAP(A,B) and QAP(*A_1, B_1*) are equivalent*" means that once we know an optimal solution to one of these problems, an optimal solution to the other one can be constructed in polynomial time.

The class of problems QAP(A,B) with matrices $A \in$ CLASS1 and $B \in$ CLASS2 is denoted by CLASS1$\times$CLASS2. For a class CLASS of matrices, we denote by NCLASS the class consisting of matrices $-A$, for all $A \in$CLASS. As an illustrative example consider the equality A-MONGE=NMONGE. Finally, the class of all matrices is denoted by MATRIX.

A permutation $\pi \in \mathcal{S}_n$ will be often given as a sequence of the images $\pi(i)$, $1 \leq i \leq n$, and denoted by $\langle \pi(1), \pi(2), \ldots, \pi(n) \rangle$. This notation will be called *sequential representation* of π.

The independent-QAP (I-QAP)

We complete this section by introducing a relaxation of the QAP, the so-called *independent-QAP*, where the rows and the columns of the coefficient matrix A may be permuted by two independent permutations $\pi, \varphi \in S_n$. Namely,

$$\min_{\phi,\pi \in S_n} \sum_{i=1}^{n} \sum_{j=1}^{n} a_{\phi(i)\pi(j)} b_{ij} \tag{4.5}$$

We denote this problem by I-QAP. The notations I-QAP(A,B) $Z(A, B, \phi, \pi)$, *optimal value* and *optimal solution* are defined similarly as for the QAP. Clearly, I-QAP is a relaxation of the QAP, i.e. the following inequality holds:

$$\min_{\pi \in S_n} \sum_{i=1}^{n} \sum_{j=1}^{n} a_{\pi(i)\pi(j)} b_{ij} \geq \min_{\phi,\pi \in S_n} \sum_{i=1}^{n} \sum_{j=1}^{n} a_{\phi(i)\pi(j)} b_{ij} \tag{4.6}$$

Thus, the optimal value of QAP(A,B) is larger than or equal to the optimal value of I-QAP(A,B). If a pair of permutations (π, π) is an optimal solution to I-QAP(A,B), then π is an optimal solution to QAP(A,B). This relationship between I-QAP(A,B) and QAP(A,B) is exploited in several proofs in this chapter.

4.2 THE CONES OF MONGE-LIKE MATRICES

Before investigating the combinatorial structure of the classes of Monge-like matrices introduced in the previous section, let us formalize the easily seen fact that the objective function of the QAP(A,B) is linear with respect to each of the coefficients matrices A, B.

Observation 4.1 *Let the matrices A and B be linear combinations of arbitrarily given matrices A_i, B_i, and constants μ_i, v_i, $1 \leq i \leq k$, $A = \sum_{i=1}^{k} \mu_i A_i$ and $B = \sum_{i=1}^{k} v_i B_i$. Then*

$$Z(A, B, \pi) = \sum_{i=1}^{k} \sum_{j=1}^{k} \mu_i v_j Z(A_i, B_j, \pi) . \qquad \square$$

Further, notice that without loss of generality we can assume the coefficient matrices of the QAP to be nonnegative, i.e. have nonnegative entries. Indeed, given a $QAP(A, B)$ of size n with coefficient matrices which may contain also negative entries, we can add s constant matrix with an enough large constant κ to A and B, and obtain a new problem $QAP(\bar{A}, \bar{B})$, where

$$\bar{A} = (\bar{a}_{ij}), \quad \bar{a}_{ij} = a_{ij} + \kappa, \quad \bar{B} = (\bar{b}_{ij}), \quad \bar{b}_{ij} = b_{ij} + \kappa.$$

QAP(A, B) and QAP$(\bar{A}, \bar{B})$ are equivalent because for all $\pi \in \mathcal{S}_n$ we have

$$Z(\bar{A}, \bar{B}, \pi) = Z(A, B, \pi) + \kappa Z(E_n, B, \pi) + \kappa Z(A, E_n, pi) + \kappa^2 Z(E_n, E_n, \pi),$$

where E_n is the $n \times n$ matrix of all ones. Then, for all $\pi \in \mathcal{S}_n$ we have

$$Z(E_n, B, \pi) = \sum_{i=1}^{n}\sum_{j=1}^{n} b_{ij} \quad Z(A, E_n, \pi) = \sum_{i=1}^{n}\sum_{j=1}^{n} a_{ij}$$

$$Z(E_n, E_n, \pi) = n^2$$

The above equalities show that $QAP(A, B)$ and $QAP(\bar{A}, \bar{B})$ have the same optimal solutions, and hence, they are equivalent.

Next, we present some results on the combinatorial structure of the classes of matrices MONGE, A-MONGE and KALMANSON. It can be shown that each of these classes forms a cone. Moreover, for each of these cones, 0-1 matrices generating the corresponding extremal rays can be specified explicitly. For more details about these results and their proofs the reader is referred to [12, 33, 50, 67, 205]. Here we give only sketches of proofs. The proofs apply a general method which can be used for all classes of matrices discussed in this section. This method is based on the following characterization of polyhedral cones (see e.g. [211]).

Proposition 4.2 *Let a polyhedral cone $C = \{x \in \mathbb{R}^n : Ax \le 0\}$ be given, where A is an $m \times n$ matrix. Let $z_1, z_2, \ldots z_r$ be a minimal set of vectors generating the r-dimensional linear subspace $L = \{x \in \mathbb{R}^n : Ax = 0\}$ of C, and let G_1, G_2, ..., G_s, be the minimal proper faces of C, i.e. the $(r+1)$-dimensional faces of C. Each G_i can be written as $G_i = \{x : \langle a_i, x\rangle \le 0, A_i' x = 0\}$, where A_i' is a submatrix of A and a_i^t is a row of A such that*

$$rank \begin{pmatrix} A_i' \\ a_i^t \end{pmatrix} = n - r \tag{4.7}$$

and $L = \{x : \langle a_i, x\rangle = 0, A_i' x = 0\}$. Now, choose for each $i = 1, 2, \ldots, s$, a vector y_i from $G_i \setminus L$, i.e. a vector y_i fulfilling $\langle a_i, y\rangle < 0$ and $A_i' y_i = 0$. Then, for each $c \in C$ there exist nonnegative real numbers λ_i, $1 \le i \le s$, and arbitrary real numbers μ_j, $1 \le j \le r$, such that

$$c = \sum_{i=1}^{s} \lambda_i y_i + \sum_{j=1}^{r} \mu_j z_j \qquad \square$$

Now assume that we want 1) to prove that a given class CLASS of $n \times n$ matrices is a polyhedral cone, and 2) to find its extremal rays. Our approach consists of the following steps. First, find a representation of CLASS as CLASS $= \{X: AX \leq 0\}$, where A is an $n^2 \times n^2$ matrix of full rank and $X \in$ CLASS is thought as an n^2-dimensional vector[2]. Secondly, find the dimension $dimL$, and a basis for the linear space $L = \{X: AX = 0\}$. Finally, investigate the minimal faces of CLASS, i.e. the faces which are determined by $\{X: \langle a_i, x\rangle \leq 0, A_i' x = 0\}$, for some row a_i^t, and some submatrix A_i' of matrix A, which fulfill (4.7).

Next, let us see how this approach can be applied to different classes of Monge-like matrices.

4.2.1 The cone of Monge matrices

First, let us introduce the $n \times n$ matrices $A^{(p,q)}$, $1 \leq p, q \leq n-1$, given as follows:

$$a_{ij}^{(p,q)} = \begin{cases} 1 & \text{if } i = p, j = q \text{ or } i = p+1, j = q+1 \\ -1 & \text{if } i = p, j = q+1 \text{ or } i = p+1, j = q \\ 0 & \text{otherwise} \end{cases}$$

Clearly, there are $(n-1)^2$ matrices of this type. Denote by $\mathcal{A}$ the $(n-1)^2 \times n^2$ matrix whose rows are the matrices $A^{(p,q)}$ (considered as n^2-dimensional vectors). From the definition of Monge matrices follows immediately that

$$\text{MONGE} = \{X: \mathcal{A}X \leq 0\}$$

Moreover, by means of elementary algebra it can be shown that the matrices $A^{(p,q)}$, $1 \leq p, q \leq n-1$, form a linear independent system in $\mathbb{R}^{(n-1)^2}$. Hence, $rank\mathcal{A} = (n-1)^2$ and, consequently, $dimL = n^2 - (n-1)^2 = 2n-1$.

Further, let us show that the matrices $H^{(p)} = (h_{ij}^{(p)})$, $V^{(q)} = (v_{ij}^{(q)})$, $1 \leq p \leq n$, $2 \leq q \leq n$, defined as follows, form a basis of the linear space L.

$$h_{ij}^{(p)} = \begin{cases} 1 & \text{if } i = p \\ 0 & \text{otherwise} \end{cases} \qquad v_{ij}^{(q)} = \begin{cases} 1 & \text{if } j = q \\ 0 & \text{otherwise} \end{cases}$$

Matrix $H^{(p)}$ has a row of 1-entries, namely the row with index p, and the rest of its entries are equal to 0. Similarly, matrix $V^{(q)}$ has a column of 1-entries, namely the column with index q, and the rest of its entries are equal to 0. It is easy to see that these matrices which are in fact SUM matrices, fulfill the Monge inequalities (4.1) with equality, or equivalently, $\mathcal{A}H^{(p)} = 0$, $\mathcal{A}V^{(q)} = 0$. Hence, $H^{(p)}$, $V^{(q)}$, belong

[2]Throughout this section we make no difference between an $n \times n$ matrix X and the n^2-dimensional vector obtained by a top-to-bottom row-wise ordering of the entries of X.

to L. It is not difficult to see that the rank of this system of matrices is $2n-1$, and that gives us a basis for L. Indeed, $\sum_{p=1}^{n} H^{(p)} = \sum_{q=1}^{n} V^{(q)}$, and $H^{(p)}$, $V^{(q)}$, $1 \leq p \leq n$, $2 \leq q \leq n$, are linearly independent. Summarizing, we obtain the following lemma.

Lemma 4.3 *The Monge matrices form a cone represented as*

$$\text{MONGE} = \{X: \mathcal{A}X \leq 0\}$$

where $\mathcal{A}$ is as above and $rank\mathcal{A} = (n-1)^2$. The dimension of the corresponding linear subspace $L = \{X: \mathcal{A}X = 0\}$ equals $2n-1$ and the set of matrices $H^{(p)}$, $V^{(q)}$, $1 \leq p \leq n$, $2 \leq q \leq n$, is a basis of L.

Next we show that each of the rows of $\mathcal{A}$ defines a minimal face of the cone MONGE. Indeed, for each of the rows $A^{(p,q)}$, consider the submatrix $\mathcal{A}^{(p,q)}$ of $\mathcal{A}$ obtained from $\mathcal{A}$ by deleting the row $A^{(p,q)}$. Clearly

$$rank \begin{pmatrix} A^{(p,q)} \\ \mathcal{A}^{(p,q)} \end{pmatrix} = rank\mathcal{A} = (n-1)^2 = n^2 - dimL \tag{4.8}$$

Let $B^{(p,q)}$ be the n^2-dimensional vector obtained as sum of all rows of $\mathcal{A}$ but the row $A^{(p,q)}$. Then, the inequality $\langle B^{(p,q)}, X\rangle \leq 0$ is fulfilled by all $X \in$ MONGE, and $\langle B^{(p,q)}, X\rangle = 0$ only if $\mathcal{A}^{(p,q)}X = 0$. Hence, the face defined by the equality $\langle B^{(p,q)}, X\rangle = 0$ equals $G^{(p,q)} = \{X: \langle A^{(p,q)}, X\rangle \leq 0, \mathcal{A}^{(p,q)}X = 0\}$, and $G^{(p,q)}$ is a minimal face due to equality (4.8). Now, for each pair of indices (p,q) let us introduce a matrix which belongs to the corresponding face $G^{(p,q)}$ but does not belong to the linear space L. Consider the matrices $L^{(p,q)} = (l_{ij}^{(p,q)})$, $2 \leq p \leq n$, $1 \leq q \leq n-1$ specified as below:

$$l_{ij}^{(p,q)} = \begin{cases} 1 & \text{if } i \geq n-p+1,\, j \leq q \\ 0 & \text{otherwise} \end{cases}$$

Matrix $L^{(p,q)}$ has a $p \times q$ submatrix with 1-entries in the left-lower corner; the rest of its entries are equal to 0. It is not difficult to check that $L^{(p,q)} \in G^{(p-1,q)} \setminus L$, for all $2 \leq p \leq n$, $1 \leq q \leq n-1$. Summarizing we get the following lemma:

Lemma 4.4 *The sets $G^{(p,q)}$, $1 \leq p, q \leq n-1$, are the minimal faces of the cone* MONGE. *Moreover, $L^{(p,q)} \in G^{(p-1,q)} \setminus L$ for $2 \leq p \leq n$, $1 \leq q \leq n-1$.*

Now Proposition 4.2 can be applied and the following characterization of the Monge matrices is immediately obtained.

Theorem 4.5 *For each $n \times n$ Monge matrix M, there exist real numbers μ_i, $1 \le i \le n$, and λ_j, $2 \le j \le n$ and nonnegative numbers κ_{ij} for $2 \le i \le n$, $1 \le j \le n-1$ such that*

$$M = \sum_{i=1}^{n} \mu_i H^{(i)} + \sum_{i=2}^{n} \lambda_i V^{(i)} + \sum_{i=2}^{n} \sum_{j=1}^{n-1} \kappa_{ij} L^{(i,j)} \qquad \square$$

The same result was obtained by Rudolf and Woeginger [205] in a direct proof. Moreover, Rudolf et al. considered the cone of *nonnegative* Monge matrices and identified its extremal rays. To this end they define the matrices $R^{(p,q)} = (r_{ij}^{(p,q)})$, $1 \le p, q \le n$, as follows:

$$r_{ij}^{(p,q)} = \begin{cases} 1 & \text{if } i \le p,\ j \ge n-q+1 \\ 0 & \text{otherwise} \end{cases}$$

Matrix $R^{(p,q)}$ has a $p \times q$ submatrix with 1-entries in the right-upper corner; the rest of its entries are equal to 0. Consider as an illustrative example the 3×3 matrices $H^{(2)}$, $V^{(3)}$, $L^{(1,2)}$ and $R^{(2,2)}$ represented below:

$$H^{(2)} = \begin{pmatrix} 0 & 0 & 0 \\ 1 & 1 & 1 \\ 0 & 0 & 0 \end{pmatrix} \qquad V^{(3)} = \begin{pmatrix} 0 & 0 & 1 \\ 0 & 0 & 1 \\ 0 & 0 & 1 \end{pmatrix}$$

$$L^{(1,2)} = \begin{pmatrix} 0 & 0 & 0 \\ 0 & 0 & 0 \\ 1 & 1 & 0 \end{pmatrix} \qquad R^{(2,2)} = \begin{pmatrix} 0 & 1 & 1 \\ 0 & 1 & 1 \\ 0 & 0 & 0 \end{pmatrix}$$

By making use of elementary means of algebra, Rudolf et al. derive the following result.

Proposition 4.6 (Rudolf and Woeginger 1995, [205])
The nonnegative $n \times n$ Monge matrices form a cone. Its extremal rays are generated by the 0-1 matrices $H^{(i)}$, $V^{(i)}$, for $1 \le i \le n$, and $L^{(i,j)}$, $R^{(i,j)}$, for $1 \le i, j \le n$. $\square$

Sometimes we will restrict our investigations to *symmetric* Monge matrices which obviously form a cone. Similarly as for the cone of nonnegative Monge matrices, Rudolf et al. identify the extremal rays of the cone of nonnegative symmetric Monge matrices. To this end the matrices $S^{(p)} = H^{(p)} + V^{(p)}$, $1 \le p \le n$, and $T^{(p,q)} =$

$L^{(p,q)} + R^{(q,p)}$, $1 \le p, q \le n$, $p+q \le n$, are introduced. For example, consider the 3×3 matrices $S^{(2)}$ and $T^{(1,2)}$ given below:

$$S^{(2)} = \begin{pmatrix} 0 & 1 & 0 \\ 1 & 2 & 1 \\ 0 & 1 & 0 \end{pmatrix} \qquad T^{(1,2)} = \begin{pmatrix} 0 & 0 & 1 \\ 0 & 0 & 1 \\ 1 & 1 & 0 \end{pmatrix}$$

In terms of matrices $S^{(p)}$ and $T^{(p,q)}$, the cone of nonnegative symmetric Monge matrices is characterized as follows.

Proposition 4.7 (Rudolf and Woeginger 1995, [205])
The nonnegative symmetric $n \times n$ Monge matrices form a cone. Its extremal rays are generated by the matrices $S^{(i)}$, $1 \le i \le n$, and $T^{(i,j)}$, $1 \le i, j \le n$, $i+j \le n$. Thus, for every nonnegative $n \times n$ matrix A in MONGE(SYM), *there exist nonnegative numbers κ_i, $1 \le i \le n$, and μ_{ij}, $1 \le i, j \le n$, $i+j \le n$, such that:*

$$A = \sum_{i=1}^{n} \kappa_i S^{(i)} + \sum_{i=1}^{n-1} \sum_{j=1}^{n-i} \mu_{ij} T^{(i,j)} \qquad \square$$

4.2.2 The cone of graded Anti-Monge matrices

Some of our polynomiality results on QAPs with Monge or Anti-Monge matrices will additionally require the Monge-like matrix to be graded, i.e. with monotone columns and rows. Analogously to the nonnegative Monge matrices, the nonnegative left-higher graded Anti-Monge matrices and the nonnegative right-lower graded Monge matrices form also cones. The following simple observation helps to describe the structure. of these matrix classes.

Observation 4.8 *(a) A matrix A is Anti-Monge if and only if*

$$\Delta_{ij} := a_{ij} - a_{i,j-1} - a_{i-1,j} + a_{i-1,j-1} \ge 0, \text{ for } 2 \le i, j \le n. \tag{4.9}$$

(b) An Anti-Monge matrix is left-higher graded, if its first row and its first column are non-decreasing, i.e. if

$$\Delta_{i1} := a_{i1} - a_{i-1,1} \ge 0, \quad \text{for } 1 < i \le n, \text{ and} \tag{4.10}$$

$$\Delta_{1j} := a_{1j} - a_{1,j-1} \ge 0, \quad \text{for } 1 < j \le n. \tag{4.11}$$

(c) A left-higher graded Anti-Monge matrix is nonnegative if

$$\Delta_{11} := a_{11} \geq 0. \tag{4.12}$$

(d) A matrix A is completely determined by the n^2 values Δ_{ij}, $1 \leq i, j \leq n$. □

Note that a matrix A is a nonnegative left-higher graded Anti-Monge matrix, if and only if the $(n+1) \times (n+1)$ matrix obtained by bordering A with an additional top row of zeros and an additional left column of zeros is an Anti-Monge matrix. In this way, inequalities (4.10)–(4.12) become special cases of (4.9), and the additional requirements of monotonicity and non-negativity appear quite natural for Anti-Monge matrices.

For the characterization of the cone of nonnegative matrices in A-MONGE(LHG) we can use again Proposition 4.2. First, we derive a description of A-MONGE of the form $\{X: \mathcal{A}X \leq 0\}$. Here $\mathcal{A}$ will be an $n^2 \times n^2$ matrix with rows determined by the matrices $A^{(p,q)} = (a_{ij}^{(p,q)})$, $1 \leq p, q \leq n$, given as follows.

$$a_{ij}^{(1,1)} = \begin{cases} -1 & \text{if } i = j = 1 \\ 0 & \text{otherwise} \end{cases} \qquad a_{ij}^{(1,q)} = \begin{cases} 1 & \text{if } i = 1,\ j = q-1 \\ -1 & \text{if } i = 1,\ j = q \\ 0 & \text{otherwise} \end{cases} \qquad \text{for } q \geq 2$$

$$a_{ij}^{(p,1)} = \begin{cases} 1 & \text{if } i = p-1,\ j = 1 \\ -1 & \text{if } i = p,\ j = 1 \\ 0 & \text{otherwise} \end{cases} \qquad \text{for } p \geq 2$$

$$a_{ij}^{(p,q)} = \begin{cases} 1 & \text{if } i = p-1, j = q \text{ or } i = p, j = q-1 \\ -1 & \text{if } i = p-1, j = q-1 \text{ or } i = p, j = q \\ 0 & \text{otherwise} \end{cases} \qquad \text{for } p, q \geq 2$$

According to Observation 4.8, the set of nonnegative left-higher graded $n \times n$ Anti-Monge matrices is given as $\{X: \mathcal{A}X \leq 0\}$. Hence, these matrices forms a cone. Moreover, it is easy to see that $\mathcal{A}$ has full rank, and hence $L = \{X: \mathcal{A}X = 0\}$ consists only of the 0-matrix, i.e. the matrix with all entries equal to 0, and $dimL = 0$. For the characterization of the cone according to Proposition 4.2 we have to identify its minimal faces. Similarly as in the previous section, we introduce submatrices $\mathcal{A}^{(p,q)}$ of $\mathcal{A}$, $1 \le p,q \le n$, obtained by deleting the corresponding row $A^{(p,q)}$. Further, similarly as in the case of Monge matrices, it is not difficult to show that the sets $G^{(p,q)} = \{X: \langle A^{(p,q)}, X\rangle \le 0, \mathcal{A}^{(p,q)}X = 0\}$, for $1 \le p,q \le n$, are the minimal faces of $\{X: \mathcal{A}X \le 0\}$. Now it remains to select a matrix which belongs to $G^{(p,q)} \setminus L$, for all $1 \le p,q \le n$. To this end we introduce the matrices $C^{(p,q)} = (c_{ij}^{(p,q)})$ for $1 \le p,q \le n$

$$c_{ij}^{(p,q)} = \begin{cases} 1 & \text{if } i \ge n-p+1,\ j \ge n-q+1 \\ 0 & \text{otherwise} \end{cases}$$

$C^{(p,q)}$ is a 0-1 matrix whose ones form a lower-right block of size $p \times q$. It is easy to check that $\mathcal{A}C^{(p,q)} \le 0$ for all $1 \le p,q \le n$, and $\langle A^{(p,q)}, C^{(p',q')}\rangle = 0$ if and only if $p' = n-p+1$, $q' = n-q+1$. By applying Proposition 4.2 we get the following theorem.

Theorem 4.9 *The nonnegative left-higher graded $n \times n$ Anti-Monge matrices form a cone. For each matrix M in this cone there exist nonnegative numbers λ_{pq}, $1 \le p,q \le n$, such that*

$$M = \sum_{p=1}^{n}\sum_{q=1}^{n} \lambda_{pq} C^{(p,q)}.$$

Hence, the 0-1 matrices $C^{(p,q)}$, $1 \le p,q \le n$, are the extremal rays of this cone. □

Based on the above characterization of the nonnegative left-higher graded Anti-Monge matrices, we can characterize the nonnegative right-lower graded Monge matrices. Notice that MONGE(RLG) = NA-MONGE(LHG). Hence, the non-positive right-lower graded Monge matrices are exactly the nonnegative left-higher graded Anti-Monge matrices multipled by -1. Each nonnegative right-lower graded Monge matrix can be obtained by some non-positive right-lower graded Monge matrix by adding to the latter some constant which is sufficiently large. Considering all these facts, it is not difficult to see that the matrices $E^{(p,q)} = \left(e_{i,j}^{(p,q)}\right)$ for $1 \le p,q \le n$, defined below, will be of interest for the characterization of the nonnegative matrices in MONGE(RLG).

$$e_{i,j}^{(p,q)} = \begin{cases} 1 & \text{if } i \le p \text{ or } j \le q \\ 0 & \text{otherwise} \end{cases} \qquad 1 \le p,q \le n$$

$E^{(p,q)}$ is a 0-1 matrix whose zeroes form a lower-right block of size $p \times q$. Consider as an illustrating example the 3×3 matrices $E^{(2,1)}$ and $C^{(2,1)}$given below:

$$E^{(2,1)} = \begin{pmatrix} 1 & 1 & 1 \\ 1 & 1 & 1 \\ 1 & 0 & 0 \end{pmatrix} \qquad C^{(2,1)} = \begin{pmatrix} 0 & 0 & 0 \\ 0 & 0 & 1 \\ 0 & 0 & 1 \end{pmatrix}$$

Summarizing, we get a straightforward corollary of Theorem 4.9.

Corollary 4.10 *The nonnegative right-lower graded $n \times n$ Monge matrices form a cone. For each matrix M in this cone there exist nonnegative numbers λ_{pq}, $1 \le p, q \le n$, such that*

$$M = \sum_{p=1}^{n} \sum_{q=1}^{n} \lambda_{pq} E^{(p,q)}$$

Hence the 0-1 matrices $E^{(p,q)}$, $1 \le p, q \le n$, are the extremal rays of this cone.

4.2.3 The cone of Kalmanson matrices

For the characterization of Kalmanson matrices we exploit again Proposition 4.2. To this end we need a matrix $\mathcal{A}$ such that KALMANSON $= \{X : \mathcal{A}X \le 0\}$. We make use of the following description of Kalmanson matrices due to Deĭneko, Rudolf and Woeginger [64].

Proposition 4.11 (Deĭneko, Rudolf and Woeginger [64], 1995)
A symmetric $n \times n$ matrix $C = (c_{ij})$ is a Kalmanson matrix if and only if the following inequalities are fulfilled:

$$c_{1i} + c_{n,i+1} \le c_{1,i+1} + c_{ni}, \quad \text{for all } 2 \le i \le n-2$$

$$c_{i,j+1} + c_{i+1,j} \le c_{ij} + c_{i+1,j+1}, \quad \text{for all } 1 \le i \le n-3,\ i+2 \le j \le n-1$$

Note that the dimension of the vector space of the $n \times n$ symmetric matrices equals $n(n+1)/2$, and hence, every symmetric matrix corresponds to an $n(n+1)/2$-dimensional vector. Throughout this section, we make no difference between the matrix and the corresponding vector. Further, let the 0-1 matrices $B^{(p)} = (b_{ij}^{(p)})$, for $2 \le p \le n-2$, and $B_{ij}^{(p,q)} = (b_{ij}^{(p,q)})$, for $1 \le p \le n-3$, $p+2 \le q \le n-1$, be defined below

$$b_{ij}^{(p)} = \begin{cases} 1 & \text{if } i=1,\ j=p \text{ or } i=p+1,\ j=n \\ -1 & \text{if } i=1,\ j=p+1 \text{ or } i=p,\ j=n \\ 0 & \text{otherwise} \end{cases}$$

$$b_{ij}^{(p,q)} = \begin{cases} 1 & \text{if } i=p,\ j=q+1 \text{ or } i=p+1,\ j=q \\ -1 & \text{if } i=p,\ j=q \text{ or } i=p+1,\ j=q+1 \\ 0 & \text{otherwise} \end{cases}$$

Then Proposition 4.11 implies KALMANSON $= \{X: \mathcal{B}X \leq 0\}$, where $\mathcal{B}$ is a matrix whose rows are $n(n+1)/2$-dimensional vectors corresponding to the symmetric matrices $B^{(p)}$, $B^{(p,q)}$ defined above. Hence $\mathcal{B}$ is an $(n(n-3)/2) \times (n(n+1)/2)$ matrix. Moreover, it can be proven by means of elementary algebra that the rows of $\mathcal{B}$ are linearly independent, i.e. $rank\mathcal{B} = n(n-3)/2$ (see [67] for a proof of this fact). The last equality implies that $dimL = n(n+1)/2 - n(n-3)/2 = 2n$, where $L = \{X: \mathcal{A}X = 0\}$. Further, it can be shown that the matrices $F^{(p,p)} = (f_{ij}^{(p,p)})$ and $F^{(p)} = f_{ij}^{(p)}$, $1 \leq p \leq n$, defined below, belong to L and are linearly independent. Hence, these matrices form a basis of L.

$$f_{ij}^{(p,p)} = \begin{cases} 1 & \text{if } i = j = p \\ 0 & \text{otherwise} \end{cases} \qquad f_{ij}^{(p)} = \begin{cases} 1 & \text{if } p = i,\ q \neq i \text{ or } p \neq i, q = i \\ 0 & \text{otherwise} \end{cases}$$

Summarizing, we obtain the following lemma.

Lemma 4.12 *The Kalmanson matrices form a cone which can be described as*

$$\text{KALMANSON} = \{X: \mathcal{B}X \leq 0\},$$

where $\mathcal{B}$ is introduced above. The matrices $F^{(p,p)}$, $F^{(p)}$, $1 \leq p \leq n$, form a basis of the corresponding linear space L.

It remains to identify the minimal faces of KALMANSON and a "proper" element in each face. Then, Proposition 4.2 can be applied. Analogously as in the cases of MONGE and A-MONGE, let the submatrices $\mathcal{B}^{(p,q)}$, $1 \leq p \leq n-3$, $p+2 \leq q \leq n-1$, and $\mathcal{B}^{(p)}$, $2 \leq i \leq n-2$, of $\mathcal{B}$ be obtained from $\mathcal{B}$ by deleting row $B^{(p,q)}$, $B^{(p)}$, respectively. The analogy goes further to proving that the sets $G^{(p,q)} = \{X: \langle B^{(p,q)}, X\rangle \leq 0, \mathcal{B}^{(p,q)}X = 0\}$, $G^{(p)} = \{X: \langle A^{(p)}, X\rangle \leq 0, \mathcal{B}^{(p)}X = 0\}$ are minimal faces of KALMANSON. Now, let us introduce the so-called *cut matrices*[3] $w^{(p)} = (w^{(p)})$, $3 \leq p \leq n-1$, $W^{(p,q)} = (w_{ij}^{(p,q)})$, $2 \leq p \leq n-2$, $p+1 \leq q \leq n-1$, which will turn out to be "proper" elements of the minimal faces of KALMANSON.

$$w_{ij}^{(p)} = \begin{cases} 1 & \text{if } |\{i,j\} \cap [p,n]| = 1 \\ 0 & \text{otherwise} \end{cases} \qquad w_{ij}^{(p,q)} = \begin{cases} 1 & \text{if } |\{i,j\} \cap [p,q]| = 1 \\ 0 & \text{otherwise} \end{cases}$$

It is easy to check that $w^{(p)} \in G^{(p-1)} \setminus L$ and $w^{(p,q)} \in G^{(p-1,q)} \setminus L$. Based on these arguments and on Lemma 4.12, Proposition 4.2 can be applied to obtain a characterization of the cone KALMANSON.

Theorem 4.13 (Bandelt et al. [12], 1992, Christopher et al. [50], 1996, Demidenko et al. [67], 1995)

[3]This term was originally used by Deĭneko and Woeginger [65]

The $n \times n$ Kalmanson matrices form a cone. For each $n \times n$ matrix M in this cone there exist real numbers α_i, β_i, $1 \le i \le n$, and nonnegative numbers λ_i, $3 \le i \le n-1$, μ_{ij}, $2 \le i \le n-2$, $i+1 \le j \le n-1$, such that

$$M = \sum_{i=1}^{n} (\alpha_i F^{(i,i)} + \beta_i F^{(i)}) + \sum_{i=3}^{n-1} \lambda_i W^{(i)} + \sum_{i=2}^{n-2} \sum_{j=i+1}^{n-1} \mu_{ij} W^{(i,j)}. \qquad \square$$

As an illustrative example for the matrices $F^{(p,p)}$, $F^{(p)}$, $W^{(p)}$, $W^{(p,q)}$ which generate the cone of Kalmanson matrices consider the 5×5 matrices $F^{(3)}$, $F^{(3,3)}$, $W^{(3)}$, $W^{(2,3)}$.

$$F^{(3,3)} = \begin{pmatrix} 0 & 0 & 1 & 0 & 0 \\ 0 & 0 & 1 & 0 & 0 \\ 1 & 1 & 0 & 1 & 1 \\ 0 & 0 & 1 & 0 & 0 \\ 0 & 0 & 1 & 0 & 0 \end{pmatrix} \qquad F^{(3)} = \begin{pmatrix} 0 & 0 & 0 & 0 & 0 \\ 0 & 0 & 0 & 0 & 0 \\ 0 & 0 & 1 & 0 & 0 \\ 0 & 0 & 0 & 0 & 0 \\ 0 & 0 & 0 & 0 & 0 \end{pmatrix}$$

$$W^{(3)} = \begin{pmatrix} 0 & 0 & 1 & 1 & 1 \\ 0 & 0 & 1 & 1 & 1 \\ 1 & 1 & 0 & 0 & 0 \\ 1 & 1 & 0 & 0 & 0 \\ 1 & 1 & 0 & 0 & 0 \end{pmatrix} \qquad W^{(2,4)} = \begin{pmatrix} 0 & 1 & 1 & 0 & 0 \\ 1 & 0 & 0 & 1 & 1 \\ 1 & 0 & 0 & 1 & 1 \\ 0 & 1 & 1 & 0 & 0 \\ 0 & 1 & 1 & 0 & 0 \end{pmatrix}$$

4.2.4 The cone-structure and the QAP

After having discussed the structure of several sets of Monge-like matrices, a natural question arises: Can we exploit the special structure of these matrix classes and their properties for the identification and solution of polynomially solvable restricted versions of the QAP?

As we will see further in this chapter, the answer is yes. More specifically, assume that a restricted version of the QAP with specially structured coefficient matrices is believed to be a constant permutation QAP. Moreover, assume that some permutation ϕ_0 is conjectured to be the constant permutation. In the case that the coefficient matrices belong to matrix classes which form cones (as described in the previous sections), the coefficient matrices A, B of some instance QAP(A,B) of the considered version of the QAP can be expressed as follows:

$$A = S_1 + \sum_i \alpha_i A_i, \qquad B = S_2 + \sum_j \beta_j B_j,$$

Here, S_1 and S_2 are in most of the cases matrices with a "simple" structure. e.g. sum matrices, A_i, B_i are 0-1 matrices generating the extremal rays of the corresponding cones, and the coefficients α_i, β_j are nonnegative. According to Observation 4.1, we have

$$\begin{aligned} Z(A,B,\phi) &= Z(S_1,S_2,\phi)+\sum_j \beta_j Z(S_1,B_j,\phi)+\sum_i \alpha_i Z(A_i,S_2,\phi) \\ &\quad +\sum_{i,j} \alpha_i\beta_j Z(A_i,B_j,\phi) \end{aligned} \tag{4.13}$$

The QAP with a sum matrix is polynomially solvable, as we will see in the next section. Moreover, the QAP with a sum matrix and another specially structured coefficient matrix, e.g. a circulant matrix, is a constant QAP (see next section). So, let us assume that the problems $QAP(S_1,S_2)$, $QAP(S_1,B_j)$, $QAP(A_i,S_2)$ are constant QAPs, i.e. every permutation is an optimal solution for each of them, and concentrate our attention at the problems $QAP(A_i,B_j)$ with 0-1 coefficient matrices. Because of (4.13) and the nonnegativity of the numbers α_i, β_j, it is sufficient to prove that the conjectured constant permutation ϕ_0 is an optimal solution for each of these instances of the QAP. In most of the cases, it is of benefit to restrict our investigations to these QAP with 0-1 coefficient matrices, as it is much simpler to check the conjecture concerning the optimality of ϕ_0 for the problems $QAP(A_i,B_j)$ than for the original QAP(A,B).

Clearly, the applicability of this approach, henceforth called *reduction to extremal rays*, is quite limited, as it can be only used if the QAP at hand is a constant permutation QAP. Moreover, we would need a good guess for the eventual constant permutation. However, generally, we can try to check whether the given problem is the constant permutation QAP by investigating the combinatorial structure of matrices A_i and B_i which generate the extremal rays of the corresponding cones. Such investigations could parallelly help to get some feeling concerning the structure of the constant permutation, in the case that the latter exists.

4.3 SIMPLE POLYNOMIALLY SOLVABLE CASES OF THE QAP

The most simple polynomially solvable special cases of the QAP are either constant QAPs, in the case that the coefficient matrices have very special properties (e.g. one matrix is skew symmetric and the other one is symmetric), or "trivial" problems because of the ordering of the entries of the coefficient matrices (the coefficient matrices are "appropriately" graded). Besides such special cases, we also discuss in this section QAPs which can be reformulated as linear assignment problems (LAPs). Such QAP versions arise if the problem data imply the "linearity" of the objective function, as e.g. in the case of sum matrices.

First, let us recall Proposition 2.1 which is exploited in many proofs throughout this chapter: Given two vectors $U = (u_i)$ and $V = (v_i)$, the minimum of the scalar product of the permuted vectors $\langle U^{(\phi)}, V^{(\psi)} \rangle$ over all pairs of permutations (ϕ, ψ) is attained when $U^{(\phi)}$ and $V^{(\psi)}$ are sorted in non-increasing and non-decreasing order, respectively. The following proposition, probably the very first result on well solvable cases of the QAP, is a straightforward corollary of Proposition 2.1.

Proposition 4.14 (Krushevski 1964, [144])
Consider two $n \times n$ matrices $A = (a_{ij})$ and $B = (b_{ij})$ such that for all 4-tuples of indices (i, j, k, l) the following equivalence holds:

$$(a_{ij} \geq a_{kl}) \Longleftrightarrow (b_{ij} \leq b_{kl})$$

Then, the identity permutation id is an optimal solution to QAP(A,B). □

The following theorem states a generalization of Proposition 4.14:

Theorem 4.15 *The identity permutation id is an optimal solution to QAPs on the classes* LHG×RLG, LLG×RHG.

Proof. We prove the result only for QAPs on the class LHG×RLG. In the other case the proof is completely analogous. Let $A = (a_{ij})$ be an $n \times n$ matrix with non-decreasing rows and columns and let $B = (b_{ij})$ be an $n \times n$ matrix with non-increasing rows and columns. Consider the relaxation I-QAP(A,B) and an arbitrary pair of permutations (φ, π) in $\mathcal{S}_n$. It can be easily seen that the following inequalities hold, due to Proposition 2.1:

$$\begin{aligned} Z(A, B, \varphi, \pi) &= \sum_{i=1}^{n} \sum_{j=1}^{n} a_{\varphi(i)\pi(j)} b_{ij} \geq \sum_{i=1}^{n} \sum_{j=1}^{n} a_{i\pi(j)} b_{ij} \\ &\geq \sum_{i=1}^{n} \sum_{j=1}^{n} a_{ij} b_{ij} = Z(A, B, id, id) \end{aligned} \tag{4.14}$$

Inequalities (4.14) show that (id, id) is an optimal solution to I-QAP(A,B). Hence, *id* is an optimal solution to QAP(A,B). □

Notice that both Proposition 4.14 and Theorem 4.15 describe classes of constant permutation QAPs At this point it is probably interesting to notice that the complexity of the apparently simple problem LHG×RHG remains an open question. Next we identify another class of constant QAPs, namely the class SKEW×SYM.

Theorem 4.16 *The problem* SKEW×SYM *is a constant QAP. The value of the objective function corresponding to an arbitrary permutation equals* 0.

Proof. Let A be a skew symmetric $n \times n$ matrix and let B be a symmetric $n \times n$ matrix. In this case the objective function of QAP(A,B) can be written as

$$Z(A,B,\pi) = \sum_{i=1}^{n}\sum_{j=1}^{n} a_{\pi(i)\pi(j)}b_{ij} = \sum_{j=2}^{n}\sum_{i=1}^{j-1} a_{\pi(i)\pi(j)}b_{ij} + \sum_{i=1}^{n} a_{\pi(i)\pi(i)}b_{ii} + \sum_{i=2}^{n}\sum_{j=1}^{i-1} a_{\pi(i)\pi(j)}b_{ij}\,.$$

Due to symmetry conditions, the first and the third term on the right hand side of the above equality cancel each other. Since the diagonal elements of a skew symmetric matrix are equal to 0, the remaining term $\sum_{i=1}^{n} a_{\pi(i)\pi(i)}b_{ii}$ equals 0 and this completes the proof. □

The next theorem presents an easy-to-prove but still surprising result. Namely, the QAP with a sum matrix is polynomially solvable.

Theorem 4.17 *The problem* SUM×MATRIX *is solvable in $O(n^3)$ time, where n is the size of the problem.*

Proof. The proof basically relies on transforming QAP(A,B) with an $n \times n$ sum matrix A and an arbitrary $n \times n$ matrix B to a linear assignment problem. Let (α_i^r) and (α_i^c) be the generating vectors of matrix A. The objective function of QAP(A,B) can be written as follows:

$$Z(A,B,\pi) = \sum_{i=1}^{n}\sum_{j=1}^{n} a_{\pi(i)\pi(j)}b_{ij} = \sum_{i=1}^{n}\sum_{j=1}^{n} (\alpha_{\pi(i)}^r + \alpha_{\pi(j)}^c)b_{ij} = \sum_{i=1}^{n}\left(\alpha_{\pi(i)}^r \sum_{j=1}^{n} b_{ij}\right) + \sum_{j=1}^{n}\left(\alpha_{\pi(j)}^c \sum_{i=1}^{n} b_{ij}\right) = \sum_{j=1}^{n} \alpha_{\pi(j)}^r \beta_j^r + \sum_{j=1}^{n} \alpha_{\pi(j)}^c \beta_j^c\,, \tag{4.15}$$

where β_i^r, (β_i^c), $1 \le i \le n$, is the sum of the i-th row (i-th column) of matrix B. Now, consider an $n \times n$ matrix $D = (d_{ij})$ defined by

$$d_{ij} = \alpha_i^r \beta_j^r + \alpha_i^c \beta_j^c \quad \text{for } 1 \le i, j \le n$$

With this notation we have $Z(A, B, \pi) = \sum_{i=1}^{n} d_{\pi(i)i}$. Hence, solving QAP(A,B) is equivalent to solving the linear assignment problem $\min_{\pi \in \mathcal{S}_n} \sum_{i=1}^{n} d_{\pi(i)i}$. □

If the matrices A and (or) B in the above theorem have some "nice property" the problem SUM×MATRIX becomes even "easier". Two results of this type are presented by the following theorem.

Theorem 4.18 *(a) The problem* SUM×CIRC *is a constant QAP, i.e. every permutation $\pi \in \mathcal{S}_n$ is an optimal solutions to* SUM×CIRC *of size n.*
(b) If at least one of the matrices A, B is symmetric or skew symmetric, then QAP(A,B) with $A \in$ SUM *is solvable in $O(n^2)$ time, where n is the size of the problem.*

Proof of (a). Consider QAP(A,B) where $A = (a_{ij})$ and $B = (b_{ij})$ are two $n \times n$ matrices, with $A \in$ SUM and $B \in$ CIRC. For an arbitrary $\pi \in \mathcal{S}_n$, we have:

$$Z(A, B, \pi) = \sum_{i=1}^{n} \sum_{j=1}^{n} a_{\pi(i)\pi(j)} b_{ij} = \sum_{p=0}^{n-1} \left(c_p \sum_{i=1}^{n} a_{\pi(i)\pi(k_{p+i})} \right), \tag{4.16}$$

where $c_0, c_1, \ldots, c_{n-1}$ are real numbers as introduced in the definition of circulant matrices and $k_j := j \ (mod\ n)$, for $j \in \{1, 2 \ldots, 2n-1\} \setminus \{n\}$ and $k_n := n$. Let (α_i^r) and (α_i^c) be the generating vectors of the sum matrix A. Then the right-hand side of equality (4.16) can be written as follows:

$$Z(A, B, \pi) = \sum_{p=0}^{n-1} \left(c_p \sum_{i=1}^{n} \alpha_{\pi(i)}^r \right) + \sum_{p=0}^{n-1} \left(c_p \sum_{i=1}^{n} \alpha_{\pi(k_{p+i})}^c \right) =$$

$$\sum_{p=0}^{n-1} \left(c_p \sum_{i=1}^{n} \alpha_i^r \right) + \sum_{p=0}^{n-1} \left(c_p \sum_{j=1}^{n} \alpha_j^c \right)$$

Note that the right hand side of the last equality does not depend on permutation π. This completes the proof in this case.

Proof of (b). Consider QAP(A,B) with $A \in$ SUM and $B \in$ MATRIX(SKEW). (The three remaining combinations SUM×MATRIX(SYM), SUM(SYM)×MATRIX, SUM(SKEW)×MATRIX,can be discussed analogously.) Denote by β_i^r (β_i^c), $1 \le i \le n$, the sum of the i-th row (i-th column) of matrix B, respectively. As B is skew symmetric, $\beta_i^r = -\beta_i^c$, $1 \le i \le n$. Then, for every $\pi \in \mathcal{S}_n$, the following equalities holds:

$$Z(A,B,\pi) = \sum_{i=1}^{n}\sum_{j=1}^{n} a_{\pi(i)\pi(j)}b_{ij} = \sum_{i=1}^{n} \alpha_{\pi(i)}^r \beta_i^r + \sum_{i=1}^{n} \alpha_{\pi(i)}^c \beta_i^c$$

$$= \sum_{i=1}^{n} \left(\alpha_{\pi(i)}^r - \alpha_{\pi(i)}^c\right) \beta_i^r$$

Thus, solving QAP(A,B) is equivalent to solving the following minimization problem

$$\min_{\pi \in \mathcal{S}_n} \sum_{i=1}^{n} \left(\alpha_{\pi(i)}^r - \alpha_{\pi(i)}^c\right) \beta_i^r$$

According to Proposition 2.1, this minimization problem can be solved by sorting the n-dimensional vectors $(\alpha_i^r - \alpha_i^c)$ and (β_i). This takes $O(n \log n)$ time, whereas the vector (β_i^r) can be computed in $O(n^2)$ time. Here we assume that the generating vectors of matrix A, (α_i^c) and (α_i^r), are given. □

The next proposition is another early result due to Krushevski [145]. It discusses a polynomially solvable case of the QAP where the algebraic structure of the coefficient matrices A and B allows us to write the objective function as a quadratic form. For more details on this special case the reader is referred to the original reference in Russian and to Rendl [193].

Proposition 4.19 (Krushevski 1965, [145])
Let real numbers $u_1 \le u_2 \le \cdots \le u_n$, $v_1 \ge v_2 \ge \cdots \ge v_n$, $u = \sum_{i=1}^n u_i$, $v = \sum_{i=1}^n v_i$, and α_1, α_2, β_1, β_2 be given. Let the $n \times n$ matrices $A = (a_{ij})$ and $B = (b_{ij})$ be defined by $a_{ij} = \beta_1 v_i + \beta_2 v_j + v_i v_j$ and $b_{ij} = \alpha_1 u_i + \alpha_2 u_j + u_i u_j$, respectively. If $\sum_{i=1}^n u_i v_i \ge -\Delta/2$ holds, where

$$\Delta = (\alpha_1 + \alpha_2)v + (\beta_1 + \beta_2)u + n(\alpha_1\beta_1 + \alpha_2\beta_2)\ ,$$

then the identity permutation yields an optimal solution to QAP(A,B).

Proof. For any fixed $\pi \in S_n$ consider the scalar product $\langle V^\pi, U\rangle$, where $U = (u_i)$ and $V = (v_i)$. The objective function of QAP(A,B) can be written as given below

$$Z(A,B,\pi) = \sum_{i=1}^{n}\sum_{j=1}^{n} a_{\pi(i)\pi(j)}b_{ij} = \langle V^\pi, U\rangle^2 + \Delta\langle V^\pi, U\rangle + (\alpha_1\beta_2 + \alpha_2\beta_1)uv \quad (4.17)$$

The function $f(x) = x^2 + \Delta x + b$ increases for $x \geq -\Delta/2$. Since by assumption, $-\Delta/2 \leq \langle V, U\rangle \leq \langle V^\pi, U\rangle$ holds for each $\pi \in \mathcal{S}_n$, the expression in the right hand side of (4.17) is minimized by $\pi = id$. □

Remark. Notice that, in general, Proposition 4.19 cannot be derived as a corollary from our previous results, although the structure of the matrices A, B is quite simple. Namely, $A = A_1 + A_2$ and $B = B_1 + B_2$, where $A_1 = (\alpha_1 v_i + \alpha_2 v_j)$, $A_2 = (v_i v_j)$, $B_1 = (\beta_1 u_i + \beta_2 u_j)$, $B_2 = (u_i u_j)$, for $1 \leq i,j \leq n$. Thus, A is a sum of two matrices $A_1 \in$ SUM and $A_2 \in$ PROD(SYM) and similarly, $B = B_1 + B_2$, where $B_1 \in$ SUM and $B_2 \in$ PROD(SYM). One could think of writing the objective function $Z(A,B,\pi)$ as sum of the objective functions of four other QAPs with a "nicer", already investigated, structure. Namely, $Z(A,B,\pi) = Z(A_1,B_1,\pi) + Z(A_1,B_2,\pi) + Z(A_2,B_1,\pi) + Z(A_2,B_2,\pi)$, for all $\pi \in \mathcal{S}_n$. We can minimize the first three summands of the form $Z(A_k,B_l,\pi)$, $k,l = 1,2$, in the above equation, by applying point (b) of Theorem 4.18. The fourth summand $Z(A_2,B_2,\pi)$ is equal to $(\sum_{i=1}^n u_{\pi(i)}v_i)^2$. In the case that all u_i, v_i, $1 \leq i \leq n$ have the same sign, the latter expression can be minimized by applying Proposition 2.1. Otherwise, the problem is NP-hard as it will be shown in Theorem 4.24 in the next section. However, even if we would have minimized each of the summands, we would be lucky to have also minimized $Z(A,B,\pi)$ only if problems $QAP(A_k,B_l)$, $k,l = 1,2$, have a common optimal solution. But, as described in the proof of Theorem 4.18, this depends on the monotonicity of matrices A_k and B_k, $k = 1,2$, i.e. on the algebraic signs of coefficients α_i and β_i, $i = 1,2$, and u_i, v_i, $1 \leq i \leq n$. In the case that α_i, β_i, $i = 1,2$, and u_i, v_i, $i = 1,2,\ldots,n$, have all the same sign, the problem is obviously solved by applying Theorem 4.15, even when condition $\sum_{i=1}^n u_i v_i \geq -\Delta/2$ is dropped. Let us formulate this as a corollary:

Corollary 4.20 *Let $u_1 \leq u_2 \leq \cdots \leq u_n$, $v_1 \geq v_2 \geq \cdots \geq v_n$, and α_1, α_2, β_1, β_2 be real numbers. Let the $n \times n$ matrices $A = (a_{ij})$ and $B = (b_{ij})$ be defined by $a_{ij} = \beta_1 v_i + \beta_2 v_j + v_i v_j$ and $b_{ij} = \alpha_1 u_i + \alpha_2 u_j + u_i u_j$, respectively. If the numbers α_i, β_i, $i = 1,2$, and u_i, v_i, $i = 1,2,\ldots,n$, have all the same sign, then the identity permutation yields an optimal solution to QAP(A,B).*

4.4 QAPS WITH MONGE-LIKE MATRICES

It is a well known fact that many "difficult" combinatorial optimization problems become "easy" to solve when the input is restricted to Monge or Monge-like matrices (or other mathematical structures with Monge or Monge-like properties). The *traveling salesman problem* (TSP) and the *transportation problem* are two classical examples[4]. The positive impact of Monge and Monge-like properties in combinatorial optimization motivated the investigation of the computational complexity of restricted versions of the QAP where both coefficient matrices are restricted to be Monge-like matrices. It turns out that in most of the cases QAP(A,B) restricted to Monge-like matrices remains NP-hard. Anyway, if the coefficient matrices A and B fulfill also additional conditions, polynomially solvable versions of the QAP arise.

Most of the results on QAPs with Monge-like matrices concern versions of the problem where the coefficient matrices belong to one of the classes MONGE, NMONGE = A-MONGE, PROD(SYM), NPROD(SYM), CHESS and NCHESS. Notice here that the following inclusions holds:

$$\text{CHESS} \subseteq_{\pi} \text{PROD(SYM)} \subseteq_{\pi} \text{NMONGE} . \tag{4.18}$$

Equivalently, we have NCHESS$\subseteq_{\pi}$NPROD(SYM)$\subseteq_{\pi}$MONGE. There exist no general methods for dealing with such QAPs. The polynomially solvable cases identified up to now are either implicit LAPs or constant permutation QAPs. The implicit LAPs arise in the case that the coefficient matrices have a very special structure, e.g. in the case of product matrices. The constant permutation QAPs can be solved by reduction to extremal rays, as we briefly introduced in Section 4.2.4. There is only one case with coefficient matrices having very special properties (large symmetric matrices or chess-board matrices), where the corresponding QAPs are neither implicits LAPs nor constant permutation QAPs. These problems are solved by dynamic programming.

Table 4.1 summarizes the results known to date on the computational complexity of QAPs with both coefficient matrices having Monge-like properties. An entry "poly" means that the QAP with input matrices taken from the corresponding row and column is solvable in polynomial time, an entry "NP" means that the corresponding problem is NP-hard and an entry "???" means that the computational complexity of the corresponding problem is unknown. When reading the table, note that two N-s cancel each other and that the table indeed contains all information. e.g. the complexity of QAP with both input matrices in PROD(SYM) is found at the

[4]For a detailed overview on the role of Monge and Monge-like properties in combinatorial optimization the reader is referred to [38].

[5]The problem CHESS(EVEN)×MONGE(EVEN) is proven to be polynomially solvable, whereas the complexity of CHESS(ODD)×MONGE(ODD) is still an open question.

Table 4.1 The computational complexity of the QAP on Monge-like matrices

	MONGE	NPROD(SYM)	NCHESS
MONGE	NP	NP	NP
NPROD(SYM)	NP	NP	NP
NCHESS	NP	$NP_{(d)}$	$poly_{(a)}$
NMONGE	???	???	$poly_{(b)}$[5]
PROD(SYM)	???	$poly_{(c)}$	poly
CHESS	poly	poly	poly

intersection of the NPROD(SYM)-row with the NPROD(SYM)-column. Because of the inclusion relations (4.18), Table 4.1 has much additional structure. An "NP" entry in the upper half makes all other entries in the upper half that lie above or to the left of it to "NP"-entries. "poly"-entries produce other "poly"-entries below and to the right of them. An analogous statement holds for the lower half. Moreover, notice that both the upper and the lower half of Table 4.1 are symmetric with respect to the corresponding main diagonals. This is immediately seen for the upper half of the table. The symmetry of the lower half of the table is due to the equivalence of problems QAP(-A,B) and QAP(A,-B), for arbitrary matrices A and B. Due to this structure, the four results marked by subscripts would imply all the others in Table 4.1. The results marked by the subscripts (a), (b), (c) and (d) are proved in Theorems 4.21, 4.22, 4.23 and 4.24, respectively.

Theorem 4.21 *The problems* CHESS×CHESS *and* CHESS×NCHESS *are solvable in polynomial time.*

Proof. The objective function $Z(A, B, \pi)$ of the problem CHESS×CHESS can be written as

$$Z(A,B,\pi) = \sum_{i=1}^{n}\sum_{j=1}^{n}(-1)^{\pi(i)+\pi(j)}(-1)^{i+j} = \left(\sum_{i=1}^{n}(-1)^{i+\pi(i)}\right)^2 = (n-4m_\pi)^2 \ ,$$

where A, B are two $n \times n$ chess-board matrices, $\pi \in \mathcal{S}_n$, and m_π is the cardinality of the set $\{i \in \{1,2,\ldots,n\} \mid \pi(i) \text{ is odd and } i \text{ is even}\}$. It is easy to check that, for each $1 \leq m \leq \lfloor n/2 \rfloor$, there exists a permutation $\pi \in \mathcal{S}_n$ such that $m = m_\pi$. Hence, the optimal value of QAP(A,B) equals $\min\{(n-4m)^2 | 0 \leq m \leq \frac{n}{2}\}$. This optimal value equals 1 for odd n, 0 for n divisible by four and 4 for $n \equiv 2 (mod\ 4)$.

Denote by m_0 the value of m for which the above minimum is achieved. Then each permutation $\pi \in \mathcal{S}_n$ with $m_0 = m_\pi$ is an optimal solution to QAP(A,B).

For the second problem CHESS×NCHESS, the identity permutation is an optimal solution as it yields a value of the objective function equal to $-n^2$ which is obviously the optimal one. □

Next, let us consider the more general problem MONGE×CHESS. Notice that this problem is a restricted version of MONGE×PROD(SYM) (referred to as an open problem in Table 4.1). We show that the problem MONGE×CHESS of even size is a constant permutation QAP. The proof exploits the fact that the nonnegative Monge matrices form a cone. This result was implicitly proven by Pferschy, Rudolf and Woeginger [185]. These authors investigated the *maximum balanced bisection* problem for Monge matrices and showed that it is solvable in polynomial time, if the size of the problem is even. An optimal bisection does not depend on the instance of the Monge matrix; it always consists of the first half of the rows (respectively columns) versus the second half of the rows (respectively columns). Clearly, the maximum bisection problem on (even) Monge matrices is equivalent to the problem MONGE×CHESS of even size.

Theorem 4.22 *Consider QAP(A,B) where A and B are $n \times n$ matrices, $A \in$ MONGE and $B \in$ CHESS. If n is even, $n = 2m$, the permutation π_* given by*

$$\pi_*(x) = \begin{cases} i & \text{if} \ \ x = 2i-1, \ \ 1 \le i \le m \\ m+i & \text{if} \ \ x = 2i, \ \ 1 \le i \le m \end{cases}$$

is an optimal solution to QAP(A,B).

Proof. First, recall that w.l.o.g. we can assume matrix A to have nonnegative entries (see Section 4.2). Next, notice that QAP(A,B) is equivalent to QAP(A,B$^{(1)}$), where $B^{(1)} = \frac{1}{2}(B + E_n)$ and E_n is the $n \times n$ matrix of all ones. (These problems even have a common set of optimal solutions.) This follows easily from Observation 4.1 and the fact that, for fixed A and $\pi \in \mathcal{S}_n$, $Z(A, E_n, \pi)$ is a constant independent on π. So, we prove the theorem for QAP(A,B$^{(1)}$) instead of proving it for QAP(A,B), where $B_{ij}^{(1)} = \left(b_{ij}^{(1)}\right)$ and

$$b_{ij}^{(1)} = \begin{cases} 1 & \text{if} \ \ i+j \ \ \text{is even} \\ 0 & \text{otherwise} \end{cases}$$

We show that permutation π_* is an optimal solution of QAP(D,B$^{(1)}$), for all 0-1 matrices D which generate some extremal ray of the cone of nonnegative $n \times n$

Monge matrices described in Section 4.2. Then by exploiting Proposition 4.6 and Observation 4.1 this optimality result carries over to all Monge matrices. There are four QAP instances to be investigated, corresponding to the four different types of 0-1 matrices which generate the cone of nonnegative Monge matrices. In each case we show that the objective function value corresponding to permutation π_* is a lower bound for the value of the objective function, and consequently, π_* is an optimal solution.

Case 1: $D = H^{(p)}$, $1 \le p \le n$.
The following equalities show that problem QAP$(H^{(p)}, B^{(1)})$ is a constant QAP:

$$Z(H^{(p)}, B^{(1)}, \pi) = \sum_{j=1}^{n} h^{(p)}_{p,\pi(j)} b^{(1)}_{i_0,j} = \sum_{j=1}^{n} b^{(1)}_{i_0,j} = |J| = n/2 \qquad (4.19)$$

where i_0 is such that $\pi(i_0) = p$ and $J = \{j \mid j \in \{1, 2, \ldots, n\}$ and $j + i_0$ is even$\}$. (In the above equality the fact that n is even is essentially used.) Thus, each permutation π is an optimal solution of QAP$(H^{(p)}, B^{(1)})$ and so does π_*.

Case 2: $D = V^{(q)}$, $1 \le q \le n$.
Analogous to Case 1. It can be shown that QAP$(V^{(q)}, B^{(1)})$ is a constant QAP.

Case 3: $D = L^{(p,q)}$, $1 \le p, q \le n$.
In this case the objective function is given by the following simple formula for all $\pi \in \mathcal{S}_n$:

$$\begin{aligned} Z(L^{(p,q)}, B^{(1)}, \pi) &= \sum_{i=n-p+1}^{n} \sum_{j=1}^{q} b^{(1)}_{\pi^{-1}(i)\pi^{-1}(j)} \\ &= |M_1||M_2| + (p - |M_1|)(q - |M_2|) \qquad (4.20) \end{aligned}$$

where π^{-1} is the inverse permutation of π and the sets M_1, M_2 are defined below:

$$\begin{aligned} M_1 &= \Big\{x \in \{n-p+1, n-p+2, \ldots, n\} \mid \pi^{-1}(x) \text{ is even}\Big\} \\ M_2 &= \Big\{x \in \{1, 2, \ldots, q\} \mid \pi^{-1}(x) \text{ is even}\Big\} \end{aligned}$$

Let us apply formula (4.20) to calculate $Z(L^{(p,q)}, B^{(1)}, \pi_*)$. First, notice that for the permutation π_*^{-1} inverse of π_* we have:

$$\pi_*^{-1}(x) = \begin{cases} 2x - 1 & \text{if} \quad 1 \le x \le m \\ 2(x - m) & \text{if} \quad m + 1 \le x \le n \end{cases}$$

Define a new matrix $B^{(2)} = \left(b_{ij}^{(2)}\right)$ by $b_{ij}^{(2)} = b^{(1)}_{\pi_*^{-1}(i)\pi_*^{-1}(j)}$, $1 \le i, j \le n$. It is easy to see that

$$b_{ij}^{(2)} = \begin{cases} 1 & \text{if } i \le m,\ j \le m \text{ or } i > m,\ j > m \\ 0 & \text{otherwise} \end{cases}$$

Considering the structure of matrix $B^{(2)}$ and the equalities

$$Z(L^{(p,q)}, B^{(1)}, \pi_*) = Z(L^{(p,q)}, B^{(2)}, id) = \sum_{i=n-p+1}^{n} \sum_{j=1}^{q} b_{ij}^{(2)}$$

we get:

$$Z(L^{(p,q)}, B^{(1)}, \pi_*) = \begin{cases} 0 & \text{if } p \le m \ q \le m \\ (p-m)q & \text{if } p > m \ q \le m \\ (q-m)p & \text{if } p \le m \ q > m \\ (p-m)m + (q-m)m & \text{if } p > m \ q > m \end{cases} \tag{4.21}$$

Now, let us show that $Z(L^{(p,q)}, B^{(1)}, \pi) \ge Z(L^{(p,q)}, B^{(1)}, \pi_*)$, for all $\pi \in \mathcal{S}_n$. This can be done by distinguishing the following four subcases:

(a) $p \le m$, $q \le m$. As the entries of matrices $L^{(p,q)}$ and $B^{(1)}$ are nonnegative, $Z(L^{(p,q)}, B^{(1)}, \pi) \ge 0 = Z(L^{(p,q)}, B^{(1)}, \pi_*)$, for all $\pi \in \mathcal{S}_n$.

(b) $p > m$, $q \le m$. Checking that the following inequalities hold, for all $\pi \in \mathcal{S}_n$, completes the proof in this case.

$$\begin{aligned} Z(L^{(p,q)}, B^{(1)}, \pi) &= |M_1||M_2| + (p - |M_1|)(q - |M_2|) \\ &= pq - |M_1|(q - |M_2|) - |M_2|(p - |M_1|) \ge pq - m|M_2| \\ &\quad -m(q - |M_2|) = q(p-m) = Z(L^{(p,q)}, B^{(1)}, \pi_*) \end{aligned}$$

(c) $p \le m$, $q > m$. Analogously as in (b) it can be shown that $Z(L^{(p,q)}, B^{(1)}, \pi) \ge p(q-m)$, for all $\pi \in \mathcal{S}_n$.

(d) $p > m$, $q > m$. We show that $Z(L^{(p,q)}, B^{(1)}, \pi) \ge (p-m)m + (q-m)m$, for all $\pi \in \mathcal{S}_n$. First notice that $p - m \le |M_1| \le m$ and $q - m \le |M_2| \le m$. Now it is elementary to see that the function f defined as

$$f \colon \{p-m, p-m+1, \ldots, m\} \times \{q-m, q-m+1, \ldots, m\} \to \mathbb{R}^+$$

$$(x, y) \mapsto xy + (p-x)(q-y) \ ,$$

achieves its minimum either at the point $(m, q-m)$ or at the point $(p-m, m)$. The minimum value is $f(m, q-m) = f(p-m, m) = (p-m)m + (q-m)m$. Considering equality (4.20) we have $Z(L^{(p,q)}, B^{(1)}, \pi) = f(|M_1|, |M_2|) \geq (p-m)m+(q-m)m = Z(L^{(p,q)}, B^{(1)}, \pi_*)$.

Case 4: $D = R^{(p,q)}$, $1 \leq p, q \leq n$.
In this case the proof is completely analogous to the proof in Case 3. □

The complexity of MONGE×CHESS remains open for odd n. The proof of Theorem 4.22 exploits essentially the fact that the size of the problem is even. Also the exchange argument in [185], used for solving the balanced bisection problem to optimality, fails for problems of odd size. Moreover, notice that the proof of Theorem 4.22 strongly relies on the fact that the considered problem is a constant permutation QAP. Problem MONGE×CHESS of odd size is not a constant permutation QAP. The following example shows that even the "simpler" problem QAP($L^{(p,q)}, B^{(1)}$) of odd size, with $B^{(1)}$ defined as in the proof of Theorem 4.22, is not a constant permutation QAP. That is, for different pairs (p, q), instances of QAP($L^{(p,q)}, B^{(1)}$) do not *necessarily* have a common optimal solution.

Example 4.1 Consider the 3×3 matrices $L^{(1,2)}$, $L^{(2,1)}$, $B^{(1)}$ and permutations $\pi_1, \pi_2, \pi_3, \pi_4 \in \mathcal{S}_3$, $\pi_1 = \langle 1, 3, 2\rangle$, $\pi_2 = \langle 2, 3, 1\rangle$, $\pi_3 = \langle 2, 1, 3\rangle$ and $\pi_4 = \langle 3, 1, 2\rangle$. It can be checked that

$$Z(L^{(1,2)}, B^{(1)}, \pi_1) = Z(L^{(1,2)}, B^{(1)}, \pi_2) = 0 \ ,$$

$$Z(L^{(1,2)}, B^{(1)}, \pi_3) = Z(L^{(1,2)}, B^{(1)}, \pi_4) = 1 \ ,$$

$$Z(L^{(2,1)}, B^{(1)}, \pi_3) = Z(L^{(2,1)}, B^{(1)}, \pi_4) = 0 \ ,$$

$$Z(L^{(2,2)}, B^{(1)}, \pi_1) = Z(L^{(2,1)}, B^{(1)}, \pi_2) = 1 \ .$$

Moreover, it can be easily seen that π_1, π_2 are the only optimal solutions to QAP($L^{(1,2)}, B^{(1)}$), and analogously, π_3, π_4 are the only optimal solutions to the problem QAP($L^{(2,1)}, B^{(1)}$). Thus, these two QAPs do *not* have any common optimal solution. □

Next we formulate and prove the result marked by the subscript (c) in Table 4.1.

Theorem 4.23 *The problem* PROD(SYM)×NPROD(SYM) *is solvable in polynomial time.*

Proof. Consider the $n \times n$ matrices A, B, where $A = (\alpha_i\alpha_j) \in$ PROD(SYM), $B = (-\beta_i\beta_j) \in$ NPROD(SYM). The QAP(A,B) is equivalent to the following maximization problem

$$\max_{\pi \in \mathcal{S}_n} \{(\sum_{i=1}^{n} \alpha_{\pi(i)}\beta_i)^2\} .$$

By applying Proposition 2.1, the maximum and the minimum of $\sum_{i=1}^{n} \alpha_{\pi(i)}\beta_i$ over all permutations $\pi \in S_n$ is determined. The maximum of the two corresponding squared values is the optimal value of the given QAP and the corresponding permutation is an optimal solution to it. □

As for the problem PROD(SYM)×PROD(SYM), even though this problem seems to be quite similar to PROD(SYM)×NPROD(SYM) and its coefficient matrices have the same simple structure, it turns out that this problem is NP-hard. Actually, even a "more restricted" version of the QAP, namely CHESS×PROD(SYM), is NP-hard, as shown by the next theorem. The proof of the NP-hardness is done by a reduction from the NP-complete EQUIPARTITION problem (cf. Garey and Johnson [88]).

> EQUIPARTITION
>
> Input: A set $\{x_1, \ldots, x_{2n}\}$ of positive integers.
>
> Question: Does there exist a subset $I \subset \{1, 2, \ldots, 2n\}$, $|I| = n$, such that $\sum_{i \in I} x_i = \sum_{i \notin I} x_i$ holds?

Theorem 4.24 *The problem* CHESS×PROD(SYM) *is NP-hard.*

Proof. We start with an instance of EQUIPARTITION and define a $2n \times 2n$ symmetric product matrix $B = (\beta_i \beta_j)$ by $\beta_i = x_i$, $1 \le i \le 2n$. A is a chess-board matrix, i.e. $A = ((-1)^{i+j})$. Clearly, QAP(A,B) is an instance of CHESS×PROD(SYM). For a given $\pi \in \mathcal{S}_{2n}$ denote $I_\pi = \{\pi(i) : i \text{ is even}\}$. Similarly as in the proof of the last theorem, $Z(A, B, \pi)$ can be written as follows:

$$Z(A, B, \pi) = \left(\sum_{i \in I_\pi} x_i - \sum_{i \notin I_\pi} x_i \right)^2, \quad \text{for all } \pi \in \mathcal{S}_{2n}. \tag{4.22}$$

Thus, QAP(A,B) is equivalent to the problem $\min\limits_{\pi \in \mathcal{S}_{2n}} \{(\sum_{i \in I_\pi} x_i - \sum_{i \notin I_\pi} x_i)^2\}$. We show that the answer to the given instance of the EQUIPARTITION problem is "YES" if and only if the optimal value of the above defined instance of QAP(A,B) is 0, and this completes the proof.

(if) Assume there exists an $I \subset \{1, 2, \ldots, 2n\}$, $|I| = n$, such that $\sum_{i \in I} x_i = \sum_{i \notin I} x_i$. Consider a permutation $\pi_0 \in \mathcal{S}_{2n}$ which permutes the numbers $2i$, $1 \le$

$i \le n$, into elements of I, i.e. the equality $I_{\pi_0} = I$ holds. Then, (4.22) implies $Z(A, B, \pi_0) = 0$.

(only if) Assume that the optimal value of QAP(A,B) is 0. Thus, there exists a permutation $\pi_0 \in \mathcal{S}_{2n}$ such that $Z(A, B, \pi_0) = 0$. Equality (4.22) implies $(\sum_{i \in I_{\pi_0}} x_i - \sum_{i \notin I_{\pi_0}} x_i)^2 = 0$. As $|I_{\pi_0}| = n$, I_{π_0} solves the instance of the EQUIPARTITION problem by "YES". □

Remark. Because of the inclusion CHESS ⊂ PROD(SYM), Theorem 4.24 implies that the problem PROD(SYM)×PROD(SYM) is NP-hard. In other words, for two vectors (α_i) and (β_i), minimizing $\left(\sum_{i=1}^{n} \alpha_{\pi(i)}\beta_i\right)^2$ over all permutations $\pi \in \mathcal{S}_n$ is NP-hard, whereas maximizing it is polynomially solvable, as shown by Theorem 4.23. The complexity of the problems MONGE × PROD(SYM) and MONGE×NMONGE remains an open question. Obviously, proving either the polynomiality for the latter problem, or the NP-hardness for the first one, would answer both open questions and complete Table 4.1.

The following result is a straightforward corollary of Theorem 4.24:

Corollary 4.25 MONGE(SYM, RLG)×MONGE(SYM) *is NP-hard.*

Proof. By Theorem 4.24, problem CHESS×PROD(SYM) is NP-hard. Moreover, it is easy to check that the following inclusions hold

NPROD(SYM, POS)$\subseteq_\pi$MONGE(SYM, RLG), NCHESS$\subseteq_\pi$MONGE(SYM) □

From Theorem 4.24 it follows also that the problem MONGE(SYM)×NCHESS is NP-hard. However, it is possible to derive better results for restricted versions of MONGE(SYM)×NCHESS, where the Monge matrix is somehow "specially structured". For example, the problem LARGE(SYM)×NCHESS can be solved polynomially by a dynamic programming approach. Also LARGE(SYM)×CHESS can be solved by an analogous dynamic programming scheme. Recall here that the problem LARGE(SYM)×CHESS is a special case of MONGE(SYM)×CHESS, and the complexity of odd sized instances of the latter problem is still an open question.

Theorem 4.26 *The problems* LARGE(SYM)×NCHESS, LARGE(SYM)×CHESS *are solvable in* $O(n^2)$ *time by dynamic programming, where* n *is the size of the problem.*

Proof. We prove the result only for the problem LARGE(SYM)×NCHESS. A completely analogous proof can be given for the problem LARGE(SYM)×CHESS. Consider QAP(A,B) of size n with $A \in$ LARGE(SYM)glossary∈ and $B \in$ NCHESS. First we make two important remarks.

1. W.l.o.g. we can assume the large symmetric matrix A to be left-higher graded, because LARGE(SYM)$\subseteq_\pi$LARGE(SYM,LHG). Indeed, if the generating vector of the large symmetric matrix is not sorted in increasing order, we can reorder it increasingly. This would result in permuting the rows and the columns of matrix A by the same permutation, say ψ, to get a matrix $A^\psi = (a_{\psi(i),\psi(j)})$. Obviously, solving QAP(A,B), is equivalent to solving QAP(A^ψ,B) because of the equality $Z(A,B,\pi) = Z(A^\psi, B, \psi^{-1} \circ \pi)$ which holds for all $\pi \in \mathcal{S}_n$.

2. Analogously as in the proof of Theorem 4.22, instead of solving QAP(A,B) we may equivalently solve QAP(A,B′), where $B' = (b'_{ij})$ and

$$b'_{ij} = \begin{cases} 0 & \text{if} \quad i+j \text{ is even} \\ 1 & \qquad \text{otherwise} \end{cases}$$

Next, we construct a dynamic programming scheme for solving QAP(A,B) with $A \in$ LARGE(SYM,LHG) and $B = B'$. Let (α_i) be the generating vector of A, $\alpha_i \le \alpha_{i+1}$, $1 \le i \le n-1$. Due to the equalities $a_{ij} = a_{ji} = \alpha_i$ for $i \ge j$, the objective function of QAP(A,B) can be rewritten as follows:

$$Z(A,B,\pi) = \sum_{i=1}^{n}\sum_{j=1}^{n} a_{ij} b_{\pi^{-1}(i)\pi^{-1}(j)} = 2\sum_{i=2}^{n}\left(\alpha_i \sum_{j=1}^{i-1} b_{\pi^{-1}(i)\pi^{-1}(j)}\right).$$

Considering the last equality and the structure of matrix B, it is easy to see that the parity of $\pi^{-1}(i)$, $1 \le i \le n$, determines the corresponding value $Z(A,B,\pi)$ of the objective function. Thus, if for two permutations $\pi_1, \pi_2 \in \mathcal{S}_n$ the equality $I_{\pi_1} = I_{\pi_2}$ holds, then $Z(A,B,\pi_1) = Z(A,B,\pi_2)$. Here I_π is defined as in the proof of Theorem 4.24, i.e. $I_\pi = \{\pi(i) : i \text{ is even}\}$, for all $\pi \in \mathcal{S}_n$. Hence, in order to solve QAP(A,B), it is sufficient to specify the subset I_{π_0} for an optimal solution π_0 rather than determining π_0 itself. In other words, we are looking for an *appropriate* set $I_0 \subset \{1,2,\ldots,n\}$, $|I_0| = \frac{n}{2}$. Once such a set I_0 is found, we can construct a permutation $\pi_0 \in \mathcal{S}_n$ such that $I_{\pi_0} = I_0$. π_0 would then be an optimal solution to QAP(A,B)[6]. We show that the required set I_0 can be constructed by a dynamic programming approach.

[6]Given some I_0, we can construct more than just one permutation π having the property $I_\pi = I_0$. Namely, there are $\binom{n}{\lfloor n/2 \rfloor} \lfloor n/2 \rfloor!$ such permutations. Clearly, all of them are optimal solutions to QAP(A,B).

For $2 \le k \le n$, denote by I_π^k the intersection $I_\pi \cap \{1, 2, ldots, k\}$. Further, for $2 \le k \le n$, $0 \le s \le \min\{k, \lfloor \frac{n}{2} \rfloor\}$, $0 \le k - s \le \min\{k, \lceil \frac{n}{2} \rceil\}$, consider the auxiliary problem $P(k,s)$ defined as follows:

$$\mathbf{P(k,s)}: \quad \min_{\pi \in \mathcal{S}_n : |I_\pi^k| = s} \left\{ 2 \sum_{i=2}^{k} \left(\alpha_i \cdot \left| \left\{ j : 1 \le j \le i-1,\ \pi^{-1}(i) + \pi^{-1}(j) \text{ is odd} \right\} \right| \right) \right\}$$

Problem $P(k,s)$ is a kind of restricted QAP, where the objective function is the sum of *a part* of the entries $a_{\pi(i)\pi(j)}$ with $1 \le \pi(i), \pi(j) \le k$, and the set of feasible solution is a subset of $\mathcal{S}_n$ consisting of those permutations which map exactly s even numbers from $\{1, 2, \ldots, n\}$ to $\{1, 2, \ldots, k\}$. Hence, QAP(A,B) is exactly P(n,$\lfloor \frac{n}{2} \rfloor$). Denote by Z(k,s) the optimal value of problem P(k,s). Then, Z(n,$\lfloor \frac{n}{2} \rfloor$) is the optimal value of QAP(A,B).

Next we introduce our dynamic programming scheme. For a feasible pair (k, s) with $s < \min\{k, \lfloor n/2 \rfloor\}$ and $k \ge 3$, an optimal solution of problem P(k,s+1) maps either an odd number or an even number to k. It is easily seen that the following equality holds for all feasible pairs (k, s):

$$Z(k, s+1) = \min\left\{ 2\alpha_k(s+1) + Z(k-1, s+1), 2\alpha_k(k-1-s) + Z(k-1, s) \right\} \quad (4.23)$$

Further, notice that $Z(2,0) = Z(2,2) = 0$, $Z(2,1) = 2\alpha_2$ and $Z(k,0) = 0$, for $2 \le k \le \lceil n/2 \rceil$. Moreover, set $Z(k,s) := +\infty$ for all unfeasible pairs (k,s), $1 \le k, s \le n$. Assume that we know the value of $Z(k-1, s)$ for all $1 \le s \le n$. As $Z(k-1, 0)$ is also known, we can consecutively calculate $Z(k,1), Z(k,2), \ldots$ by applying equality (4.23). Obviously, this can be done in $O(n)$ steps which require $O(1)$ time each. Thus, starting with $Z(2,s)$ $(k-1 = 2)$ and applying the above scheme, the computation of $Z(n, \lfloor n/2 \rfloor)$ takes $O(n^2)$ elementary operations. Then, the set I_0 can be specified by backtracking the minima in (4.23) and setting $k \in I_0$ if and only if the corresponding minimum is achieved at the second term in the right-hand side of (4.23). □

According to Table 4.1, the problem A-MONGE×CHESS is NP-hard, whereas the complexity of A-MONGE×NCHESS of odd size is unknown. The following straightforward corollary of Theorem 4.26 shows that solvable special cases of the above problems arise if the Anti-Monge matrix has some additional "nice structure".

Corollary 4.27 *The problems* SMALL(SYM)×CHESS *and* SMALL(SYM)×NCHESS *are solvable in $O(n^2)$ time by dynamic programming, where n is the size of the problem.*

Proof. According to Proposition 4.1, QAP(A,B) is equivalent to QAP(-A,-B). Next, notice that NSMALL(SYM) = LARGE(SYM). Thus, SMALL(SYM)×CHESS and SMALL(SYM)×NCHESS are equivalent to the problems LARGE(SYM)×NCHESS and LARGE(SYM)×CHESS, respectively. □

4.5 QAPS WITH CIRCULANT AND SMALL BANDWIDTH MATRICES

This section and the next one deal with QAPs restricted to diagonally structured matrices. There are two main motives for the investigation of QAPs on diagonally structured matrices.

First, several "difficult" combinatorial optimization problems, similar to QAPs with respect to their combinatorial structure, become "easy" when restricted to diagonally structured matrices. The *Hamiltonian path problem* on circulant matrices and the *traveling salesman problem (TSP)* on matrices with small bandwidth are two typical examples. It is worthy to notice here that the TSP on circulant matrices is still a challenging open problem. (For a more detailed discussion on these topics see [153]). Some positive results concerning polynomially solvable special cases of the QAP with diagonally structured matrices, more concretely *Toeplitz* matrices, were obtained in the 1960's and 1970's [41, 165, 187, 223].

The second motive concerns a large number of problems of theoretical and practical nature, which can be formulated as QAPs with diagonally structured coefficient matrices. Among applications of practical relevance let us mention the so-called *turbine problem* and different versions of *placement problems*. Some of these applications will be discussed in detail later on in this chapter. The TSP, the *taxonomy problem*, and the problem of *finding a spanning set of cycles of fixed length*, are some theoretical applications. As such problems are NP-hard in their general formulation, the identification of polynomially solvable restricted versions of them becomes an interesting question.

In the current section we consider QAPs with one of the coefficient matrices either being a circulant or having a small bandwidth. As we will see, these restrictions on the coefficient matrices are too weak to guarantee the polynomiality of the corresponding versions of the QAP. Therefore, we assume the other coefficient matrix to have Monge or Monge-like properties. It will turn out that under these conditions some polynomially solvable cases arise. The polynomially solvable versions are all constant permutation QAPs and the proofs are done by reduction to extremal rays (see Section 4.2.4).

4.5.1 Two "easy" QAPs with circulant and bandwidth-k matrices

As a straightforward corollary of Theorem 4.24 it can be shown that CIRC×MONGE is a "difficult" problem.

Corollary 4.28 *The problem* CIRC×MONGE *is NP-hard.*

Proof. Recall that in the proof of Theorem 4.24 we have shown the NP-hardness of the problem PROD(SYM)×CHESS(EVEN). Clearly, this problem is equivalent to NPROD(SYM)×NCHESS(EVEN) and therefore, the latter is NP-hard too. The proof is completed by considering the following inclusions:

$$\text{NPROD(SYM)} \subseteq_\pi \text{MONGE}\,, \qquad \text{NCHESS(EVEN)} \subseteq_\pi \text{CIRC} \qquad \square$$

We conjecture that the QAP with two circulant matrices, CIRC×CIRC, is NP-hard. Notice that this problem contains as a special case another celebrated open problem, namely, the TSP on circulant matrices. However, in the case that both circulant matrices A and B of QAP(A,B) have very special properties, polynomially solvable cases may arise. As an example consider the following theorem which is a straightforward corollary of a result on the Hamiltonicity of the so-called *circulant digraphs with two stripes*. For more details on this topic the reader is referred to [230].

Theorem 4.29 (Yang, Burkard, Çela and Woeginger 1995, [230])
Consider a QAP(A,B) with circulant $n \times n$ matrices A and B. Assume that two of the numbers c_i, $i = 1, 2, \ldots, n-1$, in the definition of the circulant matrix A are zero and the others are strictly positive. Additionally, assume that $b_{ij} > 0$ if $i - j \equiv 1(\text{mod}\, n)$ and $b_{ij} = 0$ otherwise. Then, it can be decided in $O(\log^4 n)$ time (in the unit time model) whether the optimal value of QAP(A,B) equals zero. If this is the case, an optimal solution can be founded in $O(n)$ time. $\square$

The next theorem gives a result on a polynomially solvable version of the QAP from the class MONGE(SYM)×BAND-k. The considered problem arises as a QAP formulation of the *spanning set of k-cycles* problem. In this problem we are given a weighted graph $G = (V, E)$ with vertex set V and edge set E. Let $w: E \to \mathbb{R}$ be a weight function which joins a weight w_{ij} to each edge $(i, j) \in E$. We can assume w.l.o.g. that G is complete. This is done by assigning a weight equal to ∞ to all non-existent edges. Moreover, let $|V| = k \cdot m$, $m > 1$, $k > 1$. Consider m subgraphs $G_i = (V_i, E_i)$, $1 \leq i \leq m$ of graph G. Assume that for all $1 \leq i \leq m$ G_i is a cycle of length k. Moreover, assume that $\cup_{i=1}^{k} V_i = V$ and $V_i \cap V_j = \emptyset$, $i \neq j$. Such a set of subgraphs $\{G_i: 1 \leq i \leq m\}$ is called a *spanning set of k-cycles* in G and is denoted by $\mathcal{C}_k$. Its weight is given by

$$w(\mathcal{C}_k) = \sum_{i=1}^{m} \sum_{e \in E_i} w(e)$$

For a given weighted graph $G = (V, E)$, with $|V| = mk$, $k > 1$, $m > 1$, the *spanning set of k-cycles* problem, shortly *S-k-CP*, is the problem of finding a spanning set of k-cycles with minimum weight in G .

This problem can easily be formulated as a QAP. Indeed, denote by A the weighted adjacency matrix of graph G, $A = (w_{ij})$. Let $B^{(k,m)}$ be the adjacency matrix of a graph G' on V which consists of m vertex disjoint cycles of length k each, denoted by $G_i' = (V_i', E_i')$, $1 \leq i \leq m$, where

$$V_i' = \{v_{(i-1)k+j}: j = 1, \ldots, k\} \text{ and}$$

$$E_i' = \{(v_{(i-1)k+j}, v_{(i-1)k+j+1}): j = 1, \ldots, k-1\} \cup \{(v_{ik}, v_{(i-1)k+1})\}.$$

This definition of the $km \times km$ matrix $B^{(k,m)}$ is valid throughout the rest of this section. Obviously, $B^{(k,m)}$ is a bandwidth-k matrix. As an example consider the matrix $B^{(4,3)}$ with $k = 4$ and $m = 3$ in Figure 4.1.

Clearly, with these definitions and notations, the S-k-CP can be formulated as a QAP, namely $QAP(A, B^{(k,m)})$. We show that if A is an $n \times n$ symmetric Monge matrix, $QAP(A, B^{(k,m)})$ is a constant permutation QAP ($n = km$, k, m are integers and $k > 1$, $m > 1$). Let us denote the corresponding constant permutation by $\rho^{(k,m)}$. Thus, $\rho^{(k,m)}$ is an optimal solution to all instances of $\text{MONGE(SYM)} \times B^{(k,m)}$ with fixed k and m. Together with $\rho^{(k,m)} \in \mathcal{S}_{mk}$, $k > 1$, $m > 1$, let us introduce another permutation π^* which will play a special role throughout the rest of this chapter.

Definition 4.5 a) *For $n \in \mathbb{N}$, the permutation $\pi^* \in S_n$ is defined by $\pi^*(i) = 2i-1$ for $1 \leq i \leq \lceil \frac{n}{2} \rceil$, and $\pi^*(n+1-i) = 2i$ for $1 \leq i \leq \lfloor \frac{n}{2} \rfloor$.*

$$
B^{(4,3)} = \begin{pmatrix}
0 & 1 & 0 & 0 & 0 & 0 & 0 & 0 & 0 & 0 & 0 & 0 \\
0 & 0 & 1 & 0 & 0 & 0 & 0 & 0 & 0 & 0 & 0 & 0 \\
0 & 0 & 0 & 1 & 0 & 0 & 0 & 0 & 0 & 0 & 0 & 0 \\
1 & 0 & 0 & 0 & 0 & 0 & 0 & 0 & 0 & 0 & 0 & 0 \\
0 & 0 & 0 & 0 & 0 & 1 & 0 & 0 & 0 & 0 & 0 & 0 \\
0 & 0 & 0 & 0 & 0 & 0 & 1 & 0 & 0 & 0 & 0 & 0 \\
0 & 0 & 0 & 0 & 0 & 0 & 0 & 1 & 0 & 0 & 0 & 0 \\
0 & 0 & 0 & 0 & 1 & 0 & 0 & 0 & 0 & 0 & 0 & 0 \\
0 & 0 & 0 & 0 & 0 & 0 & 0 & 0 & 0 & 1 & 0 & 0 \\
0 & 0 & 0 & 0 & 0 & 0 & 0 & 0 & 0 & 0 & 1 & 0 \\
0 & 0 & 0 & 0 & 0 & 0 & 0 & 0 & 0 & 0 & 0 & 1 \\
0 & 0 & 0 & 0 & 0 & 0 & 0 & 0 & 1 & 0 & 0 & 0
\end{pmatrix}
$$

Figure 4.1 The matrix $B^{(k,m)}$ with $k = 4$ and $m = 3$

b) *For fixed $k > 1$, $m > 1$, the permutation $\rho^{(k,m)} \in \mathcal{S}_{km}$ is given as follows:*

$$\rho^{(k,m)}((i-1)k + j) = (i-1)k + \pi^*(j)\ , \quad 1 \le i \le m,\ \ 1 \le j \le k\ , \text{ where } \pi^* \in \mathcal{S}_k.$$

In the sequential representation of permutation $\rho^{(k,m)}$ the numbers $1, 2, \ldots, km$ are divided into m groups, which are then ordered consecutively. The first group contains the numbers $1, 2, \ldots, k$, the next one contains the numbers $k+1, k+2, \ldots, 2k$ and so on, ending up with the last group which contains the numbers $(m-1)k+1, \ldots, km$. The elements of each group are sorted following the rule defined by permutation π^*: first, sort increasingly the elements placed in odd positions within the group, and then sort decreasingly the remaining elements. Consider $\rho^{(4,3)} = \langle 1, 3, 4, 2, 5, 7, 8, 6, 9, 11, 12, 10 \rangle$ as an illustrative example. Here the groups are $1, 2, 3, 4,\ \ 5, 6, 7, 8,\ \ 9, 10, 11, 12$ and $13, 14, 15, 16$.

The next theorem formulates our polynomiality result.

Theorem 4.30 *The permutation $\rho^{(k,m)}$ is an optimal solution to $QAP(A, B^{(k,m)})$ of size $n = km$, where matrix $A \in$ MONGE(SYM), and $k > 1$, $m > 1$ are integers.*

The proof is carried through by four lemmas. First, it is shown that the theorem holds for $m = 2$. Then, this result is generalized for an arbitrary m. The proof of the theorem in the case $m = 2$ is itself reduced into two simpler subcases. Namely, it is proven that $\rho^{(k,2)}$ is an optimal solution to $QAP(A, B^{(k,2)})$ where matrix A generates some extremal ray of the cone of nonnegative symmetric Monge matrices of size $2k$. The subcases correspond to the two classes of matrices, denoted by $S^{(p)}$ and $T^{(p,q)}$ which generate the extremal rays of this cone (see in Section 4.2.1). Then, by applying Observation 4.1, this results carries over to all matrices in this cone.

Lemma 4.31 *Consider the problem $QAP(S^{(l)}, B^{(k,2)})$ of size $2k$, $k > 1$, where matrix $S^{(l)}$, $1 \le l \le 2k$, generates some extremal ray of the cone of the nonnegative symmetric Monge matrices and is defined in Section 4.2.1. Permutation $\rho^{(k,2)}$ is an optimal solution to this problem.*

Proof. Consider an instance $QAP(S^{(l)}, B^{(k,2)})$ defined by some fixed integers $kandl$, where $k > 1$ and $1 \le l \le 2k$. We show that this problem is a constant QAP, i.e. the objective function value $Z(S^{(l)}, B^{(k,2)}, \pi)$ does not depend on the permutation π. Indeed, for each $\pi \in \mathcal{S}_{2k}$ the objective function can be written as follows.

$$\begin{aligned} Z(S^{(l)}, B^{(k,2)}, \pi) &= \sum_{j=1}^{k-1} s^{(l)}_{\pi(j)\pi(j+1)} + \sum_{j=k+1}^{2k-1} s^{(l)}_{\pi(j)\pi(j+1)} \\ &\quad + s^{(l)}_{\pi(k)\pi(1)} + s^{(l)}_{\pi(2k)\pi(k+1)} \end{aligned} \tag{4.24}$$

where $S^{(l)} = \left(s^{(l)}_{ij}\right)$. Clearly, there are only two terms in (4.24) with row index or (exclusive) column index equal to l. Thus, $Z(S^{(i)}, B^{(k,2)}, \pi) = 2$. □Before proving the claim for $QAP(T^{(p,q)}, B^{(k,2)}$ let us recall a result of Supnick [216] concerning the TSP on symmetric Monge matrices. We will use this result in the proof of the next lemma.

Proposition 4.32 (Supnick [216],) *Let D be an $n \times n$ symmetric Monge matrix. The traveling salesman problem with D as a distance matrix is solved to optimality by the tour $\pi^*(1) \to \pi^*(2) \to \ldots \to \pi^*(n) \to \pi^*(1)$.*

Clearly, this result applies also to the matrices $T^{(p,q)}$ which generate some extremal rays of the cone of nonnegative symmetric Monge matrices. Moreover, for the TSP with $T^{(p,q)}$ as a distance matrix, also the optimal value of the objective function, i.e. the length of an optimal tour, can be computed explicitly.

Observation 4.33 *Consider a $n \times n$ matrix $T^{(p,q)}$ with $1 \le p, q \le n$ and for $p + q \le n$, defined in Section 4.2.1. Then the length of an optimal tour of the TSP* *with* $T^{(p,q)}$ *as distance matrix is equal to* 0, 1 *or* 2 *if* $p+q<n-1$, $p+q=n-1$ *or* $p+q=n$, *respectively.*

Proof. According to Proposition 4.32, the permutation π^* defines an optimal tour for the considered TSP. By considering that $B^{(n,1)}$ is the adjacency matrix of a cycle of length n, it is easily seen that the length of this tour is equal to $Z(T^{(p,q)},B^{(n,1)},\pi^*)$. Let us assume for simplicity that $p \le q$. (Everything works analogously in the symmetric case $s>r$.) Denote by $T_*^{(p,q)}$ the matrix obtained by $T^{(p,q)} = \left(t_{ij}^{(p,q)}\right)$ when permuting its rows and columns by permutation π^*. Thus, $T_*^{(p,q)} = \left(t_{\pi^*(i)\pi^*(j)}^{(p,q)}\right)$. It is a matter of elementary calculations to see that matrix $T_*^{(p,q)}$ looks as in Figure 4.2. The 0 (1) entry in the interior of a polygone means that all entries of $T_*^{(s,r)}$ falling inside this polygone are equal to 0 (1). By

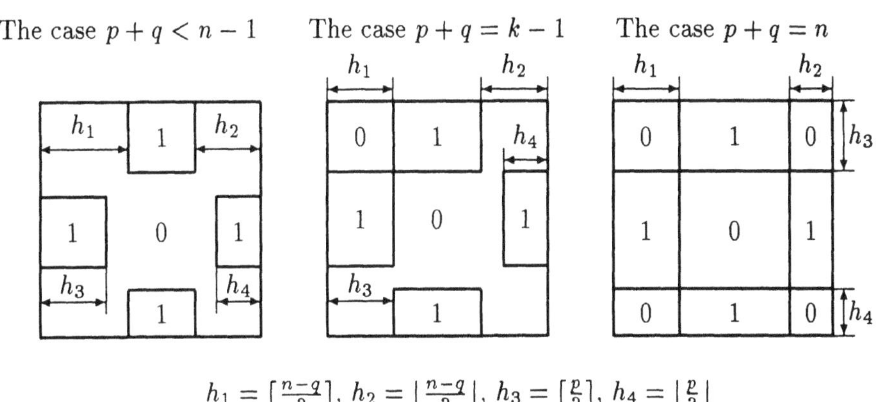

$$h_1 = \lceil \tfrac{n-q}{2} \rceil,\ h_2 = \lfloor \tfrac{n-q}{2} \rfloor,\ h_3 = \lceil \tfrac{p}{2} \rceil,\ h_4 = \lfloor \tfrac{p}{2} \rfloor$$

Figure 4.2 The $n \times n$ matrix $T_*^{(p,q)}$ for $p<q$, $p+q \le n$.

considering Figure 4.2, it is easy to see that $Z(T_*^{(p,q)},B^{(n,1)},id)$ equals $0,1,2$ in the case that $p+q$ is smaller than $n-1$, equal to $n-1$, or equal to n, respectively. Recalling that $Z(T^{(p,q)},B^{(n,1)},\pi^*) = Z(T_*^{(p,q)},B^{(n,1)},id)$ completes the proof. □

Now, we can prove the claimed result for $QAP(T^{(p,q)},B^{(k,2)})$.

Lemma 4.34 *Consider the problem* $QAP(T^{(p,q)},B^{(k,2)})$ *of size* $2k$, $k>1$, *where* $T^{(p,q)}$, $1\le p,q\le 2k$, $p+q\le 2k$, *generates some extremal ray of the cone of non-negative symmetric Monge matrices and is defined in Section 4.2.1. Permutation* $\rho^{(k,2)}$ *is an optimal solution to this problem.*

Proof. Consider an instance $QAP(T^{(p,q)}, B^{(k,2)})$ defined by some fixed integers k, p, q, where $k > 1$ and $1 \leq p, q \leq 2k$, $p + q \leq 2k$. For the sake of convenience we distinguish two cases: $p \leq q$ and $p > q$.

Case 1. $p \leq q$
First, if $q \leq k$, then obviously $Z(T^{(p,q)}, B^{(k,2)}, \rho^{(k,2)}) = 0$. Since the coefficient matrices of the considered QAP are nonnegative, we have $Z(T^{(p,q)}, B^{(k,2)}, \pi) \geq 0$ for any $\pi \in \mathcal{S}_{2k}$, and hence, $\rho^{(k,2)}$ is an optimal solution. Thus, we may assume w.l.o.g. that $q > k$. Clearly, $p + q \leq 2k$ implies then $p < k$. Thus, the non-zero entries of $T^{(p,q)}$ have either the row index or the column index smaller than k, but not both. Notice moreover, that the permutation $\rho^{(k,2)}$ permutes the elements of the set $\{k+1, \ldots, 2k\}$ to elements of the same set $\{k+1, \ldots, 2k\}$. Therefore, the above remark about the indices of the non-zero entries of matrix $T^{(p,q)}$ holds also for the matrix obtained after permuting the rows and the columns of $T^{(p,q)}$ by $\rho^{(k,2)}$. The non-zero entries of matrix $B^{(k,2)}$ may either have both row and column indices larger than k or both row and column indices smaller than or equal to k. Summarizing, only products of entries with row and column indices larger than or equal to k may contribute with positive values to the objective function $Z(T^{(p,q)}, B^{(k,2)}, \rho^{(k,2)})$. Moreover, for all $x \in \{1, 2, \ldots, k\}$ we have $\rho^{(k,2)}(k + x) = k + \pi^*(x)$, where $\pi^* \in \mathcal{S}_k$. Therefore, the following equality holds:

$$Z(T^{(p,q)}, B^{(k,2)}, \rho^{(k,2)}) = Z(T^{(p,q-k)}, B^{(k,1)}, \pi^*) , \tag{4.25}$$

where $B^{(k,1)}$ is the $k \times k$ matrix obtained from matrix $B^{(k,2)}$ by deleting the first k rows and the first k columns. Now it is easy to see that $QAP(T^{(p,q-k)}, B^{(k,1)})$ is the traveling salesman problem (TSP) with the $k \times k$ symmetric Monge matrix $T^{(p,q-k)}$ as distance matrix. ($T^{(p,q-k)}$ is an extremal ray of the cone of nonnegative symmetric Monge $k \times k$ matrices.) According to Proposition 4.32, π^* is an optimal solution to $QAP(T^{(p,q-k)}, B^{(k,1)})$. By applying Observation 4.33 with $q - k$ and k instead of q and n, respectively, we get:

$$Z(T^{(p,q)}, B^{(k,2)}, \rho^{(k,2)}) = \begin{cases} 0 & \text{if} \quad p + q < 2k - 1 \\ 1 & \text{if} \quad p + q = 2k - 1 \\ 2 & \text{if} \quad p + q = 2k \end{cases} \tag{4.26}$$

Consequently, $\rho^{(k,2)}$ is an optimal solution to $QAP(T^{(p,q)}, B^{(k,2)})$ for matrices $T^{(p,q)}$ with $p + q < 2k - 1$. It remains to prove the claim in the case that $2k - 1 \leq p + q \leq 2k$. We show that $Z(T^{(p,q)}, B^{(k,2)}, \pi) \geq Z(T^{(p,q)}, B^{(k,2)}, \rho^{(k,2)})$ for any $\pi \in \mathcal{S}_{2k}$, and this completes the proof in the two remaining cases. Consider an arbitrary $\pi \in \mathcal{S}_{2k}$. Let $z_1 \leq z_2 \leq \ldots \leq z_k$ such that $\{\pi(z_i): 1 \leq i \leq k\} = \{1, 2, \ldots, k\}$ and $y_1 \leq y_2 \leq \ldots \leq y_k$ such that $\{\pi(y_i): 1 \leq i \leq k\} = \{k+1, k+2, \ldots, 2k\}$. Denote by $T^{(z)}$ ($T^{(y)}$) the matrix obtained from $T^{(p,q)}$ by deleting the rows and columns with indices y_i, $1 \leq i \leq k$ (z_i, $1 \leq z \leq k$). Clearly, $T^{(z)}$ and $T^{(y)}$ are symmetric Monge matrices[7]. Due to the structure of matrix $B^{(k,2)}$, the following equality holds:

$$Z(T^{(p,q)},B^{(k,2)},\pi)=Z(T^{(z)},B^{(k,1)},\pi_z)+Z(T^{(y)},B^{(k,1)},\pi_y)\ ,$$

where π_z and π_y are defined by the equalities $\pi_z(i)=\pi(z_i)$ and $\pi_y(i)=\pi(y_i)-k$, for $1\le i\le k$. By the construction, the matrices $T^{(z)}$ and $T^{(y)}$ are of the form $T^{(p_z,q_z)}$ and $T^{(p_y,q_y)}$ with $p_z+p_y=p$ and $q_z+q_y=q$, $p_z+q_z\le k$, $p_y+q_y\le k$. Then, the inequality $p+q\ge 2k-1$ implies that either $p_z+q_z=k$, $p_y+q_y\ge k-1$ or $p_z+q_z\ge k-1$, $p_y+q_y=k$. Now, similarly as above, we can apply the result of Supnick for the TSP on symmetric Monge matrices. Moreover, in the case that the four coefficients p_z,p_y,q_z,q_y are strictly positive we can apply Observation 4.33 and obtain:

$$Z(T^{(p,q)},B^{(k,2)},\pi)=Z(T^{(p_z,q_z)},B^{(k,1)},\pi_z)+Z(T^{(p_y,q_y)},B^{(k,1)},\pi_y)\ge$$

$$Z(T^{(p_z,q_z)},B^{(k,1)},\pi^*)+Z(T^{(p_y,q_y)},B^{(k,1)},\pi^*)\ge 3\ge Z(T^{(p,q)},B^{(k,2)},\ \rho^{(k,2)})$$

It remains to consider the case where one of the four coefficients p_z, p_y, q_z, q_y equals 0. Assume w.l.o.g. that $p_z=0$. If $p+q=2k-1$, we have $p_y=p$ and $q_z\le k$, which imply $k-1\le p_y+q_y\le k$. Thus:

$$Z(T^{(p,q)},B^{(k,2)},\pi)=Z(T^{(p_y,q_y)},B^{(k,1)},\pi_y)\ge Z(T^{(p_y,q_y)},B^{(k,1)},\pi^*)\ge 1=$$

$$Z(T^{(p,q)},B^{(k,2)},\rho^{(k,2)})\ .$$

In the remaining case, $p+q=2k$, we again have $p_y=p$ and $q_z\le k$, which imply $p_y+q_y=k$. This yields

$$Z(T^{(p,q)},B^{(k,2)},\pi)=Z(T^{(p_y,q_y)},B^{(k,1)},\pi_y)\ge Z(T^{(p_y,q_y)},B^{(k,1)},\pi^*)=2=$$

$$Z(T^{(p,q)},B^{(k,2)},\rho^{(k,2)})\ .$$

Case 2. $p>q$
The proof is completely analogous to the proof of Case 1. □

Lemmas 4.31 and 4.34 imply the following results.

Lemma 4.35 *Let a matrix $A\in$ MONGE(SYM) be given. Permutation $\rho^{(k,2)}$ is an optimal solution of $QAP(A,B^{(k,2)})$ of size $2k$.*

[7] It is well known and easily seen that submatrices obtained from a symmetric Monge matrix by deleting rows and columns with the same set of indices are again symmetric Monge matrices. See, eg. [38].

Proof. As already argued several times in this chapter, we may assume w.l.o.g. that A is a nonnegative matrix. Under this assumption, the proof follows immediately (see Section 4.2.4). □

Finally, let us show that this result carries over to instances of size mk for $m \geq 3$.

Lemma 4.36 *Let a matrix $A \in$ MONGE(SYM) be given. Permutation $\rho^{(k,m)}$ is an optimal solution to $QAP(A, B^{(k,m)})$ of size mk.*

Proof. Assume that matrix A has nonnegative entries. We show that the following inequality holds:

$$Z(A, B^{(k,m)}, \pi) \geq Z(A, B^{(k,m)}, \rho^{(k,m)}) , \tag{4.27}$$

for an arbitrary $\pi \in \mathcal{S}_{km}$, and this completes the proof. To this end the objective function is transformed in such a way that Lemma 4.35 can be used. Let $x_1^{(j)} < x_2^{(j)} < \ldots < x_k^{(j)}$ be such that $\{\pi(x_1^{(j)}), \pi(x_2^{(j)}), \ldots, \pi(x_k^{(j)})\} = \{(j-1)k+1, (j-1)k+2, \ldots, jk\}$, for $1 \leq j \leq m$. Denote $X^{(j)} = \{x_1^{(j)}, \ldots, x_k^{(j)}\}$, for $1 \leq j \leq m$. Denote by $A^{(j)}$ the $k \times k$ matrix obtained from A by deleting all rows and columns except for those with indices in $X^{(j)}$. Define additionally the permutations $\pi^{(j)} \in \mathcal{S}_k$ by $\pi^{(j)}(i) = \pi(x_i^{(j)}) - (j-1)k$, $1 \leq i \leq k$. Clearly, the permutation π is uniquely determined by the permutations $\pi^{(j)}$ and the sets $X^{(j)}$, $1 \leq j \leq m$. The sets $X^{(j)}$ and the permutations $\pi^{(j)}$ related to a given permutation π are termed as *partial sets* and *partial permutations* of π throughout the rest of the proof. Moreover, the matrices $A^{(j)}$ and the problems $QAP(A^{(j)}, B^{(k,1)})$ are called *partial matrices* and *partial problems* of $QAP(A, B^{(k,m)})$ with respect to π, respectively. Notice that the objective function of $QAP(A, B^{(k,m)})$ is given as a sum of the objective functions of its partial problems:

$$Z(A, B^{(k,m)}, \pi) = \sum_{j=1}^{m} Z(A^{(j)}, B^{(k,1)}, \pi^{(j)}) \tag{4.28}$$

Notice moreover that the partial set $X_*^{(j)}$ of the claimed optimal permutation $\rho^{(k,m)}$ are given by

$$X_*^{(j)} = \{(j-1)k+1, (j-1)k+2, \ldots, jk\}, \qquad 1 \leq j \leq m$$

and the partial permutations of $\rho^{(k,m)}$ are equal to π^*. Next, we introduce a procedure which transforms step by step a given permutation π to permutation $\rho^{(k,m)}$ by transforming the corresponding partial sets and partial permutations. In each step of this procedure the current permutation ϕ is replaced by a permutation ϕ' with $Z(A, B^{(k,m)}, \phi) \geq Z(A, B^{(k,m)}, \phi')$.

Next let us describe a generic step of the above mentioned procedure which transforms the current permutation ϕ to a new permutation ϕ'. Such a step will be called *change*. Consider two partial problems $QAP(A^{(j)}, B^{(k,1)})$, $QAP(A^{(l)}, B^{(k,1)})$, with $j < l$. Consider additionally the symmetric Monge matrix $A^{(j,l)}$ of size $2k \times 2k$, obtained from A by deleting all rows and columns except for those with indices in $X^{(j)} \cup X^{(l)}$. Let the permutation $\phi^{(j,l)} \in \mathcal{S}_{2k}$ be defined by

$$\phi^{(j,l)}(i) = \begin{cases} \phi^{(j)}(i) & \text{if} \quad 1 \le i \le k \\ \phi^{(l)}(i) + k & \text{if} \quad k+1 \le i \le 2k \end{cases}$$

For the problem $QAP(A^{(j,l)}, B^{(k,2)})$ of size $2k$ we have:

$$Z(A^{(j,l)}, B^{(k,2)}, \phi^{(j,l)}) = Z(A^{(j)}, B^{(k,1)}, \phi^{(j)}) + Z(A^{(l)}, B^{(k,1)}, \phi^{(l)}) .$$

On the other hand, Lemma 4.35 implies that the permutation $\rho^{(k,2)}$ is an optimal solution to $QAP(A^{(j,l)}, B^{(k,2)})$. Now, it is easily seen that one of the partial sets of $\rho^{(k,2)}$, say $\bar{X}^{(j)}$, consists of the k smallest elements of $X^{(j)} \cup X^{(l)}$, and the other partial set $\bar{X}^{(l)}$ consists of the remaining elements in $X^{(j)} \cup X^{(l)}$. Let $\bar{A}^{(j)}$ and $\bar{A}^{(l)}$ be the partial matrices of $QAP(A^{(j,l)}, B^{(k,2)})$ with respect to $\rho^{(k,2)}$. (These matrices are obtained from A by deleting all rows and the columns except for those with indices in $\bar{X}^{(j)}$ and $\bar{X}^{(l)}$, respectively.) By applying equality (4.28) we get

$$Z(A^{(j,l)}, B^{(k,2)}, \rho^{(k,2)}) = Z(\bar{A}^{(j)}, B^{(k,1)}, \pi^*) + Z(\bar{A}^{(l)}, B^{(k,1)}, \pi^*)$$

Combining the two last equalities we obtain

$$\begin{aligned} Z(A^{(j)}, B^{(k,1)}, \phi^{(j)}) + Z(A^{(l)}, B^{(k,1)}, \phi^{(l)}) \ \ge \ & Z(\bar{A}^{(j)}, B^{(k,1)}, \pi^*) \\ & + Z(\bar{A}^{(l)}, B^{(k,1)}, \pi^*) \quad (4.29) \end{aligned}$$

Now, *change* replaces $X^{(j)}$ and $X^{(l)}$ by $\bar{X}^{(j)}$ and $\bar{X}^{(l)}$, respectively. It also replaces both $\pi^{(j)}$ and $\pi^{(l)}$ by π^*, whereas the other partial sets and partial permutations remain unchanged. Let ϕ' be the permutation determined by the new partial sets and the new partial permutations. Equations (4.29) and (4.28) imply $Z(A, B, \phi') \le Z(A, B, \phi)$.

In order to transform the given π to $\rho^{(k,m)}$ we apply a series of changes with $(j,l) = (1,2), (1,3), (1,4), \ldots, (1,m)$. Due to the definition of a change, after these $m-1$ changes all current partial permutations are equal to π^*, and the current partial set $X^{(1)}$ equals $\{1, 2, \ldots, k\}$. Let π' be the current partial permutation. By applying $k-2$ consecutive changes with $(j,l) = (2,3), (2,4), \ldots, (2,m)$ to π' the latter is transformed into a new permutation with the same partial permutations π^*, the same partial set $X^{(1)}$, and an (eventually) new partial set $X^{(2)} = \{k+1, \ldots, 2k\}$. This process is continued by applying other consecutive changes with

$(j,l) = (3,4),(3,5),\ldots,(3,m),\ldots,(m-1,m)$, resulting to $m(m-1)/2$ changes altogether. After this, all partial permutations equal π^* and the partial sets are given by

$$X^{(j)} = \{(j-1)k+1,(j-1)k+2,\ldots,jk\}, \qquad 1 \le j \le m$$

Thus, the claimed optimal permutation $\rho^{(k,m)}$ is obtained and this completes the proof. □

4.5.2 Special cases of the taxonomy problem

The *general taxonomy problem* was introduced by Rubinstein [201]. It belongs to a large class of problems dealing with optimal groupings of related objects. The problem is very complex in its general formulation. For a detailed description and related topics the reader is referred to [202]. Here we only describe a simple version of the general taxonomy problem which can be formulated as a QAP. We call it *taxonomy problem* (TP), and all results presented in this section hold for the problem TP introduced below.

Consider a group of n objects to be partitioned into a given number of subgroups with prespecified sizes. Assume that the given objects are identified with indices from 1 up to n. A *communication matrix* $A = (a_{ij})$ is given, where the entry a_{ij} represents a communication rate between object i and object j, $1 \le i,j \le n$. The goal is to find a partition of the given objects into groups which minimizes the weighted sum of communication rates between pairs of objects belonging to the same group, where the weights are scaling factors which depend proportionally on the size of the group.

The taxonomy problem (TP) is similar to a version of the *set partitioning problem* defined by Grötschel in [103]. Consider the given objects as a ground set E in a partitioning problem. Assign a cost c_S to each subset S of E: $c_S = |S| \sum_{i,j\in S} a_{ij}$. The above mentioned set partitioning problem consists of finding a partition of the ground set E into a prespecified number of subsets such that the overall cost of the subsets involved in the partition is minimized. In the taxonomy problem it is additionally required that the subsets of the partitioning have prespecified cardinalities. Moreover, the scaling factors $|S|$ are generally replaced by some function of $|S|$.

The TP can easily be modeled as a QAP. Let the set $\{1,2,\ldots,n\}$ represent the n objects and let n_1, n_2, $\ldots$, n_k, $\sum_{i=1}^{k} n_i = n$, be the sizes of the k subgroups.

Moreover, let $f\colon \mathbb{N} \to \mathbb{R}^+$ be a function which models the scaling factor for the communication rates within the subgroups. Finally, let $A = (a_{ij})$ be the communication matrix, i.e. a_{ij} represents the measure of communication of object i with object j, $1 \le i, j \le n$. Before formulating TP as a QAP let us introduce one more class of matrices, the definition of so-called *taxonomy* matrices.

Definition 4.6 *Let a function $f\colon \mathbb{N} \to \mathbb{R}$ and the integers n_i, $1 \le i \le k$, with $\sum_{i=1}^{k} n_i = n$, be given. An $n \times n$ matrix B is called a* taxonomy matrix *of type $(f, n_1, n_2, \ldots, n_k)$ if it is a block-diagonal matrix consisting of k blocks of non-zero elements along the diagonal, where for each i the i-th diagonal block is of size $n_i \times n_i$ and all entries within this block are equal to $f(n_i)$.*

As an example consider a 6×6 taxonomy matrix B of type $(f, 1, 3, 2)$, where f is given by the formula $f(x) = 2x$:

$$B = \begin{pmatrix} 2 & 0 & 0 & 0 & 0 & 0 \\ 0 & 6 & 6 & 6 & 0 & 0 \\ 0 & 6 & 6 & 6 & 0 & 0 \\ 0 & 6 & 6 & 6 & 0 & 0 \\ 0 & 0 & 0 & 0 & 4 & 4 \\ 0 & 0 & 0 & 0 & 4 & 4 \end{pmatrix}$$

Now it is clear that the above described TP is equivalent to QAP(A,B), where A is the communication matrix and B is a taxonomy matrix of type $(f,\ n_1,\ n_2,\ \ldots,\ n_k)$. The following theorem, formulated by Rubinstein in [203], describes three polynomially solvable cases of this version of the QAP. Our proof exploits the fact that the classes of matrices MONGE(NNEG) and MONGE(RLG,NNEG) form cones, and works with the 0-1 matrices which generate the extremal rays of these cones.

Theorem 4.37 (Rubinstein 1994, [203])
Let A be a Monge matrix and B be a taxonomy matrix of type $(f, n_1, n_2, \ldots, n_k)$ where f is a nonnegative function. Consider moreover the three following conditions on the problem input:

(a) $n_1 = n_2 = \cdots = n_k$

(b) A is a right-lower graded matrix, $n_1 \le n_2 \le \ldots \le n_k$, and f is a non-decreasing function.

(c) A is a left-higher graded matrix, $n_1 \le n_2 \le \ldots \le n_k$, and f is a non-decreasing function.

In the cases (a) and (b) the identity permutation id is an optimal solution to QAP(A,B). In the case (c) the permutation $\phi = \langle n, n-1, \ldots, 2, 1\rangle$ is an optimal solution to QAP(A,B).

Proof. We assume w.l.o.g. that matrix A is nonnegative. Recall that according to Proposition 4.6, the extremal rays of the cone of $n \times n$ nonnegative Monge matrices are generated by the 0-1 matrices $H^{(p)}$, $V^{(q)}$, $1 \leq p \leq n$, and $L^{(p,q)}$, $R^{(p,q)}$, $1 \leq p, q \leq n$, introduced in Section 4.2.1. Further, recall that according to Corollary 4.10, the extremal rays of the cone of right-lower graded nonnegative Monge matrices are generated by the 0-1 matrices $E^{(p,q)}$, $1 \leq p, q \leq n$, introduced in Section 4.2.2. Hence, in case (a) it is sufficient to prove the theorem for matrices A coinciding with any of the matrices $H^{(p)}$, $V^{(q)}$, $L^{(p,q)}$ and $R^{(p,q)}$, $1 \leq p, q \leq n$, (cf. Section 4.2.4). Analogously, in case (b) it is sufficient to prove the theorem for matrices A coinciding with any of the matrices $E^{(p,q)}$, $1 \leq p, q \leq n$. Then, by applying Observation 4.1 the results carry over to all matrices in the respective cones. The claim of the theorem in case (c) follows immediately from the claim in case (b).

Proof of (a). Consider a QAP(A,B) of size $n = k \cdot n_1$, where B is a taxonomy matrix of type $(f, n_1, n_2, \ldots, n_k)$, $n_1 = n_2 = \ldots = n_k$, and f is a nonnegative function. Let $s_0 = 0$ and $s_t = \sum_{i=1}^{t} n_i$, for $1 \leq t \leq k$. Then the objective function $Z(A, B, \pi)$ can be written as follows:

$$Z(A, B, \pi) = f(n_1) \sum_{t=1}^{k} \Big(\sum_{i=s_{t-1}+1}^{s_t} \sum_{j=s_{t-1}+1}^{s_t} a_{\pi(i)\pi(j)} \Big) \tag{4.30}$$

We consider separately the following four cases:

Case 1. $A = V^{(p)}$, $1 \leq p \leq n$.
Let $\pi \in \mathcal{S}_n$ and let s_h be the unique element among $s_0, \ldots, s_{k-1}$ for which $s_h + 1 \leq \pi^{-1}(p) \leq s_{h+1}$, where π^{-1} is the inverse permutation of π. By plugging $A = V^{(p)}$ in (4.30) and get:

$$Z(V^{(p)}, B, \pi) = f(n_1) \sum_{j=s_h+1}^{s_{h+1}} v^{(p)}_{p\pi(j)} = n_1 f(n_1) ,$$

where $V^{(p)} = \left(v^{(p)}_{ij}\right)$. Thus, $QAP(V^{(p)}, B)$ is a constant QAP and therefore *id* is an optimal solution to it.

Case 2. $A = H^{(p)}$, $1 \leq p \leq n$.
Analogously as in Case 1 it can be shown that $QAP(H^{(p)}, B)$ is a constant QAP.

Case 3. $A = R^{(p,q)}$, $1 \le p, q \le n$.
For a given $\pi \in \mathcal{S}_n$, define two k-dimensional vectors $\left(x_t^{(\pi)}\right)$, $\left(y_t^{(\pi)}\right)$, $1 \le t \le k$, as follows:

$$\begin{aligned} x_t^{(\pi)} &= |\{s_{t-1}+1 \le i \le s_t \ : \ 1 \le \pi(i) \le p\}| \\ y_t^{(\pi)} &= |\{s_{t-1}+1 \le j \le s_t \ : \ n-q+1 \le \pi(j) \le n\}| \end{aligned}$$

Then, the objective function in (4.30) can be written as

$$Z(R^{(p,q)}, B, \pi) = f(n_1) \sum_{t=1}^{k} x_t^{(\pi)} y_t^{(\pi)}$$

Thus, $QAP(R^{(p,q)}, B)$ is equivalent to the following minimization problem:

$$\min_{\pi \in \mathcal{S}_n} \sum_{t=1}^{k} x_t^{(\pi)} y_t^{(\pi)} \tag{4.31}$$

By definition, the vectors $\left(x_t^{(\pi)}\right)$, $\left(y_t^{(\pi)}\right)$ have nonnegative integer entries and fulfill the following conditions:

(i) $\sum_{t=1}^{k} x_t^{(\pi)} = p$, $\quad \sum_{t=1}^{k} y_t^{(\pi)} = q$

(ii) $0 \le x_t^{(\pi)}, y_t^{(\pi)} \le n_t$, $1 \le t \le k$

Relaxing the condition that the vectors $\left(x_t^{(\pi)}\right)$, $\left(y_t^{(\pi)}\right)$ are related to some permutation $\pi \in \mathcal{S}_n$ as described above, we get the following relaxation P1 of the minimization problem (4.31)

(P1)

$$\begin{aligned} \text{minimize} \quad & \sum_{t=1}^{k} x_t y_t \\ \text{such that} \quad & \\ & \sum_{t=1}^{k} x_t = p \\ & \sum_{t=1}^{k} y_t = q \\ & 0 \le x_t, y_t \le n_t, \quad 1 \le t \le k \\ & x_t, y_t \text{ integer}, \quad 1 \le t \le k \end{aligned}$$

Obviously, the optimal value of problem P1 is a lower bound for the optimal value of $QAP(R^{(p,q)}, B)$. We show that vectors $\left(x_t^{(id)}\right)$ and $\left(y_t^{(id)}\right)$ corresponding to the identity permutation *id* yield an optimal solution to P1. This implies then that the identity permutation *id* is an optimal solution to $QAP(R^{(p,q)}, B)$. Indeed, vectors $\left(x_t^{(id)}\right)$ and $\left(y_t^{(id)}\right)$ are given as follows:

$$x_t^{(id)} = \min\Big\{n_t, \max\{p - s_{t-1}, 0\}\Big\}, \quad y_t^{(id)} = \min\Big\{n_t, \max\{q - n + s_t, 0\}\Big\}.$$

It can be easily seen that these vectors are derived by starting with $t = 1$ and setting the entries $x_t^{(id)}$, $y_{k-t+1}^{(id)}$ as large as possible, and then iteratively increasing t by 1. This procedure is repeated until $t = k$. As an easy consequence of Proposition 2.1, this construction yields an optimal solution to problem P1.

Case 4. $A = L^{(p,q)}$, $1 \leq p, q \leq n$.
Analogous to the proof in Case 3.

Proof of (b). Consider a QAP(A,B) of size $n = \sum_{t=1}^{k} n_t$, where B is a taxonomy matrix of type $(f, n_1, n_2, \ldots, n_k)$, f is a non-decreasing function and $A = E^{(p,q)}$. For any $\pi \in \mathcal{S}_n$ the objective function $Z(A, B, \pi)$ is given as follows:

$$Z(A, B, \pi) = \sum_{t=1}^{k} \left(f(n_t) \cdot \sum_{i=s_{t-1}+1}^{s_t} \sum_{j=s_{t-1}+1}^{s_t} a_{\pi(i)\pi(j)} \right) \tag{4.32}$$

where s_t, $0 \leq t \leq k$, are defined as in the proof of (a). Set $p' = n - p$ and $q' = n - q$ and consider additionally the matrix $C^{(p',q')}$, as defined in Section 4.2.2. Clearly, $E^{(p,q)} + C^{(p',q')} = E_n$, where E_n is the $n \times n$ matrix with all entries equal to 1. From Observation 4.1 it follows that minimizing $Z(E^{(p,q)}, B, \pi)$ over all permutations $\pi \in \mathcal{S}_n$ is equivalent to maximizing $Z(C^{(p',q')}, B, \pi)$ over $\pi \in \mathcal{S}_n$ in the sense that these problems have the same set of optimal solutions. For an arbitrary $\pi \in \mathcal{S}_n$, $Z(C^{(p',q')}, B, \pi)$ can be written as follows:

$$Z(C^{(p',q')}, B, \pi) = \sum_{t=1}^{k} \left(f(n_t) \cdot \sum_{i=s_{t-1}+1}^{s_t} \sum_{j=s_{t-1}+1}^{s_t} c_{\pi(i)\pi(j)}^{(p',q')} \right) = \sum_{t=1}^{k} f(n_t) x_t^{(\pi)} y_t^{(\pi)},$$

where the vectors $\left(x_t^{(\pi)}\right)$, $\left(y_t^{(\pi)}\right)$, $1 \leq t \leq k$, are defined as follows:

$$\begin{aligned} x_t^{(\pi)} &= |\{s_{t-1} + 1 \leq i \leq s_t \ : \ p + 1 \leq \pi(i) \leq n\}| \\ y_t^{(\pi)} &= |\{s_{t-1} + 1 \leq j \leq s_t \ : \ q + 1 \leq \pi(j) \leq n\}| \end{aligned}$$

Thus, $QAP(E^{(p,q)}, B)$ is equivalent to the following maximization problem

$$\max_{\pi \in \mathcal{S}_n} \sum_{t=1}^{k} f(n_t) x_t^{(\pi)} y_t^{(\pi)} \tag{4.33}$$

By definition, $\left(x_t^{(\pi)}\right)$, $\left(y_t^{(\pi)}\right)$ have nonnegative integer entries and fulfill the following conditions:

(i) $\sum\limits_{t=1}^{k} x_t^{(\pi)} = p', \qquad \sum\limits_{t=1}^{k} y_t^{(\pi)} = q'$

(ii) $0 \le x_t^{(\pi)}, y_t^{(\pi)} \le n_t$, $1 \le t \le k$

Similarly as in the proof of Case 3 of (a), the maximization problem (4.33) can be relaxed to get the following problem P2:

$$(\mathbf{P2}) \qquad \begin{array}{ll} \text{maximize} & \sum_{t=1}^{k} f(n_t) x_t y_t \\ \text{subject to} & \\ & \sum_{t=1}^{k} x_t = p' \\ & \sum_{t=1}^{k} y_t = q' \\ & 0 \le x_t, y_t \le n_t \quad 1 \le t \le k \\ & x_t, y_t \text{ integer} \quad 1 \le t \le k \end{array}$$

Obviously, the optimal value of problem P2 is an upper bound for the maximum value of $Z(C^{(p',q')}, B, \pi)$ over $\pi \in \mathcal{S}_n$. We show that the vectors $\left(x_t^{(id)}\right)$, $\left(y_t^{(id)}\right)$, corresponding to the identity permutation *id* yield an optimal solution to P2. Clearly, this implies that *id* is an optimal solution to (4.33) and, equivalently, to $QAP(E^{(p,q)}, B)$. The vectors $\left(x_t^{(id)}\right)$, $\left(y_t^{(id)}\right)$ are given as follows:

$$x_t^{(id)} = \min\left\{n_t, \max\{p' - n + s_t, 0\}\right\}, \quad y_t^{(id)} = \min\left\{n_t, \max\{q' - n + s_t, 0\}\right\}.$$

Proposition 2.1 can be easily generalized to the case of the maximization of the "scalar products of three vectors". Burkard, Rudolf and Woeginger [42] show that given three non-decreasing n-dimensional vectors (u_i), (v_i), (w_i), the sum $\sum_{i=1}^{n} u_i v_{\phi(i)} w_{\psi(i)}$ where $\phi, \psi \in \mathcal{S}_n$, is maximized by $\phi = \psi = id$. As the function f and the vectors $\left(x_t^{(id)}\right)$, $\left(y_t^{(id)}\right)$ are non-decreasing the above mentioned result of

Burkard et al. can be applied to show that $\left(x_t^{(id)}\right)$, $\left(y_t^{(id)}\right)$ yield an optimal solution to problem P2. The proof of this claim relies on a simple exchange argument and is omitted because of its simplicity.

Proof of (c). Consider QAP(A,B) of size n with $A \in$ MONGE(LHG), B being a taxonomy matrix of type $(f, n_1, n_2, \ldots, n_k)$, and f being a non-decreasing function. Consider, moreover, the permuted matrix $A^\phi = (a_{\phi(i)\phi(j)})$, where $\phi = \langle n, n-1, \ldots, 2, 1\rangle$. The following inequality which holds for $1 \le i, j \le n-1$, shows that A^ϕ is a Monge matrix:

$$\begin{aligned} a_{\phi(i)\phi(j)} + a_{\phi(i+1)\phi(j+1)} &= a_{n-i+1,n-j+1} + a_{n-i,n-j} \le \\ a_{n-i+1,n-j} + a_{n-i,n-j+1} &= a_{\phi(i+1)\phi(j)} + a_{\phi(i)\phi(j+1)} \end{aligned}$$

Moreover, A^ϕ is a right-lower graded matrix, as shown by the following inequalities:

$$a_{\phi(i)\phi(j)} = a_{n-i+1,n-j+1} \ge a_{n-i,n-j+1} = a_{\phi(i+1)\phi(j)}, \quad 1 \le i \le n-1\,,\ 1 \le j \le n$$

$$a_{\phi(i)\phi(j)} = a_{n-i+1,n-j+1} \ge a_{n-i+1,n-j} = a_{\phi(i)\phi(j+1)}, \quad 1 \le i \le n\,,\ 1 \le j \le n-1$$

Thus, $A^\phi \in$ MONGE(RLG) and therefore, permutation *id* is an optimal solution to $QAP(A^\phi, B)$, as proven in (b). The claim follows by considering that the following equality holds:

$$Z(A, B, \phi \circ \pi) = Z(A^\phi, B, \pi)\,, \quad \text{for all } \pi \in \mathcal{S}_n$$ □

Remark 1. The first result of Theorem 4.37 was implicitly proven by Pferschy, Rudolf and Woeginger in [185]. The authors consider the so-called *balanced k-cut problem*, a special case of the TP where all subgroups have the same size (k is the number of subgroups). It is shown that the special case of the balanced k-cut problem where the communication matrix A is a Monge matrix, is solved to optimality by the identity permutation. This is exactly what Theorem 4.37 states in case (a).

Remark 2. It can be shown that both conditions, f being non-decreasing and A being a lower-right graded matrix, are essential for the claim of Theorem 4.37 in case (b). That is, there exist examples where the violation of one of these conditions is sufficient to make the identity permutation a non-optimal solution of the corresponding QAP.

Example 4.2 The conditions of Theorem 4.37 are essential.
Let $n = 3$ and let the matrices A_1, A_2, B_1, B_2 be given as:

$$A_1 = \begin{pmatrix} 3 & 3 & 3 \\ 3 & 2 & 2 \\ 3 & 2 & 1 \end{pmatrix} B_1 = \begin{pmatrix} 3 & 0 & 0 \\ 0 & 1 & 1 \\ 0 & 1 & 1 \end{pmatrix} A_2 = \begin{pmatrix} 1 & 2 & 3 \\ 2 & 2 & 3 \\ 3 & 3 & 3 \end{pmatrix} B_2 = \begin{pmatrix} 1 & 0 & 0 \\ 0 & 3 & 3 \\ 0 & 3 & 3 \end{pmatrix}$$

B_1 is a taxonomy matrix of type $(f, 1, 2)$, where $f(1) = 3$, $f(2) = 1$. Thus, f is a decreasing function. Similarly, B_2 is a taxonomy matrix and the corresponding function is increasing. Further, A_1 is a right-lower graded Monge matrix and A_2 is a left-higher graded Monge matrix. One can check that $Z(A_1, B_1, id) = 16$, $Z(A_1, B_1, \phi) = 14$, $Z(A_2, B_2, id) = 34$ and $Z(A_2, B_2, \phi) = 24$, where $\phi = \langle n, n-1, \ldots, 2, 1\rangle$. Thus, id is not an optimal solution for $QAP(A_1, B_1)$ and $QAP(A_2, B_2)$.
□

Remark 3. Finally, notice that the condition $n_1 \leq n_2 \leq \ldots \leq n_k$ in Theorem 4.37 is not essential, in the sense that for $A \in$ MONGE(LHG) or $A \in$ MONGE(RLG) and a taxonomy matrix B of type $(f, n_1, n_2, \ldots, n_k)$ with non-decreasing function f, QAP(A,B) is polynomially solvable, even if the vector (n_i) is not increasingly sorted. Indeed, given an arbitrary taxonomy matrix B of type $(f, n_1, n_2, \ldots, n_k)$, there exists a permutation ψ such that the permuted matrix B^{ψ} is again a taxonomy matrix of type $(f, n'_1, n'_2, \ldots, n'_k)$, with $n'_1 \leq n'_2 \leq \ldots \leq n'_k$ and $\{n_1, \ldots, n_k\} = \{n'_1, \ldots, n'_k\}$. For example, for the 6×6 taxonomy matrix B of type $(f, 1, 3, 2)$ with $f(x) = 2x$, introduced after Definition 4.6 in this section, the permutation $\psi = \langle 1, 5, 6, 2, 3, 4\rangle$ does the job. So, we get the following straightforward corollary:

Corollary 4.38 *Let A be a Monge matrix and let B be a taxonomy matrix of type $(f, n_1, n_2, \ldots, n_k)$. Assume additionally that A is either a left-higher or a right-lower graded matrix and that f is a non-decreasing function. Then QAP(A,B) is a constant permutation QAP. In the case that $A \in$* MONGE(RLG) *the constant permutation is ψ and in the case that $A \in$* MONGE(LHG) *the constant permutation is $\phi \circ \psi$, where $\phi = \langle n, n-1, \ldots, 2, 1\rangle$ and ψ is determined as described above.*

Proof. From Theorem 4.37 follows that $QAP(A, B^{\psi})$ is a constant permutation QAP. Its constant permutation is id or ϕ in the cases that $A \in$ MONGE(RLG) or $A \in$ MONGE(LHG), respectively. The proof is completed by considering that the equality

$$Z(A, B, \pi \circ \psi^{-1}) = Z(A, B^{\psi}, \pi)$$

holds for all $\pi \in \mathcal{S}_n$, $n = \sum_{i=1}^{k} n_i$. □

5

TWO MORE RESTRICTED VERSIONS OF THE QAP

In this chapter which is a continuation of the previous one, we consider two restricted version of the QAP: the Anti-Monge–Toeplitz QAP and the Kalmanson-Toeplitz QAP. The Anti-Monge–Toeplitz QAP is a restricted version of the QAP, where one of the coefficient matrices is a left-higher graded Anti-Monge matrix and the other one is a symmetric Toeplitz matrix. The interest in this problem is motivated by a number of practical applications, e.g. the *turbine problem* and the *data arrangement problem*, some of which will be considered in detail in the second section of this chapter. Moreover, the Anti-Monge–Toeplitz QAP contains the TSP on symmetric Monge matrices as a special case. Despite the very special structure of the Anti-Monge–Toeplitz QAP, i.e., the quite restrictive conditions to be fulfilled by its coefficient matrices, the problem is NP-hard. Namely, the turbine problem which is a special case of the Anti-Monge–Toeplitz QAP is NP-hard, as shown by Burkard, Çela, Rote and Woeginger [33]. However, for Toeplitz matrices satisfying some additional conditions, (e.g. *benevolent*, *k-benevolent*, or bandwidth-2 matrices), the Anti-Monge Toeplitz QAP becomes polynomially solvable. These polynomially solvables special cases of the problem which are constant permutation QAPs, will be described in detail in the first section of this chapter. Then, a polynomially solvable version of the Kalmanson-Toeplitz QAP, which i.e. a QAP with a *Kalmanson* and a Toeplitz matrix, is described. Further, so-called *permuted polynomially solvable cases* of the QAP are briefly discussed. The concluding section summarizes the results on special cases of the QAP obtained in the current and in the previous chapter. Here we also outline some open problems which could be object of further research.

5.1 POLYNOMIALLY SOLVABLE CASES OF THE ANTI-MONGE–TOEPLITZ QAP

The proofs of these results are based on two ideas: relaxation and reduction to extremal rays. First, a relaxation of the QAP at hand, namely the corresponding independent-QAP (I-QAP), is investigated. It is shown that the permutation claimed to be optimal for the QAP solves also the corresponding relaxation, i.e., the independent QAP. The claim is proven by exploiting the cone structure of the involved coefficient matrices. In case of benevolent matrices (to be introduced in the next section) the I-QAP over the corresponding extremal rays can be formulated equivalently as a selection problem. The latter can then be solved by elementary means. Here, we will provide only sketches of the proofs, whereas the complete proofs can be found in [33].

5.1.1 Benevolent Toeplitz matrices

Let us first give the definition of so-called *benevolent functions* and *benevolent matrices*.

Definition 5.1 *A function* $f: \{-n+1, \ldots, n-1\} \to \mathbb{R}$ *is called* benevolent *if it fulfills the following three properties.*

(BEN1) $f(-i) = f(i)$ *for all* $1 \leq i \leq n-1$.

(BEN2) $f(i) \leq f(i+1)$ *for all* $1 \leq i \leq \lfloor \frac{n}{2} \rfloor - 1$.

(BEN3) $f(i) \leq f(n-i)$ *for all* $1 \leq i \leq \lceil \frac{n}{2} \rceil - 1$.

An $n \times n$ *matrix* $B = (b_{ij})$ *is called a* benevolent matrix *if it is a Toeplitz matrix* generated *by a benevolent function as above, i.e.,* $b_{ij} = f(j-i)$ *for all* $1 \leq i, j \leq n$.

Example 5.1 The function $f: \{-7, -6, \ldots, 0, 1, \ldots 7\} \to \mathbb{R}$ defined by

$$f(0) = 3, \quad f(1) = 1, \quad f(2) = 2, \quad f(3) = 3, \quad f(4) = 4,$$

$$f(5) = 5, \quad f(6) = 5, \quad f(7) = 3$$

is benevolent. The graph of this function is shown in Figure 5.1. Note that the graph of this function for $x \in \{5, 6, 7\} \cup \{-5, -6, -7\}$ lies above the thin lines,

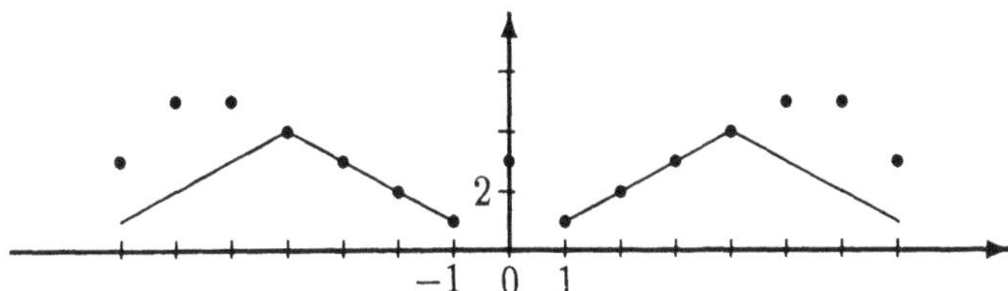

Figure 5.1 The graph of the benevolent function f

this being the geometric equivalent of property (BEN3). The benevolent matrix A generated by f is given below:

$$A = \begin{pmatrix} 3 & 1 & 2 & 3 & 4 & 5 & 5 & 3 \\ 1 & 3 & 1 & 2 & 3 & 4 & 5 & 5 \\ 2 & 1 & 3 & 1 & 2 & 3 & 4 & 5 \\ 3 & 2 & 1 & 3 & 1 & 2 & 3 & 4 \\ 4 & 3 & 2 & 1 & 3 & 1 & 2 & 3 \\ 5 & 4 & 3 & 2 & 1 & 3 & 1 & 2 \\ 5 & 5 & 4 & 3 & 2 & 1 & 3 & 1 \\ 3 & 5 & 5 & 4 & 3 & 2 & 1 & 3 \end{pmatrix}$$

Notice that by property (BEN1), a benevolent Toeplitz matrix is symmetric. □

It is easy to see that the nonnegative benevolent matrices form a cone. (These matrices are defined by certain *homogeneous* linear equalities and inequalities to be fulfilled by their entries.) Which are the extremal rays of this cone? We show that all nonnegative benevolent functions can be represented as nonnegative linear combinations of special benevolent functions, which take only 0-1 values. Hence, these 0-1 functions generate the extremal rays of the cone of nonnegative benevolent functions. These special 0-1 benevolent functions are introduced below,

Definition 5.2 *For* $\lfloor \frac{n}{2} \rfloor + 1 \le \alpha \le n-1$, *define a function* $g^\alpha\colon \{-n+1, \ldots, n-1\} \to \{0,1\}$ *by*

$$g^\alpha(x) = \begin{cases} 1 & \text{for } x \in \{-\alpha, \alpha\} \\ 0 & \text{for } x \notin \{-\alpha, \alpha\}. \end{cases}$$

For $1 \le \beta \le \lfloor \frac{n}{2} \rfloor$, *define a function* $h^\beta\colon \{-n+1, \ldots, n-1\} \to \{0,1\}$ *by*

$$h^\beta(x) = \begin{cases} 1 & \text{for } \beta \le |x| \le n-\beta \\ 0 & \text{otherwise.} \end{cases}$$

Example 5.2 For $n = 8$ and $\alpha = 5$ the function $g^5: \{-7, -6, \ldots, 0, 1, \ldots, 7\} \to \{0, 1\}$ is defined by $g^5(5) = g^5(-5) = 1$ and $g^5(x) = 0$, for $x \notin \{-5, 5\}$.
For $n = 8$ and $\beta = 2$ the function $h^2: \{-7, -6, \ldots, 0, 1, \ldots, 7\} \to \{0, 1\}$ is defined by $h^2(0) = h^2(1) = h^2(-1) = 0$, $h^2(7) = h^2(-7) = 0$ and $h^2(x) = 1$, for $x \in \{2, 3, 4, 5, 6\} \cup \{-2, -3, -4, -5, -6\}$. The Toeplitz matrices generated by these functions are

$$B(g^5) = \begin{pmatrix} 0&0&0&0&0&1&0&0 \\ 0&0&0&0&0&0&1&0 \\ 0&0&0&0&0&0&0&1 \\ 0&0&0&0&0&0&0&0 \\ 0&0&0&0&0&0&0&0 \\ 1&0&0&0&0&0&0&0 \\ 0&1&0&0&0&0&0&0 \\ 0&0&1&0&0&0&0&0 \end{pmatrix} \qquad B(h^2) = \begin{pmatrix} 0&0&1&1&1&1&1&0 \\ 0&0&0&1&1&1&1&1 \\ 1&0&0&0&1&1&1&1 \\ 1&1&0&0&0&1&1&1 \\ 1&1&1&0&0&0&1&1 \\ 1&1&1&1&0&0&0&1 \\ 1&1&1&1&1&0&0&0 \\ 0&1&1&1&1&1&0&0 \end{pmatrix}$$

Observe the circular structure of the second matrix, which is typical of the functions h^β: Going from one row to the next corresponds to a circular right shift, and this remains true when returning from the last row to the first row. The same holds for the columns. □

Consider the set of real functions defined on $\{-n+1, \ldots, n-1\}$ with the traditional operations "addition" and "scalar product". That is, if f and g are two functions which map $\{-n+1, \ldots, n-1\}$ to $\mathbb{R}$ and $\alpha \in \mathbb{R}$, then $f + g$ and αf are defined by $(f + g)(x) = f(x) + g(x)$ and $(\alpha f)(x) = \alpha f(x)$, for $x \in \{-n+1, \ldots, n-1\}$. This set of functions with these two operations forms a linear space. The following lemma states that the subset consisting of the benevolent functions which map 0 to 0 forms a cone in this space.

Lemma 5.1 *The nonnegative benevolent functions $f: \{-n+1, \ldots, n-1\} \to \mathbb{R}$ with $f(0) = 0$ form a cone. The extremal rays of this cone are generated by the functions g^α and h^β.*

Proof. Since nonnegative benevolent functions are defined by linear equations and inequalities with right-hand side equal to zero, they form a cone. The functions g^α and h^β are clearly benevolent and hence, belong to this cone. In fact, each of these functions satisfies precisely one of the characterizing inequalities (BEN2), (BEN3), and $f(1) \geq 0$ as a strict inequality and the remaining ones with equality. Further, we show that an arbitrary nonnegative benevolent function f with $f(0) = 0$ is a nonnegative linear combination of functions g^α and h^β. For this purpose we define two auxiliary functions which map $\{-n+1, \ldots, 0, \ldots, n-1\}$ to $\mathbb{R}$ and are related to f in the following way:

$$f_1(i) = \begin{cases} f(i) & \text{for} \quad |i| \le \lfloor \frac{n}{2} \rfloor \\ f(n-i) & \text{for} \quad i > \lfloor \frac{n}{2} \rfloor \\ f(-n+i) & \text{for} \quad i < -\lfloor \frac{n}{2} \rfloor \end{cases}$$

and

$$f_2(i) = \begin{cases} 0 & \text{for} \quad |i| \le \lfloor \frac{n}{2} \rfloor \\ f(i) - f(n-i) & \text{for} \quad i > \lfloor \frac{n}{2} \rfloor \\ f(i) - f(-n+i) & \text{for} \quad i < -\lfloor \frac{n}{2} \rfloor \end{cases}$$

It is easily seen that $f(i) = f_1(i) + f_2(i)$ holds for all i in $\{-n+1, \ldots, n-1\}$. Finally, observe that

$$f_1 = \sum_{i=1}^{\lfloor n/2 \rfloor} [f(i) - f(i-1)] \cdot h^i \quad f_2 = \sum_{i=\lfloor n/2 \rfloor + 1}^{n-1} [f(i) - f(n-i)] \cdot g^i \qquad (5.1)$$

and apply conditions (BEN2) and (BEN3) to see that all coefficients in these expressions are nonnegative. Hence, both f_1 and f_2 are nonnegative linear combinations of functions g^α and h^β and this completes the proof. □

5.1.2 The Anti-Monge–Toeplitz QAP with a benevolent matrix

In this section we will consider the Anti-Monge–Toeplitz QAP(A,B) where the Toeplitz matrix B is benevolent. We show that this problem is a constant permutation QAP and π^* is an optimal solution (i.e. the constant permutation). Recall that $\pi^* \in \mathcal{S}_n$ is defined by $\pi^*(i) = 2i - 1$ for $1 \le i \le \lceil \frac{n}{2} \rceil$ and $\pi^*(n+1-i) = 2i$ for $1 \le i \le \lfloor \frac{n}{2} \rfloor$ (cf. Section 4.5.1). Summarizing we prove the following theorem which is also the main result in this section.

Theorem 5.2 *The permutation π^* solves QAP(A, B) where A is a left-higher graded Anti-Monge matrix and B is a symmetric Toeplitz matrix generated by a benevolent function.*

Since the structure of a left-higher graded Anti-Monge matrix and that of a benevolent matrix do not change when adding a constant to all entries of the corresponding matrices, we can assume w.l.o.g. that both coefficient matrices A, B are nonnegative. Moreover, for the sake of convenience and in order to avoid trivialities, all matrices in this section will be of dimension $n \times n$, for some fixed $n \ge 3$. If all diagonal entries of matrix B are equal to a certain constant, the value of this constant does not influence the optimal solution of QAP(A, B) but only its optimal value. Thus, we may assume w.l.o.g. that all Toeplitz matrices in this section have 0-entries on the diagonal. Consequently, $f(0) = 0$ holds for any function f generating some Toeplitz matrix in this section.

In order to prove that π^* solves QAP(A, B), where A and B are specially structured matrices as discussed above, we will show an even stronger statement. Consider the relaxation I-QAP(A,B) of QAP(A,B), where the columns and the rows of matrix A may be permuted by different permutations *independently of each other* (see Section 4.1):

(I-QAP(A,B)) $$\min_{\pi,\psi\in S_n} \sum_{i=1}^{n}\sum_{j=1}^{n} a_{\pi(i)\psi(j)}b_{ij}$$

It will turn out that the objective function of I-QAP(A,B) is minimized by $\pi = \psi = \pi^*$. This guarantees that π^* is also an optimal solution of QAP(A, B). Thus, instead of proving Theorem 5.2 we prove the following theorem:

Theorem 5.3 *The pair of permutations (π^*, π^*) solves I-QAP(A,B) where A is a nonnegative left-higher graded Anti-Monge matrix and B is a symmetric Toeplitz matrix generated by a nonnegative benevolent function f with $f(0) = 0$.*

We have already seen that the set of nonnegative left-higher graded Anti-Monge matrices A is a cone. Moreover, the nonnegative benevolent Toeplitz matrices B with zeros on the diagonal form a cone, too. Then, clearly, (π^*, π^*) must solve I-QAP(A,B) with A and B being generating matrices for the extremal rays of the respective cones. On the other hand, it is easy to see that the general result (Theorem 5.3) follows from the optimality of (π^*, π^*) over the extremal rays, and hence, only such I-QAPs need to be considered. Indeed, this is a straightforward application of the following observation.

Observation 5.4 *Assume that I-QAP(A_1, B) and I-QAP(A_2, B) are both solved by the pair of permutations (π_0, ψ_0). Then for any two reals $k_1, k_2 \geq 0$, the problem I-QAP$(k_1A_1 + k_2A_2, B)$ is also solved by (π_0, ψ_0).*

Proof. The proof follows directly from the equality

$$Z(k_1A_1 + k_2A_2, B, \phi, \psi) = k_1Z(A_1, B, \phi, \psi) + k_2Z(A_2, B, \phi, \psi). \qquad \square$$

Similarly, an analogous statement holds for the linear combinations of the second matrix B.

To complete the proof of Theorem 5.3 and hence of Theorem 5.2, we have to show that (π^*, π^*) solves I-QAP(A,B), where A is one of the matrices $A = C^{(p,q)}$ generating the cone of left-higher graded nonnegative Anti-Monge matrices (see Section 4.2.2), and B is a Toeplitz matrix generated by one of the functions g^α or h^β. Let us first specialize A to one of the matrices $C^{(p,q)}$ and keep B as a general Toeplitz matrix. Then, I-QAP($C^{(p,q)}, B$) can be formulated as a selection problem which is more convenient to handle. Namely, we have to select p rows and q columns of matrix B such that the total sum of all pq selected entries is minimized. The rows to be selected are those which are mapped to the last p rows $n-p+1, \ldots, n$, of A by permutation π, and the selected columns are those which are mapped to the last q columns $n-q+1, \ldots, n$, of A by permutation ψ. The solution of this selection problem in case of benevolent matrices is given by the following lemma.

Lemma 5.5 (The optimal selection lemma) *Let $1 \le p, q \le n$. Let B be a nonnegative Toeplitz matrix generated by a benevolent function f with $f(0) = 0$. Suppose that p rows and q columns of matrix B have to be selected such that the total sum of all pq selected entries is minimized. Then it is optimal to select the last p elements of the sequence*

$$1, n, 2, n-1, 3, \ldots \tag{5.2}$$

as row indices and the last q elements of this sequence as column indices. $\square$

To see that Theorem 5.3 (and hence Theorem 5.2) follows from Lemma 5.5 note that row i of $A = C^{(p,q)}$ is mapped to row $\pi^{-1}(i)$ of B. Thus, the rows with the 1-entries in A are mapped to the last p elements of the sequence

$$\pi^{-1}(1), \pi^{-1}(2), \ldots, \pi^{-1}(n).$$

For $\pi = \pi^*$ this sequence coincides with the sequence given in (5.2) and hence, under the assumption that Lemma 5.5 holds, permuting the rows according to π^* in I-QAP(A,B) will be optimal. The same argument applies to the columns. By Observation 5.4, this result carries over from $C^{(p,q)}$ to all left-higher nonnegative Anti-Monge matrices A and this concludes the proof of Theorem 5.3 (hence of Theorem 5.2).

For definiteness, let us write down the rows and columns to be selected according to the above lemma. The claimed optimal solution selects the p rows from

$$p_1 := \left\lceil \frac{n-p}{2} \right\rceil + 1 \quad \text{to} \quad p_2 := n - \left\lfloor \frac{n-p}{2} \right\rfloor \tag{5.3}$$

and the q columns from

$$q_1 := \left\lceil \frac{n-q}{2} \right\rceil + 1 \quad \text{to} \quad q_2 := n - \left\lfloor \frac{n-q}{2} \right\rfloor . \tag{5.4}$$

The optimal selection lemma (Lemma 5.5) is proven separately for Toeplitz matrices B generated by the functions g^α (Lemma 5.6) and h^β (Lemma 5.8). Since the above formulation as a selection problem for rows and columns is merely another formulation of the I-QAP, Observation 5.4 can then be used to obtain Lemma 5.5.

Lemma 5.6 *The pair of permutations (π^*, π^*) solves I-QAP(A,B) with $A = C^{(p,q)}$ and a symmetric Toeplitz matrix B generated by g^α, for any $1 \le p, q \le n$ and any $\lfloor \frac{n}{2} \rfloor + 1 \le \alpha \le n-1$. In other words, selecting p rows and q columns from B according to Lemma 5.5 will minimize the sum of the selected elements.*

Sketch of the proof. First, it is shown that $p+q-2\alpha$ is a lower bound for the value of the objective function. Indeed, instead of selecting p rows and q columns of the matrix B, let us take the opposite view and delete $n-p$ rows and $n-q$ columns from B. The matrix B contains originally $2(n-\alpha)$ 1-entries and no two of them are in the same row or in the same column. Thus, by deleting $n-p$ rows and $n-q$ columns we may delete at most $(n-p)+(n-q)$ 1-entries, which gives a simple lower bound of $2(n-\alpha) - \big((n-p)+(n-q)\big) = p+q-2\alpha$ for the remaining number of 1-entries.

Secondly, it is shown by elementary means that the claimed solution achieves the value $\max\{0, p+q-2\alpha\}$, and this completes the proof. A full version of the proof can be found in [33]. □

Now let us turn to the functions h^β which generate the other extremal rays of the cone of benevolent functions f with $f(0) = 0$. For the proof it is convenient to work with a certain quantity $Q(b, m)$ whose definition is given below together with some properties. The simple and elementary proofs of these properties can be found in [33].

Definition 5.3 *For two integers $b, m \geq 0$, the quantity $Q(b, m)$ denotes the sum of the first b terms of the following sum:*

$$\min\{1, m\} + \min\{1, m\} + \min\{2, m\} + \min\{2, m\} + \min\{3, m\} + \min\{3, m\} + \cdots.$$

Lemma 5.7 *For fixed m, the difference $Q(b+1, m) - Q(b, m)$ increases with b. Therefore the function $Q(b, m)$ is a convex function in b with $Q(0, m) = 0$, and so we have in particular*

$$Q(b_1, m) + Q(b_2, m) \leq Q(b_1 + b_2, m). \tag{5.5}$$

We also have

$$Q(b+1, m) \leq Q(b, m) + m. \tag{5.6}$$

An explicit representation of $Q(b, m)$ is

$$Q(b, m) = \begin{cases} \left\lfloor \left(\frac{b+1}{2}\right)^2 \right\rfloor & \text{for } b < 2m-2 \\ m(b+1-m) & \text{for } b \geq 2m-2 \end{cases}$$ □

Now we can give the solution of the selection problem in the case of nonnegative benevolent matrices B generated by functions h^β.

Lemma 5.8 *The pair of permutations (π^*, π^*) solves I-QAP(A,B) with $A = C^{(p,q)}$ and B being a symmetric Toeplitz matrix generated by h^β, for any $1 \leq p, q \leq n$ and any $1 \leq \beta \leq \lfloor \frac{n}{2} \rfloor$. In other words, selecting p rows and q columns from B according to Lemma 5.5 will minimize the sum of the selected elements.*

Sketch of the proof: Suppose we have selected a certain set of q columns. Let x_i, $i = 1, \ldots, n$, denote the sum of the entries of the i-th row lying on the q selected columns. Clearly, in order to minimize the sum of the entries, we have to select the p rows with the smallest x_i values. We shall state some simple conditions on the sequence of numbers (x_i), and from this we will derive lower bounds for the sum of the p smallest numbers x_i. The lower bounds hold for any choice of q columns, and thus they provide a lower bound on the optimum value z^* of the objective function. This lower bound matches the objective function value of our proposed solution, and therefore this solution is optimal.

Each column of B has a very simple structure. If we arrange the row indices in a circular sequence $1, 2, \ldots, n, 1, 2, \ldots$, the 1-entries in column j form a single circular interval of length $\gamma := n - 2\beta + 1$. This circular interval of 1-entries starts at row $j + \beta$ and ends at row $j + n - \beta$, wrapping around from row n to row 1, if necessary. (In this section all indices are taken modulo n.) It is easy to see that for each set of q selected columns, the row sums x_i fulfill the following conditions:

$$0 \leq \quad x_i \quad \leq q \quad 1 \leq i \leq n \tag{5.7}$$

$$q + \gamma - n \leq \quad x_i \quad \leq \gamma \quad 1 \leq i \leq n \tag{5.8}$$

$$\sum_{i=1}^{n} x_i \quad = \quad q\gamma \tag{5.9}$$

$$|x_i - x_{i-1}| \quad \leq \quad 1, \quad 1 \leq i \leq n \tag{5.10}$$

The constraints (5.7) and (5.8) imply

$$p \cdot \max\{0, q + \gamma - n\} \leq z^* \leq p \cdot \min\{q, \gamma\}.$$

We will first deal with the cases in which these simple bounds are sufficient. Let us look at the vector x_i corresponding to our proposed optimal selection for the columns. Figure 5.2 shows such a sequence for the case $n = 19$, $q = 5$ and $\gamma = 8$.

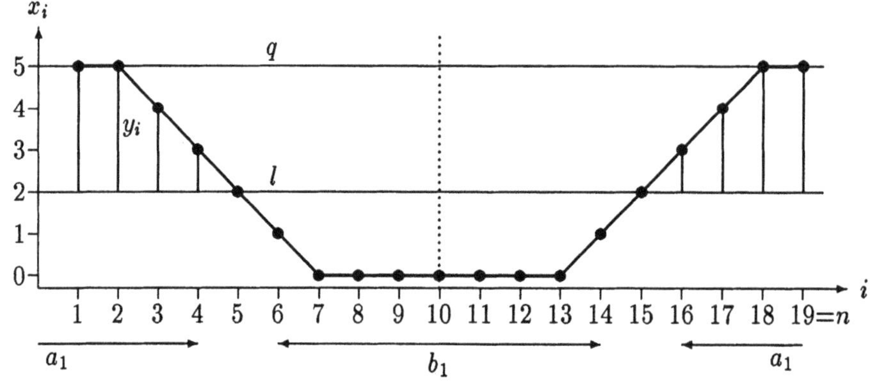

Figure 5.2 The vector (x_i) corresponding to the claimed optimal selection of columns for $n = 19$, $q = 5$, and $\gamma = 8$ ($\beta = 6$). The indices of the selected columns are 8, 9, 10, 11 and 12. If $p = 10$ or $p = 11$, the p-smallest x_i has the value $l = 2$ indicated by the horizontal line.

Since we select q "adjacent" columns, the circular sequence of values x_i consists of two horizontal intervals connected by a rising and a falling interval of slope ± 1. By analyzing the situation in detail, one can make the following statements about the two horizontal pieces. The higher piece looks as follows:

- Case U1, $q \geq \gamma$: The maximum value of x_i is γ and it occurs for $q - \gamma + 1$ adjacent positions (rows).
- Case U2, $q \leq \gamma$: The maximum value of x_i is q and it occurs for $\gamma - q + 1$ adjacent positions. (This is the case in figure 5.2.)

The lower piece looks as follows:

- Case L1, $q+\gamma \leq n$: The minimum value of x_i is 0 and it occurs for $n-q-\gamma+1$ adjacent positions.(This is the case in figure 5.2.)
- Case L2, $q + \gamma \geq n$: The minimum value of x_i is $q + \gamma - n$, and it occurs for $q + \gamma - n + 1$ adjacent positions.

These maximum and minimum values of x_i coincide with the ones given by (5.7) and (5.8). One can also check that in the sequence given by (5.2), the first rows selected (i.e., the last elements of the sequence) correspond to the positions where x_i has the minima, and the last rows selected correspond to the positions where x_i has its maxima. Therefore, if p has such a small value that only the minimal positions are selected, we know that the solution is optimal. So we get the solution for the following cases (which are not mutually exclusive):

- Case L1 ($q + \gamma \leq n$): If $p \leq n - q - \gamma + 1$, then
$$z^* = 0. \tag{5.11}$$

- Case L2 ($q + \gamma \geq n$): If $p \leq q + \gamma - n + 1$, then
$$z^* = p(q + \gamma - n). \tag{5.12}$$

Similarly, if p is big enough so that all elements except possibly the largest elements are selected, the solution is guaranteed to be optimal, taking into account that the total sum of all entries is constant, by (5.9). So we get the solutions for the following additional cases:

- Case U1 ($q \geq \gamma$): If $p \geq n - (q - \gamma + 1)$, then
$$z^* = q\gamma - (n - p)\gamma = (p + q - n)\gamma \tag{5.13}$$

- Case U2 ($q \leq \gamma$): If $p \geq n - (\gamma - q + 1)$, then

$$z^* = q\gamma - (n - p)q = (p + \gamma - n)q. \tag{5.14}$$

The remaining case that has to be considered is $|n - q - \gamma| + 1 < p < n - |q - \gamma| - 1$. Also in this case one can check that the p rows which are selected in our proposed optimal solution correspond to the p smallest x_i-values, if the q columns have been selected according to Lemma 5.5 (see [33] for more details and arguments). Consequently, the objective function value z^* corresponding to the solution claimed to be optimal can be evaluated as follows:

- Case L1 ($q + \gamma \leq n$): $z^* = (n - q - \gamma + 1) \cdot 0 + [1 + 1 + 2 + 2 + 3 + 3 + \cdots]$, where $p - (n - q - \gamma + 1)$ numbers are taken from the sum in brackets. According to Lemma 5.7 we have

$$z^* = Q(p + q + \gamma - n - 1, \infty) = \left\lfloor \left(\frac{p + q + \gamma - n}{2}\right)^2 \right\rfloor. \tag{5.15}$$

- Case L2 ($q + \gamma \geq n$): According to the above description of the vector x_i, every x_i is at least $q + \gamma - n$, and this value is taken by $q + \gamma - n + 1$ elements x_i. Summing separately the excess of x_i over $q + \gamma - n$, we can write

$$z^* = p(q + \gamma - n) + (q + \gamma - n + 1) \cdot 0 + [1 + 1 + 2 + 2 + 3 + 3 + \cdots],$$

where $p - (q + \gamma - n + 1)$ numbers are taken from the sum in brackets. This yields

$$\begin{aligned} z^* &= p(q + \gamma - n) + Q(p - (q + \gamma - n + 1), \infty) \\ &= \left\lfloor \frac{4p(q + \gamma - n) + (p - (q + \gamma - n))^2}{4} \right\rfloor = \left\lfloor \frac{(p + q + \gamma - n)^2}{4} \right\rfloor, \end{aligned}$$

which coincides with (5.15).

Further, it is shown that the value of the objective function for the claimed optimal solution is a lower bound for the number of the selected entries in an arbitrary selection, and this completes the proof. The proof of this last fact is rather technical, although elementary, and can be found in [33]. □

Remark 1. A different proof of Lemma 5.8 can be given by using a result of Çela and Woeginger (Theorem 4.1 in [46]). Translated into the language of Lemma 5.5, the authors show that there always exists an optimal selection consisting of a block of p (cyclically) adjacent rows and of a block of q (cyclically) adjacent columns. The result is formulated in a graph-theoretic setting, and the proof is based on an exchange argument.

Remark 2. Formula (5.15) can be combined with the previous bounds (5.11)–(5.14) for the easy cases L1, L2, U1 and U2 into one closed-form expression for the optimum objective function value. Let $N := \max\{0, p+q+\gamma-n\}$ and $k := \min\{p, q, \gamma, \lfloor N/2 \rfloor\}$. Then

$$z^* = N \cdot (N-k).$$

It is perhaps astonishing that this formula is completely symmetric in p, q, and γ.

5.1.3 The Anti-Monge–Toeplitz QAP with a periodic Toeplitz matrix

In this section we consider the Anti-Monge–Toeplitz QAP with a Toeplitz matrix generated by a periodic function. Again, the problem is NP-hard despite its very special structure. However, a polynomially solvable case arises if the periodic Toeplitz matrix has some additional property, namely if it is k-*benevolent*. The resulting version of the problem is again a constant permutation QAP, and the proof of this facts exploits the result obtained in the previous section on the Anti-Monge–Toeplitz QAP with a benevolent matrix. The k-benevolent matrices are a generalization of benevolent matrices, obtained by a periodic extension.

Toeplitz matrices generated by k-benevolent functions

The notion of a benevolent function can be generalized to obtain the so-called k-*benevolent function* as a result of the periodic continuation of a benevolent function. Then, analogously to the benevolent matrices, the k-*benevolent matrices* can be introduced.

Definition 5.4 *Let $k \geq 1$ and $n = kn'$. A function $f: \{-n+1, \ldots, n-1\} \to \mathbb{R}$ is called k-*benevolent *if it fulfills the following four properties.*

(i) $f(i) \leq f(i+1)$, *for* $0 \leq i \leq \lfloor \frac{n'}{2} \rfloor - 1$.

(ii) $f(i) = f(n' - i)$, *for* $0 \leq i \leq \lceil \frac{n'}{2} \rceil - 1$.

(iii) $f(i) = f(i + jn')$, *for* $0 \leq i \leq n' - 1$, $1 \leq j \leq k - 1$.

(iv) $f(-i) = f(i)$, *for* $0 \leq i \leq n - 1$.

A matrix $B = (b_{ij})$ *is called a* k-benevolent matrix *if it is a Toeplitz matrix generated by a* k*-benevolent function* f, *i.e.,* $b_{ij} = f(j - i)$ *for all* $1 \leq i, j \leq n$.

Properties (i), (ii) and (iv) are the same as the properties of benevolent functions for the range $\{-n' + 1, \ldots, n' - 1\}$, with two exceptions: The symmetry condition (ii) requires equality, and $f(0)$ is involved in (i). Property (iv) provides the periodic continuation with period n'.

Example 5.3 Let $n = 15$, $k = 3$, $n' = 5$. Define a 3-benevolent function $f: \{-14, -13, \ldots, 0, 1, \ldots, 14\} \to \mathbb{R}$ by $f(0) = 1$, $f(1) = 2$, $f(2) = 3$. The graph of function f and the 3-benevolent matrix generated by it are presented below.

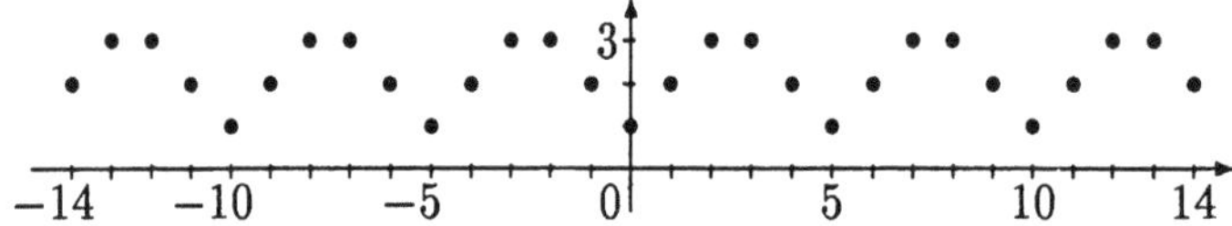

Figure 5.3 The graph of the 3-benevolent function f.

$$B = \left(\begin{array}{ccccc|ccccc|ccccc}
1 & 2 & 3 & 3 & 2 & 1 & 2 & 3 & 3 & 2 & 1 & 2 & 3 & 3 & 2 \\
2 & 1 & 2 & 3 & 3 & 2 & 1 & 2 & 3 & 3 & 2 & 1 & 2 & 3 & 3 \\
3 & 2 & 1 & 2 & 3 & 3 & 2 & 1 & 2 & 3 & 3 & 2 & 1 & 2 & 3 \\
3 & 3 & 3 & 1 & 2 & 3 & 3 & 3 & 1 & 2 & 3 & 3 & 3 & 1 & 2 \\
2 & 3 & 3 & 2 & 1 & 2 & 3 & 3 & 2 & 1 & 2 & 3 & 3 & 2 & 1 \\
\hline
1 & 2 & 3 & 3 & 2 & 1 & 2 & 3 & 3 & 2 & 1 & 2 & 3 & 3 & 2 \\
2 & 1 & 2 & 3 & 3 & 2 & 1 & 2 & 3 & 3 & 2 & 1 & 2 & 3 & 3 \\
3 & 2 & 1 & 2 & 3 & 3 & 2 & 1 & 2 & 3 & 3 & 2 & 1 & 2 & 3 \\
3 & 3 & 3 & 1 & 2 & 3 & 3 & 3 & 1 & 2 & 3 & 3 & 3 & 1 & 2 \\
2 & 3 & 3 & 2 & 1 & 2 & 3 & 3 & 2 & 1 & 2 & 3 & 3 & 2 & 1 \\
\hline
1 & 2 & 3 & 3 & 2 & 1 & 2 & 3 & 3 & 2 & 1 & 2 & 3 & 3 & 2 \\
2 & 1 & 2 & 3 & 3 & 2 & 1 & 2 & 3 & 3 & 2 & 1 & 2 & 3 & 3 \\
3 & 2 & 1 & 2 & 3 & 3 & 2 & 1 & 2 & 3 & 3 & 2 & 1 & 2 & 3 \\
3 & 3 & 3 & 1 & 2 & 3 & 3 & 3 & 1 & 2 & 3 & 3 & 3 & 1 & 2 \\
2 & 3 & 3 & 2 & 1 & 2 & 3 & 3 & 2 & 1 & 2 & 3 & 3 & 2 & 1
\end{array} \right)$$

One can clearly see that matrix B consists of $3 \times 3 = 9$ identical blocks of size $n' \times n' = 5 \times 5$ each. Two columns whose indices are congruent modulo $n' = 5$ are identical, and the same holds for the rows. □

We will show that QAP(A,B) of size $n = kn'$, with A being a left-higher graded Anti-Monge matrix and B being a k-benevolent matrix is a constant permutation QAP. The constant permutation $\pi^{(k)}$ which will be optimal for such a constant permutation QAP(A,B) is described below. Recall from Definition 4.5 that $\pi^* = \langle 1, 3, 5, 7, 9, \ldots, 8, 6, 4, 2 \rangle$. In this subsection, π^* will be regarded as a permutation of $\{1, \ldots, n'\}$. The permutation $\pi^{(k)} \in S_n$ (with $n = kn'$) is constructed in terms of $\pi^* \in S_{n'}$ as follows:

$$\pi^{(k)}((u-1)n' + i) = k\pi^*(i) - (u-1), \quad \text{for } 1 \leq u \leq k,\ 1 \leq i \leq n'. \tag{5.16}$$

Example 5.4 For $k = 3$, $n' = 5$, $n = kn' = 15$, we have

$$\pi^{(4)} = \langle \underbrace{3, 9, 15, 12, 6}, \underbrace{2, 8, 14, 11, 5}, \underbrace{1, 7, 13, 10, 5} \rangle.$$

The sequence $\langle \pi^{(k)}(1), \pi^{(k)}(2), \ldots \pi^{(k)}(20) \rangle$ is naturally divided into $k = 3$ groups with $n' = 5$ elements each. The first group, corresponding to $u = 1$ in (5.16), is obtained from $\pi^* = \langle 1, 3, 5, 4, 2 \rangle$ by multiplying every element by $k = 3$. Each successive group is obtained from the previous one by subtracting one from each entry. Thus, the numbers in the i-th group are those numbers between 1 and $n = 15$ which are congruent to $-(i-1)$ modulo k.

Next, let us divide the set of indices $\{1, 2, \ldots, n\}$ into k blocks. The set of indices belonging to the u-th block is denoted by N_u and is given below:

$$N_u := \{(u-1)n' + 1, \ldots, un'\} \qquad 1 \leq u \leq k.$$

As mentioned above, these blocks partition the row and column indices of matrix B in such a way that all submatrices B_{uv} whose rows are selected by some block N_u and whose columns are selected by some block N_v are identical.

The next theorem states our polynomiality result on the Anti-Monge–Toeplitz QAP with a k-benevolent matrix.

Theorem 5.9 *The permutation $\pi^{(k)}$ solves QAP(A, B), where A is a left-higher graded Anti-Monge matrix and B is a k-benevolent matrix.*

Similarly as for the Anti-Monge–Toeplitz QAP with a benevolent matrix, one can show that $(\pi^{(k)}, \pi^{(k)})$ solves even I-QAP(A,B), with a left-higher graded Anti-Monge matrix A, and a k-benevolent matrix B. Again, A is w.l.o.g. assumed to be nonnegative. Moreover, one can achieve $f(0) = 0$ by subtracting a constant from all values of f, where f is the function generating the k-benevolent matrix B. Clearly, this does not change the combinatorial structure of matrix B, and since $f(0)$ is the smallest value of f, the resulting matrix B will be nonnegative. Summarizing, we can prove the theorem. formulated below.

Theorem 5.10 *The pair of permutations $(\pi^{(k)}, \pi^{(k)})$ solves I-QAP(A,B), where A is a nonnegative left-higher graded Anti-Monge matrix and B is a k-benevolent matrix with zeros on the diagonal.*

As in Section 5.1.3, we can restrict our attention to the matrices $A = C^{(p,q)}$ which generate the extremal rays of the cone of nonnegative left-higher graded Anti-Monge matrices. Hence, Theorem 5.10 (and therefore, Theorem 5.9) follows from the following lemma.

Lemma 5.11 *For any $1 \leq p, q \leq n$, the pair of permutations $(\pi^{(k)}, \pi^{(k)})$ solves I-QAP(A,B), with $A = C^{(p,q)}$ and a Toeplitz matrix B generated by a k-benevolent function f with $f(0) = 0$.*

In order to prove this lemma, we could find the extremal rays of the cone of k-benevolent matrices with zeroes on the diagonal, and then proceed similarly as in the previous section. But we can also give a shorter proof which relies directly on some lemmas from Section 5.1.3.

Proof of Lemma 5.11. We know that this problem can be seen as the problem of selecting p rows and q columns from matrix B such that the total sum of all pq selected entries is minimized. Now suppose that some q columns have already been selected and we have to select the rows. Let x_i, for $i = 1, \ldots, n$, denote the sum of the selected entries in row i. Clearly, we have to select those p rows with the smallest x_i values. Since rows of B whose indices i are congruent modulo n' are identical, the numbers x_i corresponding to these rows are equal. Therefore, if $v_1 < v_2 < \ldots < v_j$ are the values taken by x_i, $1 \leq i \leq n$, and $V_t = \{x_i : x_i = v_t\}$, then $|V_t|$ is a multiple of k, for any $1 \leq t \leq j$. Moreover, the elements of V_t "are uniformly distributed in blocks", i.e., there are $|V_t|/k$ elements of V_t belonging to each block N_u of row indices. Thus, we may impose the following structure on the selected set of rows.

Claim 5.12 *There is an optimal selection of p rows, where the number p_u of selected rows in each block N_u is either $\lfloor p/k \rfloor$ or $\lceil p/k \rceil$.*

Since we can equally apply the argument to the columns once the rows are selected (in accordance with Claim 5.12), we also get:

Claim 5.13 *There is an optimal selection of p rows and q columns, where, in addition to the property of Claim 5.12, the number q_v of selected columns in each block N_v is either $\lfloor q/k \rfloor$ or $\lceil q/k \rceil$.*

The entries of B which lie in the selected rows and columns can be summed separately for each block B_{uv}, $1 \leq u, v \leq k$. Moreover, all blocks B_{uv} are identical to a certain $n' \times n'$ benevolent matrix B'. By Lemma 5.5 we know how to optimally select a given number p' of rows and a given number q' of columns from B', if we want to minimize the overall sum of the selected entries. Let us denote by $z(B', p', q')$ the optimal value of this problem, i.e., the optimal value of I-QAP$(C^{(p'q')}, B')$. Summarizing, we get the following lower bound for our problem:

$$Z(C^{(pq)}, B, \pi, \psi) \geq \sum_{u=1}^{k} \sum_{v=1}^{k} z(B', p_u, q_v)$$

Let us denote $r_p := p \bmod k$. Then, r_p of the values p_u must be equal to $\lceil p/k \rceil$ and the remaining $k - r_p$ of the values p_u are equal to $\lfloor p/k \rfloor$. Similarly, $r_q := q \bmod k$ of the values q_v are equal to $\lceil q/k \rceil$ and $k - r_q$ of them are equal to $\lfloor q/k \rfloor$. To finish the proof of the lemma, we have to show that the permutation $\pi^{(k)}$ indeed selects the optimal set of p_u rows out of each block of rows N_u and the optimal set of q_v columns out of each block of columns N_v, as specified by Lemma 5.5. This is easy to check: The indices i of selected rows and and the indices j of selected columns satisfy $\pi^{(k)}(i) > n - p$ or $\pi^{(k)}(i) > n - q$, respectively. By the way how $\pi^{(k)}$ is constructed, if we look at the indices of selected rows in each block N_u, these are precisely those indices which are mapped by π^* to the p_u largest numbers among $1, 2, \ldots, n'$:

$$\pi^{(k)}((u-1)n' + i) > n - p \quad \text{if and only if} \quad \pi^*(i) > n' - p_u,$$

for $1 \leq u \leq k$, $1 \leq i \leq n'$, where

$$p_u = \begin{cases} \lceil p/k \rceil & \text{for } u = 1, \ldots, r_p, \\ \lfloor p/k \rfloor & \text{for } u = r_p + 1, \ldots, k. \end{cases}$$

The same situation holds for the columns and this is just in accordance with Lemma 5.5. □

Toeplitz matrices generated by general periodic functions

The simplest non-trivial periodic functions f for generating a Toeplitz matrix B have period $n' = 2$ and thus, take only two values: $f(0) = f(i)$, for all even i, and $f(1) = f(i)$, for all odd i. These two values form a chess-board pattern in the matrix B. The case $f(0) \leq f(1)$ was treated above. It leads to a k-benevolent function (if n is even) and hence to a constant permutation QAP. The other case, $f(0) > f(1)$ leads to an NP-hard problem. It is no loss of generality to assume $f(0) = 1$ and $f(1) = -1$. In this case $B = (b_{ij})$ can be written as $b_{ij} = (-1)^{i+j}$.

Theorem 5.14 *QAP(A,B) is NP-hard even if A is a $(2k) \times (2k)$ nonnegative left-higher graded Anti-Monge matrix and $B = (b_{ij})$ is a $(2k) \times (2k)$ symmetric 0-1 Toeplitz matrix with $b_{ij} = (-1)^{i+j}$.*

Proof. The Toeplitz matrix B which fulfills the conditions of the theorem is a chess-board matrix. Further, Table 4.1 shows that A-MONGE×CHESS is NP-hard. Recall that this fact was proven in Theorem 4.24 and the reduction used there actually leads to a nonnegative symmetric product matrix. With an appropriate sorting of the generating vector such a product matrix is a nonnegative left-higher graded Anti-Monge matrix, and this is just what we need. Thus, *QAP*(A, B) is NP-hard. □

5.1.4 Symmetric Toeplitz matrices with small bandwidth

According to Definition 4.4 the bandwidth of a Toeplitz matrix generated by function f is the smallest i such that $f(j) = 0$, for all $j \geq i$. In the first part of this section we show that symmetric Toeplitz matrices with bandwidth equal to 2 lead to constant permutation QAPs. In the second part we give some examples which show that for any bandwidth larger than two, the resulting problem is no longer a constant permutation QAP.

Toeplitz matrices with bandwidth two

Let us investigate the Anti-Monge–Toeplitz QAP with a symmetric Toeplitz matrix with bandwidth two. It turns out that QAP(A,B) with $A \in$ A-MONGE(LHG) and a bandwidth-2 Toeplitz matrix B is a constant permutation QAP. By modifying the main diagonal we may assume that $f(0) = 0$ and hence $f(1) = f(-1)$ are the only two non-zero values of f.

The case $f(1) = f(-1) \leq 0$ leads to a benevolent function, and hence, is covered by Theorem 5.2. The remaining interesting case is $f(1) = f(-1) > 0$. By scaling we may then assume without loss of generality that $f(1) = f(-1) = 1$. The optimal permutation of this special version of the Anti-Monge–Toeplitz QAP is the so-called zig-zag permutation.

Definition 5.5 *The* zig-zag *permutation $\pi_z \in S_n$ is defined as follows:*

$$\pi_z(i) = \begin{cases} n-i & \text{if } i \leq \frac{n}{2} \text{ and } i \text{ is odd} \\ i & \text{if } i \leq \frac{n}{2} \text{ and } i \text{ is even} \\ i & \text{if } i > \frac{n}{2} \text{ and } n-i \text{ is even} \\ n-i & \text{if } i > \frac{n}{2} \text{ and } n-i \text{ is odd} \end{cases}$$

Example 5.5 For $n = 12$ we have $\pi_z = \langle \underline{11}, 2, \underline{9}, 4, \underline{7}, 6, \underline{5}, 8, \underline{3}, 10, \underline{1}, 12 \rangle$ and for $n = 13$ we have $\pi_z = \langle \underline{12}, 2, \underline{10}, 4, \underline{8}, 6, 7, \underline{5}, 9, \underline{3}, 11, \underline{1}, 13 \rangle$. The underlined entries are those which are computed as $\pi_z(i) = n - i$ in the above formula. Note that, for even n, the distinction between the cases $i \leq \frac{n}{2}$ and $i > \frac{n}{2}$ is irrelevant. □

As in the case of the Anti-Monge–Toeplitz QAP with a benevolent or a k-benevolent matrix, we can consider the relaxation I-QAP(A,B) with a left-higher graded Anti-Monge matrix A, and B being a symmetric Toeplitz matrix with bandwidth equal to 2. It can be shown that this problem is solved by the pair of permutations (π_z, π_z), and consequently, π_z is an optimal solution of $QAP(A, B)$ with A and B as specified above. The proof concerning the solution of I-QAP(A,B) is done again by reduction to extremal rays.

Theorem 5.15 *The pair of permutations (π_z, π_z) solves I-QAP(A,B), where A is a left-higher graded Anti-Monge matrix, and B is a symmetric Toeplitz matrix generated by the function $f\colon \{-n+1, \ldots, n-1\} \to \{0, 1\}$, with $f(1) = f(-1) = 1$ and $f(i) = 0$ for $i \neq \pm 1$.*

Proof. It is sufficient to show that permutation π_z is an optimal solution of $I-QAP(A, B)$ for all $A = C^{(p,q)}$ with $1 \leq p, q \leq n$. We use again the formulation of I-QAP(A,B) as a selection problem: p rows and q columns of the matrix B must be selected, such that the total sum of all pq selected entries is minimized. The claim is that it is optimal to select the rows $\pi_Z^{-1}(i)$, for $i \geq n-p+1$, and the columns $\pi_Z^{-1}(j)$, for $j \geq n-q+1$, where π_Z^{-1} is the inverse permutation of π_Z.

First, we compute the value of the objective function for the claimed optimal selection. This value depends on the quantity $h := p+q-n$ and is given below.

$$Z(C^{(p,q)}, B, \pi_Z, \pi_Z) = \begin{cases} 0 & \text{if } h \leq 0 \\ 1 & \text{if } h = 1,\ p \neq q \\ 0 & \text{if } h = 1,\ p = q = \frac{n+1}{2} \\ 2(p+q-n-1) & \text{if } h \geq 2. \end{cases} \tag{5.17}$$

Then, it will be easy to show that this expression is a lower bound for the value of the objective function of I-QAP$(C^{(p,q)}, B)$.

To compute $Z(C^{(p,q)}, B, \pi_Z, \pi_Z)$, recall that $b_{ij} = 1$ if $j = i+1$ or $j = i-1$, and $b_{ij} = 0$ otherwise. Thus, each pair (i,j) with $j = i \pm 1$ with $\pi_Z(i) \geq n-p+1$ and $\pi_Z(j) \geq n-q+1$ contributes with 1 to the sum

$$Z(C^{(pq)}, B, \pi_Z, \pi_Z) = \sum_{i=1}^{n}\sum_{j=1}^{n} C^{(p,q)}_{\pi_Z(i),\pi_Z(j)} b_{ij}$$

For a given value $u = \pi_Z(i)$ we call the values $\pi_Z(i \pm 1)$ the two *neighbors* of u. The neighbors are the two numbers which are adjacent to u in the sequence $\langle \pi_Z(1), \pi_Z(2), \ldots \rangle$ (see also Example 5.5). Of course, $\pi_Z(1) = n-1$ and $\pi_Z(n) = n$ have only one neighbor each. We distinguish two cases.

Case 1. n is even.
In this case the sum of two neighbors is always equal to $n-1$ or $n+1$, as can be checked from Definition 5.5. More precisely, we have:

- Every index $1 \leq u \leq n-2$ has two neighbors u' and u''. For one of them, the sum $u + u' = n+1$, and for the other one, $u + u'' = n-1$.
- The indices $u = n-1$ and $u = n$ have one neighbor u' each, and $u + u' = n+1$.

Now let us look at the numbers $u \geq n-p+1$ and see how many neighbors u' with $u' \geq n-q+1$ they have. An easy calculation shows that the following holds:

- If $u \leq q-2$ (and hence, $u \leq n-2$), then both neighbors $u' = n+1-u$ and $u'' = n-1-u$ fulfill $u', u'' \geq n-q+1$.

- If $u = q-1$ or $u = q$, then just one neighbor of u, namely $u' = n+1-u$, fulfills $u' \geq n-q+1$.
- If $u \geq q+1$, no neighbor u' of u fulfills $u' \geq n-q+1$.

Thus, by counting the number of neighbors $u'(u'')$ with $u' \geq n-q+1$ ($u'' \geq n-q+1$) for each $u = n-p+1, \ldots, n$, we get

$$2 \cdot \left|\{n-p+1, \ldots, n\} \cap \{1, \ldots, q-2\}\right| + \left|\{n-p+1, \ldots, n\} \cap \{q-1, q\}\right|,$$

This sum can be checked to be given by the right-hand side of (5.17). (The third case in (5.17) cannot occur when n is even.)

Case 2. n is odd.
The situation is similar as in Case 1, with the only exception that the sum of the two neighbors $\pi_Z(\frac{n-1}{2}) = \frac{n-1}{2}$ and $\pi_Z(\frac{n+1}{2}) = \frac{n+1}{2}$ equals n, and not $n \pm 1$. The situation concerning the sums of the neighbors for each index u, changes as follows:

- For $u = \frac{n-1}{2}$, we have a neighbor $u' = \frac{n+1}{2}$ with sum $u + u' = n$, instead of $u + u' = n-1$.
- On the other hand, for $u = \frac{n+1}{2}$, we have a neighbor $u'' = \frac{n-1}{2}$ with sum $u + u' = n$ instead of $u + u'' = n+1$.

The effects of these two changes cancel each other except in one special case: $p = q = \frac{n+1}{2}$. In this case, we lose one pair of neighbors and get $Z(C^{(p,q)}, B, \pi_Z, \pi_Z) = 0$, instead of the value 1 obtained for $p+q-n = 1$ in the case that n is even.

To complete the proof, let us show that (5.17) provides a lower bound for the value of the objective function. Instead of selecting p rows and q columns of the matrix B, let us take the opposite view and delete $n-p$ rows and $n-q$ columns from B. The matrix B originally contains $2n-2$ 1-entries, and each row or column contains at most two 1-entries. Thus, by deleting $n-p$ rows and $n-q$ columns we may delete at most $2(n-p)+2(n-q)$ ones, which gives an easy lower bound of $Z(C^{(pq)}, B, \pi, \psi) \geq 2n-2-\big(2(n-p)+2(n-q)\big) = 2(p+q-n-1)$ for the remaining number of 1-entries. This proves that (5.17) provides a lower bound in the case that $p+q \neq n+1$ or $p = q$. In the case that $p+q = n+1$ and $p \neq q$, if (π_Z, π_Z) is not an optimal solution of I-QAP$(C^{(p,q)}, B)$, there is a possibilty of deleting $n-p$ rows and $n-q$ columns from B so that no 1-entries remain. Since there are $2(n-1)$ 1-entries altogether, it follows that with each of the $n-p$ rows and $n-q$ columns deleted from B, we have to delete *precisely* two 1-entries. Moreover, some 1-entry which is contained in a deleted row must not be contained in a deleted column. Now, we show that there exists no such deletion strategy, and this completes the proof. Consider $b_{12} = 1$. If we delete row 1, we would delete only one 1-entry; therefore b_{12} must be deleted by deleting column 2. Similarly, $b_{21} = 1$ must be deleted by deleting row 2. Elements $b_{23} = 1$ and $b_{32} = 1$ are now also deleted. Now consider $b_{34} = 1$. Row 3 contains the already deleted element b_{32} and therefore b_{34} must be deleted by deleting column 4. Similarly, we have to delete row 4, and so on. This process can be continued until all elements are deleted, only if the number $n-p$ of deleted rows and the number $n-q$ of deleted columns are equal to $\frac{n-1}{2}$. But this is not the case because we have assumed $p \neq q$. □

Toeplitz matrices with larger bandwidth

Next, let us show that the bandwidth equal to 2 is an essential condition for Theorem 5.15. If one tries to find other 0-1 Toeplitz matrices which lead to constant permutation QAPs and have small bandwidth, the simplest matrices one can think about are those of type $T^{(1)}$ or $T^{(2)}$ which are generated by the functions $f^{(1)}, f^{(2)}\colon \{-n+1, -n+2, \ldots, 0, \ldots, n-1\} \to \{0,1\}$, respectively:

$$f^{(1)}(x) = \begin{cases} 1 & \text{if } x \in \{-2,-1,1,2\} \\ 0 & \text{otherwise} \end{cases} \qquad f^{(2)}(x) = \begin{cases} 1 & \text{if } x \in \{-2,2\} \\ 0 & \text{otherwise} \end{cases}$$

These matrices look as follows:

$$T^{(1)} = \begin{pmatrix} 0&1&1&0&0&\cdots&0&0&0 \\ 1&0&1&1&0&\cdots&0&0&0 \\ 1&1&0&1&1&\cdots&0&0&0 \\ 0&1&1&0&1&\cdots&0&0&0 \\ 0&0&1&1&0&\cdots&0&0&0 \\ \vdots&\vdots&\vdots&\vdots&\vdots&\ddots&\vdots&\vdots&\vdots \\ 0&0&0&0&0&\cdots&0&1&1 \\ 0&0&0&0&0&\cdots&1&0&1 \\ 0&0&0&0&0&\cdots&1&1&0 \end{pmatrix} \qquad T^{(2)} = \begin{pmatrix} 0&0&1&0&0&\cdots&0&0&0 \\ 0&0&0&1&0&\cdots&0&0&0 \\ 1&0&0&0&1&\cdots&0&0&0 \\ 0&1&0&0&0&\cdots&0&0&0 \\ 0&0&1&0&0&\cdots&0&0&0 \\ \vdots&\vdots&\vdots&\vdots&\vdots&\ddots&\vdots&\vdots&\vdots \\ 0&0&0&0&0&\cdots&0&0&1 \\ 0&0&0&0&0&\cdots&0&0&0 \\ 0&0&0&0&0&\cdots&1&0&0 \end{pmatrix}$$

Both $T^{(1)}$ and $T^{(2)}$ have bandwidth 3 and are 0-1 Toeplitz matrices. We give two examples showing that the Anti-Monge–Toeplitz QAPs $QAP(A, T^{(1)})$ and $QAP(A, T^{(2)})$ are *not* constant permutation QAPs.

Example 5.6 For $n = 6$ the 6×6 matrices $T^{(1)}$, $C^{(4,4)}$ and $C^{(1,4)}$ look as follows:

$$T^{(1)} = \begin{pmatrix} 0\,1\,1\,0\,0\,0 \\ 1\,0\,1\,1\,0\,0 \\ 1\,1\,0\,1\,1\,0 \\ 0\,1\,1\,0\,1\,1 \\ 0\,0\,1\,1\,0\,1 \\ 0\,0\,0\,1\,1\,0 \end{pmatrix}, \quad C^{(4,4)} = \begin{pmatrix} 0\,0\,0\,0\,0\,0 \\ 0\,0\,0\,0\,0\,0 \\ 0\,0\,1\,1\,1\,1 \\ 0\,0\,1\,1\,1\,1 \\ 0\,0\,1\,1\,1\,1 \\ 0\,0\,1\,1\,1\,1 \end{pmatrix}, \quad C^{(1,4)} = \begin{pmatrix} 0\,0\,0\,0\,0\,0 \\ 0\,0\,0\,0\,0\,0 \\ 0\,0\,0\,0\,0\,0 \\ 0\,0\,0\,0\,0\,0 \\ 0\,0\,0\,0\,0\,0 \\ 0\,0\,1\,1\,1\,1 \end{pmatrix}$$

We will show that for each permutation π_0 which is an optimal solution for problem QAP$(C^{(4,4)}, T^{(1)})$, the set $I_{\pi_0} := \{\pi_0^{-1}(3), \pi_0^{-1}(4), \pi_0^{-1}(5), \pi_0^{-1}(6)\}$ must be equal to $\{1,2,5,6\}$, whereas no optimal solution of QAP$(C^{(1,4)}, T^{(1)})$ shares this property. This implies that $QAP(A, T^{(1)})$ with a left-higher graded Anti-Monge matrix A *is not* a constant permutation QAP.

Let π be an arbitrary permutation in $\mathcal{S}_6$. $Z(C^{(4,4)}, T^{(1)}, \pi)$ is equal to the number of 1-entries in $T^{(1)}$ whose row and column indices fall in the set I_π, defined similarly as I_{π_0} above. There are $\binom{6}{2} = 15$ distinct sets I_π for all $\pi \in \mathcal{S}_6$. It can be checked that for any $I \subset \{1,2,\ldots,6\}$ with $|I| = 4$, the number of the 1-entries in $T^{(1)}$ whose row and column indices fall in I is larger than or equal to 4. Equality holds only for $I = \{1,2,5,6\}$. Thus, $I_{\pi_0} = \{1,2,5,6\}$ holds for each optimal solution π_0 to QAP$(C^{(4,4)}, T^{(1)})$. On the other hand, for all $\pi \in \mathcal{S}_6$, $Z(C^{(1,4)}, T^{(1)}, \pi)$ equals the number of 1-entries in $T^{(1)}$ lying on row $\pi^{-1}(6)$ with column indices falling in I_π. We have $I_{\pi_0} = \{1,2,5,6\}$, for each of the above optimal permutations π_0. If we select columns $\{1,2,5,6\}$ from $T^{(1)}$ we see that there exists no row with only zero entries on columns $\{1,2,5,6\}$. Thus, regardless of the value of $\pi_0^{-1}(6)$, we have $Z(C^{(1,4)}, T^{(1)}, \pi_0) \geq 1$. However, $Z(C^{(1,4)}, T^{(1)}, \psi) = 0$ for $\psi = \langle 6,1,2,3,4,5 \rangle$. Hence, π_0 cannot be an optimal solution for QAP$(C^{(1,4)}, T^{(1)})$ and the problems QAP$(C^{(4,4)}, T^{(1)})$, QAP$(C^{(1,4)}, T^{(1)})$ have no common optimal solution. This completes the proof. □

Example 5.7 For $n = 9$ the 9×9 matrices $T^{(2)}$, $C^{(4,9)}$ and $C^{(3,7)}$ look as follows:

$$T^{(2)} = \begin{pmatrix} 0\,0\,1 \cdots 0 \\ 0\,0\,0 \cdots 0 \\ 1\,0\,0 \cdots 0 \\ 0\,1\,0 \cdots 0 \\ \vdots\;\vdots\;\vdots\;\ddots\;\vdots \\ 0\,0\,0 \cdots 1 \\ 0\,0\,0 \cdots 0 \\ 0\,0\,0 \cdots 0 \end{pmatrix} \quad C^{(4,9)} = \begin{pmatrix} 0\,0 \cdots 0\,0 \\ 0\,0 \cdots 0\,0 \\ \vdots\;\vdots\;\ddots\;\vdots\;\vdots \\ 0\,0 \cdots 0\,0 \\ 1\,1 \cdots 1\,1 \\ 1\,1 \cdots 1\,1 \\ 1\,1 \cdots 1\,1 \\ 1\,1 \cdots 1\,1 \end{pmatrix} \quad C^{(3,7)} = \begin{pmatrix} 0\,0\,0 \cdots 0 \\ 0\,0\,0 \cdots 0 \\ \vdots\;\vdots\;\vdots\;\ddots\;\vdots \\ 0\,0\,0 \cdots 0 \\ 0\,0\,0 \cdots 0 \\ 0\,0\,1 \cdots 1 \\ 0\,0\,1 \cdots 1 \\ 0\,0\,1 \cdots 1 \end{pmatrix}$$

Consider the problem QAP$(C^{(4,9)}, T^{(2)})$. For any permutation $\pi \in \mathcal{S}_9$ denote $I_\pi = \{\pi^{-1}(6), \pi^{-1}(7), \pi^{-1}(8), \pi^{-1}(9)\}$. Note that $Z(C^{(4,9)}, T^{(2)}, \pi)$ is equal to the number of 1-entries in $T^{(2)}$ lying on the rows whose indices fall in I_π. From this observation it follows immediately that $Z(C^{(4,9)}, T^{(2)}, \pi) \geq 4$, for all $\pi \in \mathcal{S}_9$, and that the equality holds only for permutations π with $I_\pi = \{1,2,8,9\}$. Thus, if π_0 is an optimal solution to QAP($C^{(4,9)}, T^{(2)}$), then $I_{\pi_0} = \{1,2,8,9\}$. Now consider QAP($C^{(3,7)}, T^{(2)}$). It is easy to check that a permutation ψ yields an objective function value equal to 0, $Z(C^{(3,7)}, T^{(2)}, \psi) = 0$, if and only if the equalities $\{\psi(1), \psi(5), \psi(9)\} = \{7,8,9\}$ and $\{\psi(3), \psi(7)\} = \{1,2\}$ hold. (We have to find 3 distinct rows whose all ones lie on no more than 2 distinct columns.) Thus, for an optimal solution to QAP($C^{(3,7)}, T^{(2)}$) these equalities must hold. Clearly, $\{\psi(1), \psi(5), \psi(9)\} = \{7,8,9\}$ is not fulfilled by a permutation π_0 as above which is an optimal solution to QAP($C^{(4,9)}, T^{(2)}$). Thus, there exists no common optimal solution to problems QAP($C^{(4,9)}, T^{(2)}$) and QAP($C^{(3,7)}, T^{(2)}$). Consequently, the Anti-Monge–Toeplitz QAP $QAP(A, T^{(2)})$ *is not* a constant permutation QAP. □

Finally, let us remark that when n is a multiple of 4, $QAP(A, T^{(2)})$ is a constant permutation QAP, for $A \in$ A-MONGE(LHG). The optimal permutation is obtained by "interleaving" two "copies" of π_z, one copy permuting the even indices and the other copy permuting the odd indices. More generally, we conjecture that if $T^{(k)}$ is an $n \times n$ Toeplitz matrix generated by a function $f\colon \{-n+1, \ldots, 0, \ldots n-1\} \to \{0,1\}$ with $f(k) = f(-k) = 1$, $f(x) = 0$, $x \neq \pm k$, and A is a left-higher graded Anti-Monge matrix, $QAP(A, T^{(k)})$ is a constant permutation QAP, whenever n is a multiple of $2k$.

Before concluding this section let us summarize in a table the complexity results on the Anti-Monge–Toeplitz QAP. An entry in row i corresponds to a QAP(A,B) where matrix B belongs to the class of matrices indicated at the left-most entry of the row i. The table will have only one column as we are considering the Anti-Monge–Toeplitz QAP(A,B) and we know that $A \in$ A-MONGE(LHG). Similarly as in Table 4.1 the entries of the table will be "NP", "poly" or "???". An entry "NP" means that the corresponding QAP is NP-hard, an entry "poly" means that the corresponding QAP is polynomially solvable and an entry "???" means that the computational complexity of the corresponding QAP is unknown. We will denote by BENEVOLENT and k-BENEVOLENT the classes of benevolent and k-benevolent matrices, respectively. Further we will denote by PERIOD the property of a Toeplitz matrix which is generated by a periodic function.

Table 5.1 The computational complexity of the Anti-Monge–Toeplitz QAP

	A-MONGE(LHG)
TOEPLITZ(SYM)	NP
BENEVOLENT	poly
TOEPLITZ(SYM,PERIOD)	NP
k-BENEVOLENT	poly
TOEPLITZ(SYM) $\cap$ BAND-k	???
TOEPLITZ(SYM) $\cap$ BAND-2	poly

5.2 THE ANTI-MONGE–TOEPLITZ QAP: APPLICATIONS

In this section we describe three well known combinatorial optimization problems that can be modeled via A-MONGE(LHG)×TOEPLITZ(SYM): 1) The so-called *turbine problem*, i.e., the assignment of given masses to the vertices of a regular polygon such that the distance of the center of gravity of the resulting system to the center of the polygon is minimized. 2) The *traveling salesman problem* on symmetric Monge matrices. 3) A *data arrangement problem* concerning the arrangement of data records with given access probabilities in a linear storage medium in order to minimize the average access time.

First, we consider the turbine problem which was shown to be NP-hard [33]. This implies that the general problem A-MONGE(LHG)×TOEPLITZ(SYM) is NP-hard. Then, we investigate the two other combinatorial optimization problems and show how the polynomiality result of Theorem 5.2 generalizes and unifies several earlier results related to those problems.

5.2.1 The turbine problem

Hydraulic turbine runners as used in electricity generation consist of a cylinder around which a number of blades are welded at regular spacings. Due to inaccuracies in the manufacturing process, the weights of these blades differ slightly, and it is desirable to locate the blades around the cylinder in such a way that the distance between the center of mass of the blades and the axis of the cylinder is minimized. This problem was introduced by Mosevich [170] in 1986.

Laporte and Mercure [149] observed that this problem can be formulated as a QAP in the following way. The places at regular spacings on the cylinder are modeled by the vertices $v_1, \ldots, v_n$ of a regular n-gon on the unit circle in the Euclidean plane. Thus

$$v_i = \left(\sin(\frac{2i\pi}{n}), \cos(\frac{2i\pi}{n})\right), \quad 1 \le i \le n.$$

The masses of the n blades are given by the positive reals $0 < m_1 \le m_2 \le \cdots \le m_n$. The goal is to assign the n masses to the n vertices in such a way that the center of gravity of the resulting mass system is as close to the origin as possible, i.e., to find a permutation $\phi \in S_n$ that minimizes the Euclidean norm of the vector

$$\sum_{i=1}^{n} m_{\phi(i)} \begin{pmatrix} \sin(\frac{2i\pi}{n}) \\ \cos(\frac{2i\pi}{n}) \end{pmatrix}.$$

An easy calculation reveals that minimizing the Euclidean norm of this vector is equivalent to minimizing the expression

$$\sum_{i=1}^{n} \sum_{j=1}^{n} m_{\phi(i)} m_{\phi(j)} \cos\left(\frac{2(i-j)\pi}{n}\right). \tag{5.18}$$

But this is a quadratic assignment problem $QAP(A, B)$. Note that the matrix $A = (a_{ij})$ defined by $a_{ij} = m_i \cdot m_j$ is a symmetric product matrix and therefore a left-higher graded Anti-Monge matrix, since the masses m_i are sorted in non-decreasing order. The matrix $B = (b_{ij})$ defined by $b_{ij} = \cos\left(\frac{2(i-j)\pi}{n}\right)$ is a symmetric Toeplitz matrix. Note that the function $f(x) = \cos(2\pi x/n)$ which generates B is not benevolent, and therefore, Theorem 5.2 cannot be applied to solve the turbine problem. Laporte and Mercure [149] and Schlegel [210] proposed and tested several heuristics for this problem. However, no fast (polynomial time) algorithm to solve the turbine problem to optimality has been derived till today. This is not a coincidence: Burkard at al. in [33] show that the turbine problem is NP-hard and hence, the existence of a polynomial solution algorithm would imply $\mathcal{P} = \mathcal{NP}$.

Proposition 5.16 (Burkard, Çela, Rote and Woeginger [33], 1996)
The turbine problem, i.e., the minimization of (5.18) *over all permutations* $\phi \in S_n$*, is an NP-hard problem.*

On the other hand, notice that the function $f' = -f$ with $f'(x) = -\cos(2\pi x/n)$ which generates matrix $-B$ is benevolent. Since matrix $-B$ is benevolent, Theorem 5.2 implies that $QAP(A, -B)$ is a constant permutation QAP and π^* is its constant permutation. This problem corresponds to the *maximization* of (5.18). Hence, Theorem 5.2 can be applied to solve the *maximization version of the turbine problem*, i.e., the version of the problem where the goal is to *maximize* the distance of the center of gravity of the mass system from the axis of the cylinder.

Corollary 5.17 *The maximization version of the turbine problem, i.e., the maximization of (5.18) over all permutations $\phi \in \mathcal{S}_n$, is solved to optimality by permutation π^*.* □

5.2.2 The TSP on symmetric Monge matrices

The *traveling salesman problem* (TSP) consists on finding a shortest closed tour through a set of cities with a given distance matrix. This problem is a fundamental problem in combinatorial optimization and well-known to be NP-hard. For more information, the reader is referred to the comprehensive book edited by Lawler, Lenstra, Rinnooy Kan, and Shmoys [153]. Several restricted versions of the TSP with special combinatorial structures of the distance matrix, are known to be solvable in polynomial time. For more information on polynomially solvable special cases of the TSP the reader is referred additionally to the recent review article of Burkard, Deĭneko, Van Dal, Van der Veen and Woeginger [34]. Several special cases of the TSP are known to be solvable in polynomial time due to special combinatorial structures in the distance matrix. The following proposition states one of the first results of this flavor which was obtained by Supnick [216]. We show that this result can be easily derived as a corollary of Theorem 5.2.

Proposition 5.18 (Supnick [216], 1957)
Every instance of the TSP with a symmetric Monge distance matrix $D = (d_{ij})$ is solved by the permutation π^*.

Proof. Let $\Delta = 2\max\{|d_{ij}|: 1 \le i,j \le n\}$ and define a sum matrix $S = (s_{ij})$ by $s_{ij} = (i+j)\Delta$. Moreover, let us define a symmetric Toeplitz matrix B by its generating benevolent function $f(1) = f(-1) = f(n-1) = f(-n+1) = -1$ and $f(i) = 0$ for $i \notin \{\pm(n-1), \pm 1\}$. The proof relies on the following three simple steps.

Firstly, QAP$(S-D, B)$ is solved by π^*: Since S and $-D$ both are Anti-Monge matrices, so is $S-D$. Moreover, it is straightforward to verify that the matrix $S-D$ is left-higher graded. Since B is a symmetric Toeplitz matrix generated by the benevolent function f (defined above), Theorem 5.2 applies.

Secondly, QAP$(-S, B)$ is solved by π^*: Since $-S$ is a sum matrix and B is a circulant matrix, Theorem 4.18 implies that QAP$(-S, D)$ is solved by every permutation $\pi \in S_n$.

Finally, based on Observation 4.1, we can sum up $S - D$ and S and get that QAP$(-D, B)$ is solved by π^*. Since QAP$(-D, B)$ and QAP$(D, -B)$ are equivalent, the problem QAP$(D, -B)$ is also solved by π^*.

Now, it is easily checked that matrix $-B$ is the adjacency matrix of an undirected cycle on n vertices. Hence, QAP$(D, -B)$ is exactly the TSP with distance matrix D. □

5.2.3 Data arrangement in a linear storage medium

Consider a set of n records $r_1, \ldots, r_n$ which are referenced repetitively, where the reference is to record r_i with probability p_i, and different references are independent. Without loss of generality the records are numbered such that $p_1 \le p_2 \le \cdots \le p_n$. The goal is to place these records into a linear array of storage cells, like a magnetic tape, such that the expected distance between two consecutively referenced records is minimized, i.e., one wishes to minimize

$$\sum_{i=1}^{n} \sum_{j=1}^{n} p_{\pi(i)} p_{\pi(j)} d_{ij} , \tag{5.19}$$

where d_{ij} is the distance between the records placed at cells i and j. In the late 1960s and early 1970s, much research has been done on the special case of this problem where the distance d_{ij} is given as $d_{ij} = f(|i-j|)$, $f\colon \{0, \ldots, n-1\} \to \mathbb{R}$, i.e., d_{ij} only depends on the absolute value of the difference between i and j. The following proposition summarizes three of these results in order of increasing generality.

Proposition 5.19 *(a) If $d_{ij} = |i-j|$, then the data arrangement problem is solved by permutation π^*.* (Timofeev and Litvinov [223], 1969)

(b) If $d_{ij} = f(|i-j|)$ with a non-decreasing and convex function f, then the data arrangement problem is solved by permutation π^*. (Burkov, Rubinstein and Sokolov [41], 1969)

(c) If $d_{ij} = f(|i-j|)$ with a non-decreasing function f, then the data arrangement problem is solved by permutation π^*. (Metelski [165], Pratt [187], 1972) □

Metelski [165] and Pratt [187] realized that the above results are all contained in the following result due to Hardy, Littlewood and Pólya, here formulated in the language of the QAP and proved by applying our Theorem 5.2.

Proposition 5.20 (Hardy, Littlewood and Pólya [114], 1926)
Let $A = (a_{ij})$ be defined by $a_{ij} = x_i y_j$ for nonnegative real numbers $x_1 \leq \cdots \leq x_n$ and $y_1 \leq \cdots \leq y_n$. Let $B = (b_{ij})$ be a symmetric Toeplitz matrix generated by a function f that is non-decreasing on $\{0, \ldots, n\}$. Then QAP(A, B) is solved by π^*.

Proof. It is easy to verify that matrix A is a left-higher graded Anti-Monge matrix. Since, moreover, matrix B is generated by a benevolent function f, Theorem 5.2 can be applied. □

Motivated from the data arrangement problem, Rubinstein [201] showed that the result stated in Proposition 5.20 holds also for left-higher graded Anti-Monge matrices instead of product matrices. Obviously, this result is also contained as a special case in Theorem 5.2.

5.3 THE KALMANSON–TOEPLITZ QAP

Another restricted version of the QAP which leads to polynomially solvable cases of the problem is the so-called Kalmanson–Toeplitz QAP. In an instance QAP(A,B) of the Kalmanson–Toeplitz QAP, matrix A is a Kalmanson matrix and B is a symmetric Toeplitz matrix. The complexity of the Kalmanson–Toeplitz QAP is an open problem. In the case that the Toeplitz matrix possesses some additional properties, a polynomial solvable case arises, as shown by Deĭneko and Woeginger [65]. Namely, if A is a Kalmanson matrix and the generating function $f: \{-n+1, \ldots, -1, 0, 1, \ldots, n-1\} \to \mathbb{R}$ of an $n \times n$ Toeplitz matrix B fulfills the conditions (i)-(iii) given below, $QAP(A, B)$ is solved by the identity permutation.
(i) $f(i) = f(-i)$, for $i = 1, 2, \ldots, n-1$.
(ii) $f(i) = f(n-i)$, for $i = 1, 2, \ldots, \lfloor \frac{n}{2} \rfloor$
(iii) $f(i-1) \geq f(i)$, for $i = 1, \ldots, \lfloor \frac{n}{2} \rfloor$

Proposition 5.21 (Deĭneko and Woeginger [65], 1996)
The QAP(A,B) with a Kalmanson matrix A and a Toeplitz matrix B generated by a function f which fulfills conditions (i)-(iii) above, is solved to optimality by the identity permutation.

Sketch of the proof. The proof is done by reduction to extremal rays. Theorem 4.13 gives a description of the cone of Kalmanson matrices. Further, it can be easily shown that the nonnegative Toeplitz matrices generated by functions fulfilling conditions (i)-(iii) form a cone. The extremal rays of this cone are the Toeplitz matrices generated by the functions $f^{(\beta)}$, for $0 \le \beta \le \lfloor \frac{n}{2} \rfloor$, given by

$$f^{(\beta)}(i) = \begin{cases} 1 & \text{if } |i| \le \beta \text{ or } |i| \ge n - \beta \\ 0 & \text{otherwise} \end{cases}$$

Thus, if we denote by $D^{(\beta)}$ the Toeplitz matrix generated by $f^{(\beta)}$, every Toeplitz matrix B fulfilling the conditions of the theorem can be written as

$$B = K + \sum_{\beta=0}^{\lfloor \frac{n}{2} \rfloor} \theta_\beta D^{(\beta)}$$

where K is a constant matrix, i.e., a matrix with all entries having the same value, and θ_β are nonnegative coefficients. As argued in Section 4.2.4 and in accordance with Observation 4.1, it is sufficient to prove that the identity permutation solves the special problem $QAP(A_1, B_1)$, where A_1 is one of the matrices $F^{(p,p)}$, $F^{(p)}$, $W^{(p,q)}$, $W^{(p)}$, $1 \le p, q \le n$, which generate the cone of Kalmanson matrices, and B_1 is either a constant matrix or one of the matrices $D^{(\beta)}$, $0 \le \beta \le \lfloor \frac{n}{2} \rfloor$. A QAP with one of its coefficient matrices being a constant matrix is a constant QAP, i.e., it is solved by every permutation. $QAP(F^{(p,p)}, B)$, $1 \le p \le n$, with a Toeplitz matrix B, is also a constant QAP. Indeed, the diagonal entries of a Toeplitz matrix have all the same value and $F^{(p,p)}$ is a 0-1 matrix with only one 1-entry lying on the diagonal, on the p-th row and the p-th column. Analogously $QAP(F^{(p)}, B)$, $1 \le p \le n$, with a circulant matrix B, is a constant QAP[1]. This is due to the following two facts. Firstly, the row (column) sum of the off-diagonal elements of a circulant matrix has a constant value which does not depend on the specific row or column. Secondly, $F^{(p)}$ is a $0-1$ matrix with ones all over the p-th row and the p-th column except for the diagonal element, and zeroes elsewhere. So, what remains to be investigated are the problems $QAP(W^{(p,q)}, D^{(\beta)})$ and $QAP(W^p, D^{(\beta)})$. These problems can be formulated as a graph packing problem as follows (see also [65]). Consider the (undirected) graphs G_1, G_2, with adjacence matrices given by $W^{(p,q)}$ ($W^{(p)}$) and $D^{(\beta)}$, respectively. It is not difficult to see that G_1 is a complete bipartite graph of the form $K_{\alpha,n-\alpha}$, where $\alpha = q-p$ in the case of $W^{(p,q)}$, and $\alpha = p$ in the case of $W^{(p)}$. G_2 is a circulant graph, i.e., each node i, $i = 0, 1, \ldots, n-1$, is connected only with the nodes $i+1$, $i+2$, ..., $i+\beta$, $i-1$, $i-2$, ..., $i-\beta$, where all node indices are taken modulo n. In this setting the problem is to map the vertices

[1]Notice that a Toeplitz matrix generated by a function f fulfilling conditions (i)-(iii) is a symmetric circulant matrix.

of $G_1 = K_{\alpha,n-\alpha}$ into the the vertices $\{0,1,\ldots,n-1\}$ of G_2, so that the number of edges of G_1 which are mapped to edges of G_2 is minimized. Deĭneko et al. [65] have shown by elementary arguments that the identity permutation is the required optimal mapping. The authors give also the number of the edge coincidences in an optimal mapping, i.e., the value of the objective function of the two QAPs mentioned above, in terms of $\alpha = q-p$ and β. Namely,

$$Z(W^{(p,q)}, D^{(\beta)}, id) = \begin{cases} 2(q-p)(2\beta+1+p-q) & \text{if } q-p \leq \beta \text{ or } n-\beta \leq q-p \\ 2\beta(\beta+1) & \text{if } \beta \leq q-p \leq n-\beta \end{cases}$$

$$Z(W^{(p)}, D^{(\beta)}, id) = \begin{cases} 2p(2\beta+1-p) & \text{if } p \leq \beta \text{ or if } n-\beta \leq p \\ 2\beta(\beta+1) & \text{if } \beta \leq p \leq n-\beta \end{cases}$$

□

5.4 PERMUTED POLYNOMIALLY SOLVABLE CASES

As already mentioned in Section 4.1, two problems $QAP(A,B)$ and $QAP(A_1,B_1)$ with $A \in$Class1, $B \in$Class2, $A_1 \in$PermClass1, $B_1 \in$PermClass2 are equivalent. More concretely, if $A_1 = A^{\phi}$ and $B_1 = B^{\psi}$, the equality

$$Z(A_1,B_1,\pi) = Z(A,B,\phi\circ\pi\circ\psi^{-1})$$

holds for each permutation π, where "$\circ$" denotes the composition of permutations and ψ^{-1} is the inverse permutation of ψ. This equality implies that if π_1 is an optimal solution to $QAP(A_1,B_1)$, $\pi = \phi\circ\pi_1\circ\psi^{-1}$ is an optimal solution of QAP(A,B), and conversely, if ρ is an optimal solution of QAP(A,B), $\rho_1 = \phi^{-1}\circ\rho\circ\psi$ is an optimal solution of $QAP(A_1,B_1)$. Let us assume that the QAP Class1×Class2 is polynomially solvable. What can be said upon the complexity of the so-called *permuted problem* PermClass1×PermClass2? Clearly, if for each matrix A of PermClass$i$, $i=1,2$, a permutation π with $A^{\pi} \in$Class$i$ can be found in polynomial time, then the permuted problem is also solvable in polynomial time. Form a practical point of view, the identification of such a permutation π is strongly related to another problem, the so-called *recognition problem* for the corresponding class of matrices. For a matrix class Class the recognition problem can be formulated as follows: Given a matrix A, decide whether it belongs to PermClass, and if the answer is "yes" find a permutation π such that $A^{\pi} \in$ Class. Obviously, if Class1×Class2 is polynomially solvable and the recognition problems for Class1 and Class2 are polynomially solvable, then the problem PermClass1×PermClass2 is polynomially solvable, too.

All polynomially solvable restricted versions of the QAP presented in this chapter involve matrix classes with quite restrictive properties. Consequently, the class of polynomially solvable QAPs is still very small. However, notice that this class is a bit larger than it seems, due to the polynomial solvability of most of the corresponding permuted problems. The polynomial solvability of the permuted problem follows from the fact that the recognition problems for the involved matrix classes are polynomially solvable. Let us briefly recall these matrix classes and discuss the related recognition problems.

Monge, Anti-Monge and Kalmanson matrices. Given an $n \times n$ matrix A, it can be decided in $O(n^2)$ whether there exists a permutation π such that A^π is a Monge (Anti-Monge) matrix. The algorithm is a straightforward modification of the algorithm of Deĭneko and Filonenko [63] for the analogous recognition problem where the rows and columns are permuted by (possibly) different permutations (see also Rudolf [204]). An $n \times n$ permuted Kalmanson matrix can be recognized in $O(n^2)$ time by an algorithm of Christofer, Farach and Trick [50].

(Symmetric) Sum matrices and symmetric product matrices. Clearly, a permuted sum (product) matrix is a sum (product) matrix, i.e., if A is a sum (product) matrix, A^π is also a sum (product) matrix. Since a sum matrix is defined in terms of n^2 linear equations which involve $2n$ variables (or n variables, in the case of symmetric matrices), the recognition of a (symmetric) sum matrix can be done by solving such a system of linear equations. The recognition of symmetric product matrices is trivial: for an $n \times n$ symmetric product matrix A with generating vector (α_i) we have $a_{ii} = \alpha_i^2$, $1 \le i \le n$. Thus, $\alpha_i = \pm\sqrt{a_{ii}}$. Notice that if A is a symmetric product matrix generated by a vector (α_i), then A is also generated by $(-\alpha_i)$. Due to this property we can arbitrarily choose α_1 to be positive or negative and then argue upon the sign of the other elements α_j with $j > 1$, based on the equalities $a_{ij} = \alpha_1\alpha_j$ for $j > 1$. Then it is straightforward to check in $O(n^2)$ whether the resulting vector (α_i) indeed generates matrix A.

Large (small) symmetric matrices. A permuted large (small) symmetric matrix is again a large (small) symmetric matrix. So, the problem consists of recognizing the (large) small symmetric matrices, and this is trivial. Indeed, if $A = (a_{ij})$ is a large (small) symmetric matrix, there exists a generating vector (α_i) such that $a_{ij} = \max\{\alpha_i, \alpha_j\}$ ($a_{ij} = \min\{\alpha_i, \alpha_j\}$), from which follows that $\alpha_i = a_{ii}$, for all i.

Toeplitz matrices fulfilling conditions (i)-(iii) in the Kalmanson–Toeplitz QAP. The recognition of this kind of permuted Toeplitz matrices is discussed in [65]. The authors give an elementary $O(n^2)$ algorithm, where n is the size of the matrix to be recognized.

Benevolent and k-benevolent matrices. Without going into details, let us mention that the recognition of permuted k-benevolent matrices can be done analogously as that of Toeplitz matrices fulfilling (i)-(iii) (see [65]). As for benevolent matrices the problem is not completely solved. Permuted benevolent matrices, which fulfill BEN3 *with equality* (see Definition 5.1) can be recognized by applying the approach of Deĭneko et al. [65][2]. In the case that BEN3 is fulfilled with inequality the recognition problem remains open, to author's knowledge.

2-bandwidth symmetric Toeplitz matrices. The entries of a permuted 2-bandwidth symmetric Toeplitz matrices take only two values, 0 and $v \neq 0$. The recognition of such matrices is trivial: Check whether the graph with vertex set $\{1, 2, \ldots, n\}$ and an edge (i, j) whenever $a_{ij} = a_{ji} = v$, is a Hamiltonian path, where n is the size of the given matrix. In the affirmative case each of the two permutations describing the ordering of the vertices along the path transform the given matrix into a 2-bandwidth symmetric Toeplitz matrix.

Taxonomy matrices and (negative) chess-board matrices. Due to their block-diagonal structure, the recognition of the permuted taxonomy matrices is trivial and takes $O(n^2)$ elementary operations, where n is the size of the matrix. A necessary condition for a matrix $A = (a_{ij})$ to be a permuted taxonomy matrix is that $a_{ii} = 0$ implies $a_{ij} = a_{ji} = 0$, for all $1 \leq j \leq n$. Let us consider the following relation $\mathcal{R}$ defined on the set of indices $\{i: 1 \leq i \leq n, a_{ii} \neq 0\}$: $i\mathcal{R}j$ if and only if $a_{ij} \neq 0$ or $a_{ji} \neq 0$. It is not difficult to see that a matrix A which fulfills the above described necessary condition is a permuted taxonomy matrix if and only if $\mathcal{R}$ is an equivalence relation. The permutation π which transforms the given matrix in a taxonomy matrix can be easily constructed in terms of the classes of equivalence of $\mathcal{R}$. If $I = \{i_1, i_2, \ldots, i_k\}$ is some equivalence class of $\mathcal{R}$, then π permutes the indices $i_1, i_2, \ldots, i_k$ to a contiguous interval of indices, say $t, t+1, \ldots, t+k-1$.

As for the permuted (negative) chess-board matrices notice that by applying some linear transformation such matrices can be transformed into $0-1$ taxonomy matrices with two almost equally sized blocks of ones along the diagonal. Though we do not need to make this transformation. Given a matrix with entries equal to ± 1, we check whether the relation $\mathcal{R}'$, defined by $i\mathcal{R}'j$ if and only if $a_{ij} = 1$ or $a_{ji} = 1$ ($a_{ij} = -1$ or $a_{ji} = -1$ in the case of negative chess-board matrices), is an equivalence relation with two equivalence classes. In the affirmative case we check whether the cardinality of the two equivalence classes is equal to $\lfloor \frac{n}{2} \rfloor$ and $\lceil \frac{n}{2} \rceil$, respectively.

[2] If a benevolent matrix which fulfills BEN3 *with equality* is multiplied by -1, *all but the diagonal* entries of the resulting Toeplitz matrix fulfill properties (i)-(iii).

Graded matrices. The recognition of the permuted graded matrices is a straightforward matter. Basically, the problem we are facing here is the following: Given n n-dimensional vectors, can we find a permutation $\pi \in \mathcal{S}_n$ such that all given vectors are transformed into non-decreasing (non-increasing) ones when permuted according to π? It is straightforward to construct an $O(n^2 \log n)$ time algorithm to solve this problem. In the case of a matrix, one just needs to permute the given matrix iteratively so as to produce one more row (and/or column) with the required monotonicity in each iteration, without destroying the monotonicity of the already sorted rows (and/or columns). If this is possible, the composition of all permutations applied to the given matrix, in the order of their application, produces the required permutation which transforms the given matrix into a graded one.

5.5 OPEN PROBLEMS AND CONCLUSIONS

In this chapter we investigated restricted versions of the QAP with coefficient matrices possessing special combinatorial properties. We tried to draw a line of separation between polynomially solvable cases and NP-hard ones. In this context we first considered QAPs, where both coefficient matrices belong either to the matrix class MONGE or the matrix class A-MONGE. It is shown that this restriction does not simplify the problem in general, i.e., QAP(A,B) where both A and B are Monge (or Anti-Monge) matrices remains NP-hard. Moreover, in general, even coefficient matrices which possess more restrictive properties do not yield polynomially solvable cases of the QAP. Recall, for example, that the problems NPROD(SYM)×NPROD(SYM) and NPROD(SYM)×NCHESS are NP-hard. Only in the "very special" case where both matrices A and B are (negative) chess-board matrices, the problem QAP(A,B) is polynomially solvable.

The scenario of QAPs with a Monge matrix and an Anti-Monge matrix is more optimistic. The problem PROD(SYM)×NPROD(SYM) and the even-sized instances of CHESS×MONGE are polynomially solvable, whereas the complexity of the odd sized instances of the latter problem remains an open question. Here, an interesting open question concerns the computational complexity of the more general problem MONGE×PROD(SYM), or even MONGE×A-MONGE. Proving the NP-hardness of the first one or the polynomiality of the latter would complete Table 4.1.

Further, we have shown that the QAP with a circulant and a Monge matrix remains NP-hard, although there is at least one solvable case of it, namely the TSP on Monge matrices. We singled out several solvable cases of QAPs with one Monge matrix and the other matrix being a small bandwidth matrix. Solvable cases arising from the taxonomy problem are also classified into this group of problems. It should be noticed that in all these solvable cases the Monge matrix possesses also additional properties. All these polynomially solvable problems are constant permutation QAPs. A big open question related to QAPs with circulant matrices is the complexity of the problem CIRC×CIRC. Recall that CIRC×CIRC contains as a special case the so-called *circulant TSP* whose complexity is still a challenging open question, see [153].

Further, we investigated the Anti-Monge–Toeplitz QAP, i.e., the QAP with a left-higher graded Anti-Monge matrix and a symmetric Toeplitz matrix. It can be shown that the Anti-Monge–Toeplitz QAP is NP-hard. However, additional conditions on the Toeplitz matrix yield polynomially solvable versions of this problem which are constant permutation QAPs. Among these solvable cases, the most celebrated arises when the Toeplitz matrix is generated by a benevolent function. The other solvable cases involve k-benevolent matrices or Toeplitz matrices with bandwidth two, Generally, for Toeplitz matrices with larger bandwidth the problem A-MONGE(LHG)×TOEPLITZ is not a constant permutation QAP. A more recent result on QAPs with a Monge-like matrix and a Toeplitz matrix concerns the Kalmanson–Toeplitz QAP. Under specific conditions on the Toeplitz matrix, similar to those fulfilled by benevolent matrices, this problem becomes polynomially solvable.

Concerning the problem A-MONGE(LHG)×TOEPLITZ, we believe that more can be done in order to derive other properties of the Toeplitz matrix which yield constant permutation QAPs. The results presented here are only the first steps in this direction. However, deriving a characterization of all Toeplitz matrices which yield constant permutation problems from the class A-MONGE(LHG)×TOEPLITZ is an open problem whose complete solution is currently out of sight.

Summarizing, the results presented in this chapter bear evidence of the fact that the QAP is a very difficult problem from a theoretical point of view. Much structure on the coefficient matrices is required in order to obtain polynomially solvable special cases. Some of the solvable cases are inherently linear assignment problems (eg. PROD(SYM,NNEG)×PROD(SYM,NNEG)), whereas others are constant QAPs and thus, inherently trivial. More interesting are those solvable cases which can be solved in $O(n)$ time as *constant permutation QAPs*, where n is the size of the problem. There are only a few solvable cases which cannot be classified as members of any of these groups. In our opinion, the fact that almost all solvable cases of the problem discussed in this chapter are either constant QAPs or constant permutation QAPs is very intriguing. Is this behavior an inherent characteristic of the problem or is it due to our poor knowledge? There is no hope to answer this question without identifying other solvable and provably hard cases of the problem at hand.

The general method used to solve the constant permutation QAPs and also to prove their polynomiality, can be specified as *reduction to extremal rays*. In all cases certain (experimental) hints are exploited in order to make a guess for the constant permutation. In turn, the advantageous combinatorial structure of the problem data is exploited. Since the corresponding matrix classes form cones, the investigations can be restricted on simpler instances with 0-1 coefficients only. Proving that the guessed constant permutation is really an optimal solution for such instances is done - in most of the cases - by a simple but peculiar case-analysis or by exchange arguments. Often, the relaxation I-QAP of the problem has been considered, where the rows and the columns of the corresponding coefficient matrix are permuted independently by two different permutations, say π and ψ. Further, it is shown that the I-QAP is also a constant permutation problem with an optimal solution which permutes both rows and columns by the same permutation, say π_0. Then, clearly, π_0 is an optimal solution to the original QAP.

The constant QAPs or QAPs which inherently are linear assignment problems are usually quite easy to handle. The work to be done amounts on recognizing these properties by finding an appropriately equivalent reformulation of the problem. While analyzing the methods used in this chapter, we should notice that in one single case, namely when dealing with problems of the classLARGE(SYM)$\times$NCHESS, a dynamical programming approach is applied.

Concluding, the challenging and difficult research on polynomially solvable special cases of the QAP is still in its infancy. A lot remains to be done in the future towards drawing a borderline between easy and hard versions of the QAP.

6

QAPS ARISING AS OPTIMIZATION PROBLEMS IN GRAPHS

Many optimization problems in graphs can be formulated as QAPs with a special structure. Most of these optimization problems in graphs are NP-hard. However a number of polynomially solvable special cases have been identified for them. We believe that the general knowledge and understanding of the QAP can benefit from a closer look at these problems and the corresponding formulations as QAPs. This is the motivation for writing this chapter.

We present a number of well known results for placement problems (linear arrangement problems, in particular) , feedback arc set problems and packing problems, translated into the language of the QAP. In some cases, graph theoretical results related to the problems mentioned above lead to algorithmic approaches for solving the corresponding special cases of the QAP, e.g. in the case of packing problems. Some optimization problems in graphs lead to the so-called *generalized QAP* (GQAP) which is a straightforward generalization of the QAP. As in the QAP, also in the GQAP we are given two coefficient matrices A, B and wish to minimize the objective function $Z(A, B, \phi)$ of the related QAP. The difference to the QAP consists of the fact that here the minimization is done over a *subset* of the set $\mathcal{S}_n$ of all permutations of $\{1, 2, \ldots, n\}$, where n is the size of the problem, i.e. the size of the matrices A, B. Thus, the set of the feasibles solutions of the GQAP is a subset $\mathcal{H}_n$ of $\mathcal{S}_n$. An instance of the GQAP is specified by its coefficient matrices A, B, and its set of feasible solutions $\mathcal{H}_n \subseteq \mathcal{S}_n$, and is denoted by GQAP$(A, B, \mathcal{H}_n)$. In the case of $\mathcal{H}_n = \mathcal{S}_n$ we obtain the QAP as a special case of the GQAP. In this chapter we present some NP-hard and some polynomially solvable cases of the GQAP. Most of these GQAPs occur in the context of an optimization problem on isomorphic graphs, namely the problem of finding an isomorphism with "minimum weight" for two given isomorphic graphs. Along with polynomially solvable and provably hard special versions of the QAP, this chapter provides a number of open questions which could be attacked and hopefully solved in the future.

The chapter is organized as follows. In the first section the *placement problem* is introduced. We focus on the *linear arrangement problem* and review some well known results which lead to solvable cases of the QAP. In the second section special cases of the QAP arising from the *feedback arc set problem* are considered. A detailed analysis of QAP formulations of graph packing problems and their complexity analysis follows in Section 6.3. Finally, in the fourth section some results on polynomially solvable and provably hard cases of the GQAP are presented. We conclude by showing that the so-called *pyramidal QAP* is NP-hard, in contrast to the *pyramidal TSP* which is known to be polynomially solvable (see e.g. [153]).

6.1 QAPS RELATED TO PLACEMENT PROBLEMS

In *placement problems* we are given n modules that have to be placed in a one-dimensional array at unit intervals. The modules are pairwise connected by a number of wires. Let a_{ij} denote the number of wires connecting the modules i and j. The objective is to place the modules in such a way that the overall length of the connecting wires is minimized. Clearly, this problem can be formulated as a QAP(A,B), where the $n \times n$ symmetric matrix $A = (a_{ij})$ represents the number of the wire-connections, and $B = (b_{ij})$ is a symmetric Toeplitz matrix defined by $b_{ij} = |i-j|$, $1 \le i,j \le n$. Thus, QAP(A,B) belongs to MATRIX(SYM)×TOEPLITZ(SYM).

This problem was originally introduced in 1961 by Steinberg [215] and was called "the backboard wiring problem". The problem was reconsidered in 1972 by Hanan and Kurtzberg [110] under the name "the module placement problem". The special case of the above problem, where matrix A is the adjacency matrix of a graph is known as the *linear ordering* or the *linear arrangement* problem (or sometimes as the *minimum sum labeling* problem). In general, this problem is NP-complete (see Garey and Johnson [88]). Evan and Shiloach [75] have shown that even restricted versions like the linear arrangement problem in acyclic digraphs or in bipartite graphs are NP-complete. What does this mean in terms of QAPs? Consider a topological labeling of an acyclic digraph G on n vertices (i.e. a labeling of the vertices of the graph such that the tail of every arc has smaller label than its head). Then, if $A = (a_{ij})$ is the adjacency matrix of G, $a_{ij} = 0$ for all $i \ge j$, $1 \le i,j \le n$. Thus, the adjacency matrix A of a topologically labeled acyclic digraph is a *upper triangular matrix*, i.e. all entries of A lying on and under the diagonal are equal to 0. Now, the result of Evan and Shiloach can be formulated in the QAP language as follows.

Theorem 6.1 (Evan and Shiloach [75], 1975)
Let the $n \times n$ matrix $B = (b_{ij})$ be given by $b_{ij} = |i - j|$, $1 \leq i, j \leq n$. Consider an $n \times n$ upper triangular matrix A with entries equal to 0 or 1. Then $QAP(A, B)$ is NP-complete. □

Throughout the rest of this section we will review several polynomiality results on the linear arrangement problem, formulating them in the language of the QAP. The first result concerns (undirected) trees. In 1976 Goldberg and Klipker [98] proved that the linear arrangement problem on (undirected) trees is polynomially solvable. They proposed an $O(n^3)$ algorithm, where n is the size of the problem, i.e. the number of vertices of the tree. In 1979 Shiloach [212] derived a new algorithm improving the time complexity to $O(n^{2.2})$. Finally, in 1984, Chung [55] improved the time complexity to $O(n^{\lambda})$, where λ can be chosen to be any real which satisfies $\lambda > \log_2 3 \approx 1.58$.

Theorem 6.2 (Goldberg and Klipker [98], 1976, Shiloach [212], 1979, Chung [55], 1984)
Consider the $n \times n$ matrices A and $B = (b_{ij})$, where A is the adjacency matrix of a tree on n vertices and $b_{ij} = |i - j|$, $1 \leq i, j \leq n$. $QAP(A, B)$ is solvable in polynomial time by an $O(n^{\lambda})$ algorithm, where $\lambda > \log_2 3$. □

Both Chung and Shiloach propose recursive algorithms where the optimal linear arrangement for the given tree results as an appropriate combination of optimal linear arrangements for some suitably chosen rooted subtrees. The improvement in the time complexity results from a more efficient use of the subtrees in the recursive process rather then from some principally new idea. Note that the result of Proposition 6.2 applies only in the case that $A = (a_{ij})$ is a 0-1 matrix, i.e. the entries a_{ij} are either equal 0 or equal to 1. None of the above algorithms works in the case that A is the *weighted* adjacency matrix of a weighted tree. To our knowledge the time complexity of $QAP(A, B)$ in this case remains an open question.

A polynomially solvable version of the linear arrangement problem, where A is the *weighted* adjacency matrix of a *rooted* tree, was identified by Adolphson and Hu [3]. In this version of the problem not all permutations are considered as feasible arrangements. A feasible arrangement π has the property that for each arc (i, j) in the rooted tree, vertex i should be embedded to the left of vertex j, i.e.

$\pi(i) < \pi(j)$. Notice that because of this additional restriction this problem is a GQAP. Adolphson and Hu [3] propose a rather technical $O(n \log n)$ algorithm for this problem.

Theorem 6.3 (Adolphson and Hu [3], 1973)
Let $A = (a_{ij})$ be the $n \times n$ weighted adjacency matrix of a rooted tree, where the entries which correspond to non-existing edges are equal to 0. Let the $n \times n$ matrix $B = (b_{ij})$ be given by $b_{ij} = |i - j|$, $1 \le i, j \le n$. Denote by $\mathcal{S}_n^A$ the set of those permutations π of $\{1, 2, \ldots, n\}$ such that $\pi(i) < \pi(j)$ for each pair of indices i, j with $a_{ij} \neq 0$. Then the problem $GQAP(A, B, \mathcal{S}_n^A)$ is solvable in $O(n \log n)$ time.

$$(\mathbf{GQAP}(A, B, \mathcal{S}_n^A)) \qquad \min_{\pi \in \mathcal{S}_n^A} \sum_{i=1}^{n} \sum_{j=1}^{n} a_{\pi(i)\pi(j)} b_{ij} \qquad \square$$

Another polynomially solvable version of the linear arrangement problem occurs in the context of network flows. Assume that $A = (a_{ij})$ is the weighted adjacency matrix of a symmetric, undirected, edge weighted network. Gomory and Hu [101] defined the so-called Gomory-Hu tree that gives a concise representation of the maximum flows between any pair of vertices in the network. For any two vertices i and j in the network, the value of the maximum flow (minimum cut) from i to j equals the weight of the shortest edge on the path joining i to j in the Gomory-Hu tree. The weights on the edges of the network are the capacities involved in the related maximum flow problems. A given symmetric $n \times n$ matrix A can be considered as weighted adjacency matrix of a network with edge capacities equal to the corresponding entries in the matrix. The Gomory-Hu tree of this network is termed then as *Gomory-Hu tree associated with matrix A*. If the Gomory-hu tree associated with matrix A has a special structure, namely, in the case that this tree is a path, polynomially solvable cases of the linear arrangement problem arise.

Theorem 6.4 (Adolphson and Hu [3], 1973)
If the Gomory-Hu tree associated with a symmetric $n \times n$ matrix A is a path and matrix $B = (b_{ij})$ is defined by $b_{ij} = |i - j|$, $1 \le i, j \le n$, then $QAP(A, B)$ is polynomially solvable.

Sketch of the proof. This result is derived by Adolphson and Hu [3] as a corollary of a more general result. Let N_A be a symmetric (undirected), (edge) weighted network and let A be the symmetric matrix of the edge weights (capacities). It can be shown that the overall sum of the edge weights of the Gomory-Hu tree corresponding to N_A is a lower bound on the objective function value of $QAP(A, B)$. In the case that the Gomory-Hu tree is a *path*, it can be shown that the sum of its edge weights equals $Z(A,B,\pi)$, where $\pi(i)$ denotes the position of vertex i on the Gomory-Hu path. Thus, such a permutation π is an optimal solution to $QAP(A,B)$. (Clearly, there exist at least two optimal solutions π_1 and π_2 which are related by the equality $\pi_1(i) = \pi_2(n-i+1)$, for all $i = 1,2,\ldots,n$.) Thus, in this case, the $QAP(A,B)$ can be solved by computing the Gomory-Hu path for the network N_A. As the Gomory-Hu tree for a given symmetric, undirected, edge weighted network can be computed in polynomial time (see e.g. Gomory and Hu [101] for an $O(n^3)$ algorithm), $QAP(A,B)$ is polynomially solvable. □

It would be nice to have a characterization of networks whose Gomory-Hu tree is a path, or at least sufficient or necessary combinatorial conditions for symmetric, edge weighted networks which have this property. To our knowledge this problem has not been investigated up to now.

To conclude this section notice that in the case that A is a left-higher graded Anti-Monge matrix (see Section 4.1), the corresponding placement problem is solved by the permutation π^* defined in the previous chapter. This is a straightforward corollary of Theorem 5.2, since matrix $B = (|i-j|)$ is a Toeplitz matrix generated by a benevolent function (see Section 4.1 and Definition 5.1).

6.2 SPECIAL CASES RELATED TO THE FEEDBACK ARC SET PROBLEM

In the *minimum weight feedback arc set* problem (FAS) a weighted digraph $G = (V,E)$ with vertex set V and arc set E is given. The goal is to remove a set of arcs from E with minimum overall weight, such that all directed cycles, so-called *dicycles*, in G are destroyed and an acyclic directed subgraph remains. Clearly, the minimum weight feedback arc set problem is equivalent to the problem of finding an acyclic subgraph of G with maximum weight. It is worthy to notice here that this problem was sometimes termed as *linear ordering problem*, as for example in [190]. Notice however that the problem considered in this section is completely different from that treated in the previous section under the name linear ordering (arrangement) problem. We have adopted the terminology used (among others) by Garey and Johnson in [88] for both the linear ordering problem and the minimum weight feedback arc set problem. The unweighted version of the FAS, that is a FAS where the edge weights of the underlying digraph equal 0 or 1, is often called *the acyclic subdigraph problem* and is treated extensively by Jünger [126]. An interesting application of FAS is the so-called *triangulation of the input-output tables* which arises along with input-output analysis in economics (details and further references can be found in [190]).

The minimum weight feedback arc set problem can be formulated as a QAP. Since the vertices of an acyclic subdigraph can be labeled topologically (i.e. in such a way that the tail of every arc a has smaller label than its head), the FAS is equivalent to QAP(A,B), where A is the weighted adjacency matrix of G and matrix B equals the *lower triangular matrix* $L_n = (\ell_{ij})$, where $\ell_{ij} = -1$ if $i \leq j$ and $\ell_{ij} = 0$, otherwise.

The FAS is well known to be NP-complete (see Karp [136], Garey and Johnson [88]). Gavril [90] proved NP-completeness even for the case of an unweighted digraph where every vertex has in-degree and out-degree at most three.

Theorem 6.5 (Gavril [90], 1977)
Let A be an $n \times n$ 0-1 matrix where every row and every column contains at most three entries of value 1. *Then the problem $QAP(A, L_n)$ is NP-complete.* □

Throughout the rest of this section we review some results on polynomially solvable cases of the problem $QAP(A, L_n)$. The cornerstone for most of the polynomiality results on FAS presented in this section is a fundamental minimax theorem of Lucchesi and Younger [160], which is probably the very first result of this kind. Most of the later results, both algorithmic and theoretical ones, make use of the theorem of Lucchesi and Younger or generalize its basic idea. In order to formulate the result of Lucchesi and Younger, we need the definition of a *dicut* and the definition of a *transversal of dicuts*.

Definition 6.1 **(a)** *Consider a digraph $G = (V, E)$ with vertex set V and edge set E. A* directed cut (dicut) *in G is a set of arcs with tail in S and head in $V \setminus S$, provided that there are no arcs going from $V \setminus S$ to S and S is a strict subset of V, i.e. $S \subset V$, $S \neq \emptyset$, $S \neq V$.*

(b) *Consider a set of subsets of E, $\{E_1, E_2, \ldots, E_k\}$, $E_i \subseteq E$, $1 \leq i \leq k$, $k \in \mathbb{N}$. A* transversal of $\{E_1, E_2, \ldots, E_k\}$ *in G is a set of arcs E', $E' \subseteq E$, with the property that each of the sets E_i, $1 \leq i \leq k$, contains at least one arc from E'. In the case that $\{E_1, E_2, \ldots, E_k\}$ is the set of <u>all</u> dicycles or <u>all</u> dicuts in G, we have a transversal of dicycles or a transversal of dicuts, respectively.*

(c) *Assume that G is a weighted digraph with a weight function $w: E \to \mathbb{N}$, $e \mapsto w(e)$. The weight of a set of arcs E', $E' \subseteq E$, is given as sum of the weights of all its elements: $w(E') = \sum_{e \in E'} w(e)$.*

Proposition 6.6 (Lucchesi [159], 1976, Lucchesi and Younger [160], 1978)
Consider a digraph G. The minimum cardinality of a transversal of dicuts in G equals the maximum cardinality of a collection of pairwise disjoint dicuts. □

Let us give an illustrating example for dicuts, transversals and for Proposition 6.6.

Example 6.1 Dicuts and transversals.
Consider the digraph presented in Figure 6.1. It is easy to see that there are only three dicuts in this digraph: $D_1 = \{(1,2),(1,4)\}$, $D_2 = \{(1,4),(2,4),(2,3)\}$ and $D_3 = \{(2,3),(4,3)\}$. The set of arcs $T = \{(1,2),(2,3),(2,4)\}$ is a transversal of these dicuts, since $T \cap D_i \neq \emptyset$, for $i = 1,2,3$. Notice that $D_1 \cap D_2 \cap D_3 = \emptyset$ and $D_2 \cap D_3 = \{(2,4)\}$. Thus, the maximum cardinality of a collection of pairwise disjoint dicuts equals 2. In fact, the only collection of pairwise disjoint dicuts is $\{D_1, D_3\}$. According to Proposition 6.6 there exists a transversal of dicuts with cardinality equal to 2. Indeed, the set $\{(1,3),(2,4)\}$ is such a transversal. According to Proposition 6.6 there exist no transversal of dicuts with cardinality 1. Indeed, the existence of such a transversal would imply $D_1 \cap D_2 \cap D_3 \neq \emptyset$. □

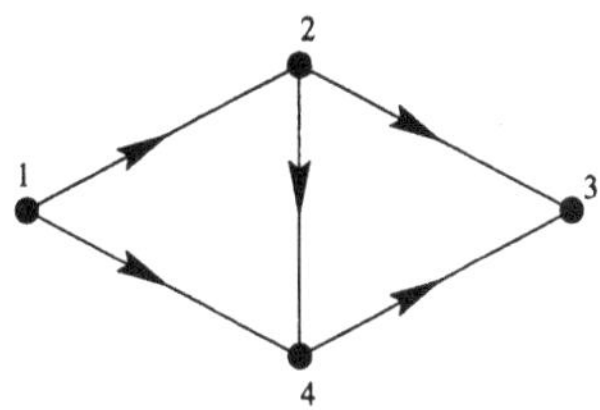

Figure 6.1 Dicuts and transversals.

Several alternative, sometimes constructive proofs for Proposition 6.6 have been given in [81, 137, 158]. Moreover, many authors have generalized this basic theorem for weighted digraphs and have obtained the following more general result.

Proposition 6.7 (Lucchesi [159], 1976, Edmonds and Giles [69], 1977, Frank [81], 1981)
Consider a weighted digraph $G = (V, E)$ with nonnegative integral weights $w(e)$ on its arcs $e \in E$. The minimum weight of a transversal of dicuts in G equals the maximum weight of a collection $\mathcal{C}$ of directed cuts with the property that e occurs in at most $w(e)$ dicuts from $\mathcal{C}$, for all $e \in E$. □

As a corollary of Proposition 6.7, the following result was implicitly proven by different authors, Lucchesi [159], Frank [81] and Gabow [85].

Theorem 6.8 (Lucchesi [159], 1976, Frank [81], 1981, Gabow [85], 1993) *Let A be the weighted adjacency matrix of a weighted* planar *digraph $G = (V, E)$. Then $QAP(A, L_n)$ is solvable in polynomial time, where L_n is a triangular matrix as defined previously in this section.*

Sketch of the proof. Let $G = (V, E)$ be a weighted planar digraph with weights $w(e)$ on its arcs $e \in E$, and let G^* be its dual digraph. With each arc e' of the dual digraph G^* associate a weight $w(e')$ equal to the weight of its corresponding arc e in G, $w(e') = w(e)$. It is well known that the one to one correspondence between arcs of G and arcs of G^* implies a one to one correspondence between directed cycles (dicycles) in G and dicuts in G^*. Clearly, there is also a one to one correspondence between transversals of dicycles in G and transversals of dicuts in G^*. Under these conditions, corresponding objects (dicuts and dicycles, transversals of dicuts and transversals of dicycles) have the same weight. Thus, as a corollary of Proposition 6.7, we get:

The minimum weight of a transversal of dicycles in a weighted planar digraph $G = (V, E)$, with nonnegative integral weights $w(e)$ on its edges $e \in E$, equals the maximum weight of a collection $\mathcal{C}$ of dicycles in G, with the property that e occurs in at most $w(e)$ dicycles from $\mathcal{C}$, for all $e \in E$.

Solving $QAP(A, L_n)$ is equivalent to finding a transversal of dicycles in G with minimum weight. Indeed, once such a transversal is found, the arcs of the transversal are removed from G and the remaining graph contains no dicycles. Hence, this graph can be sorted topologically (in polynomial time) and the corresponding labeling of vertices yields an optimal solution to $QAP(A, L_n)$. On the other side, finding a transversal of dicycles with minimum weight in G is equivalent to finding a transversal of dicuts in G^* with minimum weight G^* . Lovasz [158] has pointed out that finding a minimum weight transversal of dicuts in G^* is equivalent to finding a minimum weight set of arcs whose contraction transforms G^* in a strongly connected digraph. Frank [81] derived an $O(n^5)$ combinatorial algorithm for finding such a set of arcs in an arbitrary weighted digraph on n vertices. □

As far as we know, the best algorithm for solving $QAP(A, L_n)$, where A is the weighted adjacency matrix of a planar digraph, is due to Gabow [85]. Its time complexity is $O(n^3)$, where n is the size of the problem.

Notice that the problem $QAP(A, L_n)$ of size n with a symmetric matrix A is trivial. In this case $QAP(A, L_n)$ is a *constant* QAP (see the introduction of Chapter 4. This version of the QAP corresponds to the weighted feedback edge set problem for (undirected) graphs with a symmetric weighted adjacency matrix.

In Theorem 6.8, the condition that A is a weighted adjacency matrix of a *planar* digraph is a sufficient condition for the polynomial solvability of $QAP(A, L_n)$, but not a necessary one, as shown by polynomiality results on $QAP(A, L_n)$ to be presented next. The propositions formulated below are results concerning the FAS or its unweighted version, the *acyclic subdigraph problem*, formulated in the language of the QAP. The first of these results deals with the FAS on so-called *reducible flow graphs* which are defined as follows.

Definition 6.2 *A* flow graph *(sometimes also-called a* rooted directed graph*) is a directed graph $G = (V, E, r)$ with vertex set E, arc set E, and a distinguished vertex r, such that there is a directed path in G from r to every vertex in $V \setminus \{r\}$. A* reducible flow graph *is a flow graph which can be transformed in a single vertex by some sequence of transformations of type (i) or (ii):*

(i) If an arc $e = (v, v)$ exists in E, remove e from E.
(ii) Let v_2 be a vertex in $V \setminus \{r\}$ and let it have a single incoming arc (v_1, v_2). Replace (v_1, v_2) by a single vertex v. Predecessors of v_1 become predecessors of v. Successors of v_1 and v_2 become successors of v. There is an arc (v, v) if there was formerly an arc (v_2, v_1) or (v_1, v_1).

Theorem 6.9 (Ramachandran [189], 1988)
Let A be the weighted adjacency matrix of a reducible flow graph on n vertices. $QAP(A, L_n)$ is solvable in $O(n^2 m \log(\frac{n^2}{m}))$ steps, where m is the number of the non-zero entries of A. In the case that A is the 0-1 adjacency matrix of a reducible flow graph, $QAP(A, L_n)$ can be solved in $O(m^2)$ time.

Sketch of the proof. The proof is based on a simple idea but is however technical and goes through several steps. The proof makes use of an alternative characterization of reducible flow graphs as rooted digraphs for which the depth first search acyclic digraph (DAG) is unique (see Tarjan [221]). This characterization implies that the arcs of a reducible flow graph can be partitioned in a unique way into two sets, as DAG or *forward* and *backward* arcs. In the following we give a coarse description of the algorithm.

First, dominance relations are introduced in the set V_h of the heads of the backward arcs in the given reducible flow graph G. With the help of these relations, for each vertex $v \in V_h$ sets of so-called *dominated back arc vertices* are introduced and denoted by V_v. Then, for each $v \in V_h$ a subgraph $G_s(v)$ of G induced by the set V_v of vertices is introduced and a related *maximum flow network* $G_m(v)$ is constructed. There is a direct relationship between the cuts separating the source from the sink in $G_m(v)$ and feedback arc sets in $G_s(v)$. In order to extend this relationship to the weights of the corresponding cuts and feedback arc sets, respectively, additional networks, the so-called *minimum cost maximum flow networks* $G_{mm}(v)$, need to be introduced and constructed, for each $v \in V_h$. The minimum cost flow networks are constructed recursively, starting from vertices in V_h which are minima with respect to the dominance relation, and going towards the root which is a maximum with respect to this relation. The computation of capacities of $G_{mm}(v)$ involves solving a number of minimum cut problems in the networks $G_{mm}(vi)$ for v_i which are *immediately* dominated by v. The nice thing about the networks $G_{mm}(v)$ is that a minimum cut separating the source from the sink in $G_{mm}(v)$ represents a minimum feedback arc set in $G_s(v)$ and vice-versa. Since $G_s(r) = G$, where r is the root of the reducible flow graph, we just need to compute a minimum cut for $G_{mm}(r)$. As mentioned above the construction of $G_{mm}(v)$ involves the computation of minimum cuts in the networks $G_{mm}(v)$ for $v \in V_h$.

Summarizing, an optimal feedback arc set is found by first, constructing recursively the minimum cost maximum flow network for the root of the reducible flow graph, and secondly, solving a minimum cut problem in this network. □

The next polynomiality result concerns, $K_{3,3}$-free digraphs, i.e. digraphs which do not contain any subgraph homeomorphic to the complete bipartite graph $K_{3,3}$ or any of its divisions.

Theorem 6.10 (Penn and Nutov [175], 1994)
If A is the weighted adjacency matrix of a $K_{3,3}$-free digraph on n vertices, then $QAP(A, L_n)$ is solvable in $O(n^3)$ time.

Sketch of the proof. First, the given $K_{3,3}$-free digraph is decomposed into three-connected components. It can be shown that in such graphs the three-connected components are either planar graphs, or coincide with orientations of K_5, i.e. the complete graph on 5 vertices. Secondly, the minimum feedback arc set problem is solved in each of the connected components. As stated in Theorem 6.8, for those components which are planar graphs the FAS is polynomially solvable. Clearly, for the K_5 the FAS is solved in constant time. Finally, the optimal feedback arc sets computed separately for each component can be merged in polynomial time to obtain an optimal feedback arc set for the whole graph. The merging process is a recursive one. It involves the computation of minimum feedback arc sets of larger and larger graphs, starting with the "union" of two three-connected components, then the "union" of the latter with a third component and so on, until the whole graph is obtained. □

Theorem 6.10 can also be derived as a corollary of a stronger result of Grötschel, Jünger and Reinelt [105]. To formulate this stronger result we need one more notion from graph theory, namely the notion of a *weakly acyclic digraph* (see [105]). This definition makes use of the *acyclic subgraph polytope* which is defined below. Given a digraph $G = (V, E)$, let $\mathcal{A}(G)$ be the set of all sets $F \subseteq E$ which induce acyclic subgraphs in G. For some $F \subseteq E$ let x^F be the characteristic vector of F in E, i.e. an $|E|$-dimensional 0-1 vector, with $x_i^F = 1$ if $i \in F$ and $x_i^F = 0$, otherwise. The *acyclic subgraph polytope* $P_{AC}(G)$ is defined as the convex hull of all x^F for $F \in \mathcal{A}(G)$, i.e. $P_{AC}(G) = conv\{x^F : F \in \mathcal{A}(G)\}$. Further let us denote

$$P_C(G) := \left\{ x = (x_i) \in \mathbb{R}^{|E|} : 0 \leq x_i \leq 1, \sum_{i \in C} x_i \leq |C| - 1, \forall \text{ dicycles } C \text{ in } G \right\}$$

Definition 6.3 *A digraph $G = (V, A)$ is said to be* weakly acyclic *if $P_{AC}(G) = P_C(G)$, where $P_{AC}(G)$ and $P_C(G)$ are defined as above.*

The study of weakly acyclic digraphs is still in its infancy. The planar digraphs and the $K_{3,3}$-free digraphs are weakly acyclic digraphs (see [105] and [14], respectively). There exist also other weakly acyclic digraphs (cf. [105]), however no other "nice" classes are known. For more details on this topic the reader is referred to [14, 105].

Theorem 6.11 (Grötschel, Jünger and Reinelt [105], 1985)
If A is the weighted adjacency matrix of a weakly acyclic digraph on n vertices, then $QAP(A, L_n)$ is solvable in polynomial time. □

Remarks. 1. Grötschel et al. [105] prove that the separation problem over $P_C(G)$ is polynomially solvable. It is shown that this problem boils down to *the shortest dicycle problem*, and the latter can be solved by a modified algorithm for the shortest dipath problem. The polynomial solvability of the separation problem implies that the optimization problem over $P_C(G)$ is also polynomially solvable[1]. But, in the case of weakly acyclic digraphs, optimizing over $P_C(G)$ is equivalent to optimizing over $P_{AC}(G)$, and the latter means solving the acyclic subgraph problem. Grötschel et al. [105] solve this problem by means of an algorithm which makes use of the ellipsoid method. Notice that the algorithm derived by Penn and Nutov in [175] for $K_{3,3}$-free digraphs, a special case of weakly acyclic digraphs, is a combinatorial one.

[1]For the terminology used here the reader is referred to Grötschel, Lovász and Schrijver [106]

2. What about the recognition of the well solvable cases of the QAP described above? Obviously, the polynomial recognition algorithms for planar graphs, $K_{3,3}$-free digraphs and reducible flow graphs (see, for example, Hopcroft and Tarjan [122, 123], Tarjan [220]) can be used to recognize the corresponding properties of the coefficient matrix A for a given $QAP(A,B)$. To our knowledge, there exists no polynomial time algorithm for the recognition of weakly acyclic digraphs and the computational complexity of this problem is an open question.

Another polynomially solvable special version of the $QAP(A, L_n)$ can be obtained as a straightforward corollary of Theorem 4.15.

Corollary 6.12 *For a left-lower graded matrix A the problem $QAP(A, L_n)$ is polynomially solvable.*

Proof. The claim follows from Theorem 4.15 by considering that L_n is a right-higher graded matrix. □

The complexity of the FAS for digraphs with weighted adjacency matrices which exhibit a special combinatorial structure remains an open question. Namely, polynomial solvable cases of the problem $QAP(A, L_n)$ may arise in the case that matrix A has some "nice" properties. For example, the problems MONGE$\times L_n$, A-MONGE$\times L_n$ and their complexity might be next-to-investigate open questions .

6.3 QAPS ARISING FROM PACKING PROBLEMS IN GRAPHS

Another line of research on polynomially solvable cases of the QAP may be started from the theory of *graph packing*, cf. Bollobás [23]. Let us first introduce the graph packing problem. Consider two weighted (undirected) graphs $G_1 = (V_1, E_1)$, $G_2 = (V_2, E_2)$ with vertex sets $V_1 = \left\{v_i^{(1)} \mid 1 \le i \le n\right\}$ and $V_2 = \left\{v_i^{(2)} \mid 1 \le i \le n\right\}$ and edge sets E_1 and E_2, respectively. Denote by $A = (a_{ij})$ and $B = (b_{ij})$ the weighted adjacency matrices of G_1 and G_2, respectively. Throughout this section we assume that the weights on the edges of both graphs are positive. Moreover, we set $a_{ij} := 0$ or $b_{ij} := 0$ in the case that $\left(v_i^{(1)}, v_j^{(1)}\right) \notin E_1$ or $\left(v_i^{(2)}, v_j^{(2)}\right) \notin E_2$, respectively.

Definition 6.4 *A permutation $\pi \in \mathcal{S}_n$ is called a* packing of the graph G_2 into the graph G_1 *if* $\left(v_i^{(2)}, v_j^{(2)}\right) \in E_2$ *implies* $\left(v_{\pi(i)}^{(1)}, v_{\pi(j)}^{(1)}\right) \notin E_1$, *for* $1 \le i, j \le n$.

In other words, a packing of G_2 into G_1 is an embedding of the vertices of G_2 into the vertices of G_1 such that no pair of edges coincide. For two graphs G_1 and G_2 as above the *graph packing problem* consists of finding a packing of G_2 into G_1, if one exists, or proving that no packing exists. It is easy to see that the concept of a packing is symmetric, i.e, whenever a packing of G_2 into G_1 exists, there exists also a packing of G_1 into G_2. If this is the case, we may say that a *packing of* G_1 *and* G_2 exists.

The graph packing problem is strongly related to the QAP. Consider an arbitrary embedding π of G_2 into G_1, that is a one to one mapping of the vertices of G_1 into the vertices of G_2. Assume that there exists a pair of indices (i, j) such that $\left(v_i^{(2)}, v_j^{(2)}\right) \in E_2$ and $\left(v_{\pi(i)}^{(1)}, v_{\pi(j)}^{(1)}\right) \in E_1$. We will refer to this fact saying as the occurrence of an *edge coincidence*. For an embedding π, the number of pairwise different pairs (i, j) which yield an edge coincidence, is called *number of edge coincidences*. A packing is an embedding which causes no edge coincidences. Obviously, the objective function value of $QAP(A, B)$ corresponding to a packing π is equal to zero:

$$Z(A, B, \pi) = \sum_{i=1}^{n}\sum_{j=1}^{n} a_{\pi(i)\pi(j)} b_{ij} = 0$$

Conversely, it is clear that a permutation $\pi \in \mathcal{S}_n$ such that $Z(A, B, \pi) = 0$ is a packing of G_2 into G_1. Thus, solving the graph packing problem for two given graphs G_1, G_2 with weighted adjacency matrices A, B, respectively, is equivalent to finding the optimal value of $QAP(A, B)$ and an optimal solution of it in the case that this optimal value equals 0.

Example 6.2 Packing and edge coincidences.
Consider a connected graph $G = (V, E)$ and its complement $\bar{G} = (V, \bar{E})$, where $V = \{v_1, v_2, \ldots, v_n\}$ is the vertex set of both G and $\bar{G}$, and $\bar{E} = \{(v_i, v_j) : 1 \le i, j \le n, (v_i, v_j) \notin E\}$. Clearly, the identity permutation is a packing of $\bar{G}$ into G. In the case that G is complete or empty, i.e. the graph without edges, each embedding of the vertices of $\bar{G}$ into the vertices of G is a packing. Otherwise, let v_i be a vertex with degree smaller than $n-1$ and larger than 0 in G. Denote by v_j and v_k two vertices such that $(v_i, v_j) \in E$ and $(v_i, v_k) \notin E$. A permutation π such that $\pi(j) = k$, $\pi(k) = j$ and $\pi(t) = t$ for all $t \notin \{j, k\}$ is not a packing of $\bar{G}$ to G. Indeed, $(v_i, v_k) \in \bar{E}$ and $(v_i, v_j) = (v_{\pi(i)}, v_{\pi(k)}) \in E$. In this case we have an *edge coincidence*. □

Throughout the rest of this section we review some results on graph packing and formulate them in the language of the QAP. Most of these results lead to polynomially solvable cases of the QAP. Let $\Delta(G)$ denote the maximal degree of the vertices of a graph G throughout the rest of this section. The following theorem concerns "sparse" QAPs, i.e. QAPs whose coefficient matrices have together a "large number" of zero entries. It is intuitively clear that if the overall number of zero entries of matrices A and B is very large, then the optimal value of $QAP(A, B)$ is 0. Also by intuition, it should be "easy" to find an optimal solution to such a $QAP(A, B)$. The following theorem confirms this feeling and mathematically prescribes what "large number" of zero entries means.

Theorem 6.13 (Bollobás and Eldrige [24], 1978)
Let A and B be two symmetric $n \times n$ matrices with nonnegative entries and zeros on the diagonals. In the case that condition (i) or condition (ii) holds, and $n > 9$, the optimal value of $QAP(A, B)$ is 0 and an optimal solution can be found in $O(n^2)$ time or $O(n \log n)$ time, respectively:

(i) The matrices A and B have (collectively) at most $4n - 6$ non-zero entries and each of them has at most $n - 2$ non-zero entries per row.

(ii) The matrices A and B have (collectively) at most $3n - 3$ non-zero entries.

Clearly, for $n \leq 9$ the corresponding QAPs can be solved in constant time.

Sketch of the proof.
Proof of (i). Consider two weighted graphs $G_1 = (V_1, E_1)$ and $G_2 = (V_2, E_2)$ with weighted adjacency matrices A and B, respectively. The zero entries in A (B) correspond to non-existing edges in G_1 (G_2). Obviously, $|V_1| = |V_2| = n$, $\Delta(G_1) < n-1$, $\Delta(G_2) < n-1$ and $|E_1| + |E_2| \leq 2n-3$. Bollobás and Eldrige [24] prove that under these conditions, G_2 can be packed into G_1 (with a finite number of exceptions). The proof given in [24] is done by induction on the number of vertices of G_1 and G_2. This proof leads in a natural way to an $O(n^2)$ recursive algorithm for finding a packing. There are at most n recursive iterations and in each of them $O(n)$ constant-time graph operations are performed. Before starting the algorithm a preprocessing is needed for sorting the vertices of G_1 with respect to the degree, for computing the connected components of G_2, and for sorting these components according to their densities. The preprocessing takes $O(n \log n)$ time. Summarizing, a packing π of G_2 into G_1 exists and can be found in $O(n^2)$ time. Obviously, the permutation π is an optimal solution to $QAP(A, B)$ as $Z(A, B, \pi) = 0$ and the coefficient matrices A, B are nonnegative. Finally, each of the finitely many exceptions consists of a pair of graphs which have at most 9 vertices. Hence,

the corresponding QAPs can be solved in constant time. Clearly, in these cases the optimal value of $QAP(A, B)$ is positive.

Proof of (ii). Again, consider two weighted graphs $G_1 = (V_1, E_1)$ and $G_2 = (V_2, E_2)$ with adjacency matrices A and B, respectively, where the zero entries in A (B) correspond to non-existing edges in G_1 (G_2). Obviously, the inequality $|E_1| + |E_2| \leq \frac{3(n-1)}{2}$ holds. Bollobás and Eldrige [24] have shown that in this case G_2 can be packed into G_1. The proof is again inductive and leads naturally to a recursive algorithm. The recursive procedure is called at most n times and in each call a finite number of elementary ($O(1)$) graph operations are performed. Moreover, the same preprocessing as in (i) is needed. Summarizing, a packing, or equivalently, an optimal solution to $QAP(A, B)$ can be found in $O(n \log n)$ time. □

At this point, an interesting question is the investigation of the "complexity threshold" for the above type of problem. More precisely, let $f(n)$ denote a non-decreasing function mapping $\mathbb{N}$ to $\mathbb{N}$. Consider a restricted version of the QAP, where the input is restricted to pairs (A, B) of symmetric $n \times n$ matrices with nonnegative entries which have collectively at most $2f(n)$ non-zero entries and zeros on the diagonals. For what functions $f(n)$ is this special case of the QAP polynomially solvable and for what functions $f(n)$ is this problem NP-hard? What functions mark the complexity threshold? Theorem 6.13 gives a lower bound for this complexity threshold, namely $f(n) = 2n - 3$, for $n \in \mathbb{N}$. The following theorem, whose rather technical proof is omitted, gives an upper bound for this threshold.

Theorem 6.14 (Burkard, Çela, Metelski, Demidenko and Woeginger [31], 1997) *Consider the restricted version of $QAP(A, B)$ with symmetric nonnegative coefficients matrices A, B which collectively have at most $f(n)$ non-zero entries, where n is the size of the problem. We assume, moreover, that the matrices A and B have zeros on the diagonals. If $f(n) = \Omega(n^{1+\varepsilon})$, for some fixed real $\varepsilon > 0$, then this version of QAP is $\mathcal{NP}$-hard.* □

We believe that the above mentioned complexity threshold is much closer to the lower bound given by Theorem 6.13 than to the upper bound given by Theorem 6.14, say around $f(n) = 4n$. However, this is only a conjecture and closing the gap between the above lower and upper bounds remains an apparently difficult open problem.

The next theorem, implicitly proved by Bollobás and Eldrige, suggests that *separately* bounding the number of non-zero entries of the coefficient matrices A and

B instead of bounding the sum of these numbers, leads to another scenario for the complexity threshold.

Theorem 6.15 (Bollobás and Eldrige [24], 1978)
Consider a $QAP(A, B)$, where A and B are two symmetric $n \times n$ matrices with nonnegative entries and zeros on the diagonals. Assume that the matrices A and B have at most $2\alpha n$ and $2[(1-2\alpha)/5]n^{3/2}$ non-zero entries, respectively, for some $0 < \alpha < 1/2$. Then $QAP(A, B)$ is solvable in $O(n^{5/2})$ time and its optimal value equals zero.

Sketch of the proof. The matrices A and B can be viewed as weighted adjacency matrices of two undirected graphs $G_1 = (V_1, E_1)$ and $G_2 = (V_2, E_2)$, whose edge sets E_1, E_2 fulfill the inequalities $|E_1| \le \alpha n$ and $|E_2| \le [(1-2\alpha)/5]n^{3/2}$, for some $0 < \alpha < 1/2$. Bollobás and Eldrige have shown that under these conditions there exists a packing of G_2 into G_1. Their proof is constructive and yields immediately an algorithm for computing this packing as follows. Begin by assigning the isolated vertices of G_1 to the vertices of G_2 with largest degrees. Ties are broken arbitrarily. Then update both graphs by deleting the already assigned vertices and the edges incident to them. Consider the updated graphs, say G_1' and G_2', respectively. Assign the isolated vertices of G_2' to vertices of G_1' with largest degrees and again update G_1', G_2', in the same way as done for the graphs G_1, G_2. Denote the updated graphs by G_1'' and G_2'', respectively. Consider an arbitrary embedding of G_2'' into G_1'', that is a one to one mapping of the vertices of G_2'' into the vertices of G_1''. Notice that this embedding is not necessarily a packing. Apply iteratively an improvement procedure until a packing results. This procedure gets an arbitrary embedding as input and outputs a new embedding which has a *strictly smaller* number of edge coincidences than the embedding given as input. This procedure, which can be naturally derived from the proof in [24], runs in $O(n^{3/2})$ time. Considering that the improvement procedure is repeated at most $O(n)$ times (this is the maximum possible number of common edges) completes the proof. This algorithm involves a preprocessing phase to distinguish the isolated vertices in each of the given graphs and to sort the vertices according to non-decreasing degrees. The time complexity is dominated by the time spent on improving the embeddings, which amounts to $O(n^{5/2})$. □

The following theorems present a number of results on QAPs which arise as packing problems in certain special graphs.

Theorem 6.16 *Let A and B be the weighted symmetric $n \times n$ adjacency matrices of two graphs $G_1 = (V_1, E_1)$ and $G_2 = (V_2, E_2)$ with nonnegative weights on the edges and zeros on the diagonals. Then, $QAP(A, B)$ is solvable in linear time if one of the following conditions is fulfilled:*

(a) (Hedetniemi, Hedetniemi and Slater [117], 1982)

G_1 and G_2 are trees. None of them is isomorphic to the complete bipartite graph $K_{1,n-1}$. ($K_{1,n-1}$ is the star *on n vertices.)*

(b) (Metelski [166], 1984)

G_1 is a regular graph of degree 2 (i.e. all of its vertices have degree 2) and G_2 is a tree. G_2 is neither isomorphic to the star $K_{1,n-1}$, nor isomorphic to a tree resulting from the star $K_{1,n-2}$ when adding one *vertex x to it so as to subdivide* one *of its edges, say (u, v), into two other edges (u, x) and (x, v).*

(c) (Metelski [166], 1984)

Both G_1 and G_2 are regular graphs of degree 2.

(d) (Azarionok [8], 1987)

Both G_1 and G_2 have bandwidth 2.

Sketch of the proof.
Proof of (a). In [117], Hedetniemi, Hedetniemi and Slater prove the existence of a packing of G_2 into G_1, in the case that G_1 and G_2 are weighted graphs satisfying condition (a). First, a simple auxiliary result is shown: there exists a packing of an arbitrary tree on n vertices which is not a star, into a tree $K'_{1,n-1}$ obtained by subdividing in 2 parts an edge of the star $K_{1,n-2}$ (with $n-1$ vertices). Such a packing is given explicitly in [117]. Then, the authors consider trees which are neither stars, nor are obtained as subdivisions of stars as described above. The proof in this case is straightforward and is done by induction. It immediately leads to an $O(n)$ time algorithm for finding a packing of G_2 into G_1, or equivalently, an optimal solution to $QAP(A, B)$ with optimal value equal to 0. Such a linear time algorithm is presented by the authors in [116].

Proof of (b), (c). If the graphs G_1 and G_2 on n vertices, $n > 6$, fulfill condition (b) or (c), there exists a packing of G_2 into G_1, as shown by Metelski [166]. The proof is constructive and leads to an $O(n)$ time algorithm for the construction of such a packing. Clearly, there is a finite number of pairs of graphs with at most 6 vertices and for each of them the corresponding QAP can be solved in constant time.

Proof of (d). Azarionok [8] proves the existence of a packing of G_2 into G_1, for graphs G_1 and G_2 on n vertices, $n > 16$, fulfilling condition (d). He gives a concrete embedding of G_1 into G_2 and proves that it is a packing. This embedding can be constructed by performing $O(n)$ elementary operations. Obviously, QAPs corresponding to packing problems on graphs with at most 16 vertices can be solved in constant time. □

As confirmed by the following NP-hardness result, the fact that the weights on the edges are nonnegative is essential in Theorem 6.16 (a) .

Theorem 6.17 (Korneenko [142], 1982)
The special case of $QAP(A,B)$, where A and $-B$ are the adjacency matrices of two unweighted trees, is NP-hard. □

Remark. Given the matrices A and B, minimizing $Z(A,B,\pi)$ (over all permutations π) is equivalent to maximizing $Z(A,-B,\pi)$. The last two theorems provide one more example where the computational complexity of maximizing the objective function of a certain QAP is different from the complexity of the analogous minimization problem. We have already seen that another notorious QAP exhibits such a behavior. This is the turbine problem (see Section 5.2.1), where the maximization problem is "easy" and the minimization is "hard".

Other polynomially solvable QAPs arise as packing problems on special graphs, where the product of the maximum degrees of the vertices or the product of the number of edges is "sufficiently small".

Theorem 6.18 (Sauer and Spencer [207], 1978)
Let A and B be two symmetric $n \times n$ matrices with nonnegative entries and zeros on the diagonals. Denote by α and β the maximum number of non-zero entries per row for the matrices A and B, respectively. In the case that $2\alpha\beta < n$, the problem $QAP(A,B)$ is polynomially solvable and its optimal value equals 0.

Sketch of the proof. Consider two weighted graphs G_1 and G_2 on n vertices with weighted adjacency matrices A and B, respectively. Clearly, $2\alpha\beta < n$ implies $2\Delta(G_1)\Delta(G_2) < n$, where $\Delta(G_i)$ is the maximum degree of the vertices of G_i, $i = 1,2$. Sauer and Spencer [207] show that for G_1 and G_2 fulfilling $2\Delta(G_1)\Delta(G_2) < n$, there exists a packing π of G_2 into G_1. This is proven by considering an embedding of G_2 into G_1 with a minimum number of edge coincidences. Under the assumption that this embedding is not a packing, Sauer and Spencer derive a new embedding of G_2 into G_1 with a strictly smaller number of edge coincidences. This contradiction proves the existence of a packing of G_2 into G_1. The construction of Sauer and Spencer can be used to derive an algorithm for finding a packing of G_2 into G_1. Begin with an arbitrary embedding σ of G_2 to G_1. If σ is a packing, we are done. Otherwise, choose a common edge (u,v), i.e. (u,v) is an edge of G_1 and $(\sigma^{-1}(u),\sigma^{-1}(v))$ is an edge of G_2, and find a vertex x which fulfills the following conditions:

(1) $x \neq u$

(2) (u, x) is not a common edge

(3) For any vertex y of G_2 either $(u, \sigma(y))$ is not an edge of G_1, or $(\sigma^{-1}(x), y)$ is not an edge of G_2

(4) For any vertex y of G_1 either $(\sigma^{-1}(u), \sigma^{-1}(y))$ is not an edge of G_2, or (x, y) is not an edge of G_1

It can be show that such a vertex x exists in the case that the conditions of the theorem are fulfilled. A straightforward complexity analysis shows that a vertex x as above can be found in $O(n^2)$ time. Construct a new embedding σ' with

$$\sigma'(i) = \begin{cases} \sigma(i) & \text{if } i \notin \{\sigma^{-1}(u), \sigma^{-1}(x)\} \\ x & \text{if } i = \sigma^{-1}(u) \\ u & \text{if } i = \sigma^{-1}(x) \end{cases}$$

It is easy to show that σ' has a strictly smaller number of edge coincidences than σ. Repeat this improving step until a packing is constructed. Clearly, the number of repetitions cannot exceed $\min\{|E_1|, |E_2|\}$ which is in $O(n^{3/2})$, under the conditions of the theorem. Summarizing, we have an $O(n^3)$ algorithm for finding a packing of G_2 into G_1, or equivalently, an optimal solution to $QAP(A, B)$. □

Theorem 6.19 (Sauer and Spencer [207], 1978)
Let A and B be two symmetric $n \times n$ matrices with nonnegative entries and zeros on the diagonals. Denote by α and β the number of non-zero entries in A and B, respectively. If $\alpha\beta < 2n(n-1)$, the optimal value of $QAP(A, B)$ equals zero.

Sketch of the proof. Similarly as in the proof of the previous theorem, consider two weighted graphs $G_1 = (V_1, E_1)$, $G_2 = (V_2, E_2)$ on n vertices such that A, B are their weighted adjacency matrices, respectively. (Zero entries correspond to non-existing edges.) Clearly, $\alpha\beta < 2n(n-1)$ implies $2|E_1||E_2| < n(n-1)$. Sauer and Spencer [207] show that for graphs G_1, G_2 fulfilling this condition, there exists a packing π of G_2 into G_1. Then, obviously, $Z(A, B, \pi) = 0$, and this completes the proof. The existence of a packing is shown by the following simple probabilistic argument. Consider the uniform distribution on the set of all embeddings of V_2 into V_1. It can easily be shown that the probability that at least one edge coincidences occurs is smaller than one. This means that there exists at least one embedding which involves no edge coincidence i.e. there exists at least one packing. □

Remark. To our knowledge, the computational complexity of the QAP version considered in Theorem 6.19 is still an open problem. We know that the optimal value of such a QAP equals zero, but we have no polynomial algorithm for finding an optimal solution, that is a permutation whose corresponding value of the objective function equals 0. A straightforward use of the proof of Theorem 6.19 to derive such an algorithm seems to be impossible.

6.4 SPECIAL CASES OF THE GQAP

Most of the results in this section concern versions of the GQAP where the coefficient matrices A, B are (symmetric) weighted adjacency matrices of two *isomorphic* (undirected) graphs G_1, G_2, and the minimization occurs over *the set of isomorphisms* between these graphs. Throughout this section, *the set of isomorphisms between two isomorphic graphs* G_1 and G_2, is denoted by $\mathcal{I}(G_1, G_2)$. Clearly, $\mathcal{I}(G_1, G_2) \subseteq S_n$. This version of GQAP was initially tackled by Christofides and Gerrard [53] in 1976 and was termed as *"a graph theoretical formulation of the QAP"*. Given two arbitrary graphs G_1 and G_2, the set $\mathcal{G}(G_1, G_2)$ consists of the subgraphs of G_2 which are isomorphic to G_1. For any $G \in \mathcal{G}(G_1, G_2)$ the set of isomorphisms between G_1 and G is denoted by $\mathcal{I}(G_1, G)$. Finally, the weighted adjacency matrices of G_1 and G are denoted by $A = (a_{ij})$ and $B^G = (b_{ij}^G)$, respectively. Christofides and Gerrard considered the following minimization problem, closely related to the GQAP:

$$\min_{G \in \mathcal{G}(G_1, G_2)} \left\{ \min_{\pi \in \mathcal{I}(G_1, G)} \sum_{i=1}^{n} \sum_{j=1}^{n} a_{\pi(i)\pi(j)} b_{ij}^G \right\}, \tag{6.1}$$

where n is the cardinality of the vertex set of G_1. Hence, among all subgraphs of G_2 isomorphic to G_1 and the corresponding isomorphisms, we want to find a subgraph with a minimum weight isomorphism, where the weight of an isomorphism π is given by the double sum in (6.1). If G_1 and G_2 are isomorphic, we have $\mathcal{G}(G_1, G_2) = \{G_2\}$ and the problem formulated in (6.1) is the following GQAP:

$$\min_{\pi \in \mathcal{I}(G_1, G_2)} \sum_{i=1}^{n} \sum_{j=1}^{n} a_{\pi(i)\pi(j)} b_{ij}, \tag{6.2}$$

where $B = (b_{ij})$ is the adjacency matrix of G_2 and both G_1 and G_2 are graphs on n vertices. Christofides and Gerrard identified trivial cases of problem (6.2), where the cardinality of $\mathcal{I}(G_1, G_2)$ grows polynomially with the size n of the problem. As simple examples consider the cases when both graphs G_1 and G_2 are chains, cycles or wheels (see Figure 6.2 below).

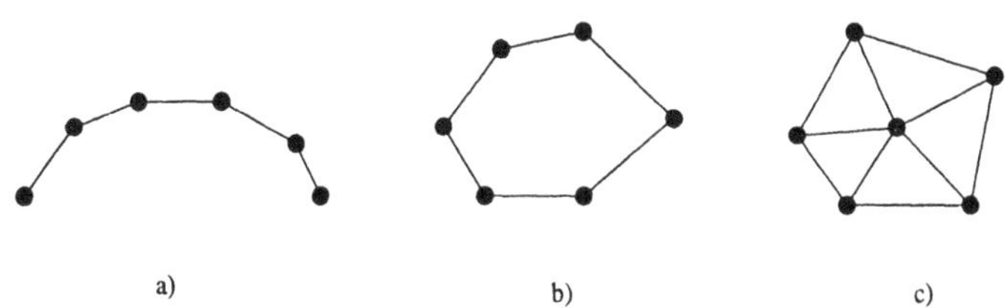

Figure 6.2 a) A chain with 6 vertices. b) A cycle with 6 vertices. c) A wheel with 6 vertices.

Observation 6.20 (Christofides and Gerrard [53], 1976)
Let G_1, G_2, be two isomorphic (a) chains, (b) cycles or (c) wheels on n vertices. In all cases the cardinality of $\mathcal{I}(G_1, G_2)$ grows polynomially with n: in case (a) $|\mathcal{I}(G_1, G_2)| = 2$, in case (b) $|\mathcal{I}(G_1, G_2)| = 2n$ and in case (c) $|\mathcal{I}(G_1, G_2)| = 2(n-1)$, for $n \geq 4$.

The strongest result of Christofides and Gerrard [53] concerning non-trivial cases of the GQAP, where $|\mathcal{I}(G_1, G_2)|$ grows exponentially with the size of the problem, is stated by the following theorem.

Theorem 6.21 (Christofides and Gerrard [53], 1976)
Let T_1 and T_2 be two isomorphic trees on n vertices with arbitrary weights on the edges, and let A and B be the weighted (symmetric) adjacency matrices of T_1 and T_2, respectively. Then the following GQAP

$$\min_{\pi \in \mathcal{I}(T_1, T_2)} \sum_{i=1}^{n} \sum_{j=1}^{n} a_{\pi(i)\pi(j)} b_{ij} \qquad (6.3)$$

is polynomially solvablo.

Sketch of the proof. The proof relies on a polynomial time dynamic programming scheme for (6.3). Actually, the authors construct a dynamic programming scheme for the special case when both trees are *multistars*. Then, they notice that this scheme can be modified in order to work also for general trees. Below we describe the dynamic programming approach for general trees T_1 and T_2.

First, we define *roots* in T_1 and T_2 and assign *levels* to the vertices of T_1 and T_2. This is done by applying a *shelling* procedure to T_1 and T_2, as follows. Assign level 0 to all leaves of the considered tree. Delete all vertices with level equal to 0 together with their incident edges. Assign level 1 to all leaves of the tree resulting from this deletion. Then, delete all vertices with level equal to 1 and the edges incident to them. Repeat this procedure while increasing the value of the level assigned to the leaves of the current tree by one in each iteration. Continue until all vertices are labeled. In the last labeling step, one vertex or two vertices joined by an edge may be left. In the first case, the unique vertex which is left is the root of the tree. In the other case subdivide the remaining edge in two new edges. This is done by adding an additional vertex which is incident with the extremities of this edge and removing the edge itself. The vertex we add is considered as the root of the tree. The level of the root is called *level of the tree*. Once a root is defined in a tree, father and son relationships may also be defined as usually. The father of vertex i is denoted by $p(i)$ and the set of sons of i is denoted by $S(i)$. From this construction follows that that each isomorphism between two isomorphic trees T_1 and T_2 preserves the level of the vertices, i.e. a vertex of T_1 with level equal to k is mapped to a vertex of T_2 with level equal to a k and vice-versa. Moreover, it is easily seen that each isomorphism between T_1, T_2 maps their roots to each other. Therefore, isomorphic trees have equal levels, say ℓ.

After the preprocessing step where levels are assigned to the vertices of T_1 and T_2, the dynamic programming scheme works as follows. For each pair of vertices i, j with level equal to 0 in T_1, T_2, respectively, compute the *cost* $c_{ij} = 2a_{ip(i)}b_{jp(j)}$ of mapping vertex i to vertex j. Assume that we have recursively computed such costs for all pairs of vertices with level smaller than k, and want to compute them for vertices with level equal to k. An isomorphism between T_1 and T_2 may map a vertex i with level equal to k in T_1 into a vertex j with level equal to k in T_2, only if $|S(i)| = |S(j)|$. Let i and j be two such vertices in T_1 and T_2, respectively. Consider an $|S(i)| \times |S(j)|$ matrix $C^{(i,j)}$ whose entries $c_{tl}^{(i,j)}$ are the costs of mapping t to l for any pair (t,l) with $t \in S(i)$, $l \in S(j)$. Solve the linear assignment problem with coefficient matrix $C^{(i,j)}$ and denote its optimal value by $\Lambda(i,j)$. Now the cost c_{ij} of mapping i to j is given by the following formula

$$c_{ij} = 2a_{ip(i)}b_{jp(j)} + \Lambda(i,j)$$

Apply this process until the cost $c_{r_1r_2}$ is calculated, where r_1, r_2 are the roots of T_1, T_2, respectively. This is the optimal value of the considered GQAP. The optimal solution corresponding to this value can be found by back-tracking in the usual dynamic programming manner.

The above algorithm runs in polynomial time. Its time complexity is dominated by the time required to solve the linear assignment problems. For computing the costs of the assignments of the level-k vertices of T_1 to the level-k vertices of T_2, $1 \le k \le \ell - 1$, at most n_k^2 linear assignment problems are solved, where n_k is the number of vertices of level k in T_1 (and also in T_2). The size of all these linear assignment problems is smaller than or equal to $n-1$. Thus, a straightforward upper bound for the time complexity of this step is $n_k^2 n^3$. Summing up the time complexity for all steps of the dynamic programming scheme, we get the following inequality which completes the proof:

$$\sum_{k=1}^{\ell} n_k^2 n^3 \leq n^3 \left(\sum_{k=1}^{\ell} n_k \right)^2 \leq n^5 \qquad \square$$

Rendl [192] investigated the possibility of generalizing the above result to vertex series-parallel digraphs (VSP). In the following we briefly introduce this class of graphs. The definition of this class is done in terms of its minimal members, the *minimal vertex series-parallel digraphs*, or shortly *MVSP digraphs* (see Valdes, Tarjan and Lawler [225]) which are defined recursively as follows:

(i) The digraph containing a single vertex and no edges is a MVSP.

(ii) If $G_i = (V_i, E_i)$, $i = 1, 2$, are two vertex disjoint MVSP digraphs, then the digraphs constructed by the following operations are also MVSP:

- **a)** Parallel composition: $G_p := (V_1 \cup V_2, E_1 \cup E_2)$,
- **b)** Serial composition: $G_s := (V_1 \cup V_2, E_1 \cup E_2 \cup (T_1 \times S_2))$, where T_1 is the set of sinks of G_1 and S_2 is the set of sources of G_2.

A *vertex series-parallel digraph* is a digraph whose transitive closure equals the transitive closure of a MVSP digraph. It is usual to represent a MVSP by a rooted binary tree called the *decomposition tree*. The leaves of this tree are the vertices of the MVSP. The parallel (serial) decomposition of two MVSP digraphs G and G' is represented by a node labeled by "P" ("S") with the roots of the decomposition trees of G and G' being its sons (cf. [225]). Contracting the dipaths which consist only of "P" ("S") vertices into a single "P" ("S") vertex, yields the so-called *canonical decomposition tree* which is no longer binary. The canonical decomposition tree is unique up to the reordering of the successors of "P" vertices. In [225] it is proven that, given a digraph with n vertices and m arcs, it can be recognized in $O(n + m)$ time whether it is a VSP (MVSP) or not. Moreover, in the case that the considered digraph is MVSP, its decomposition tree can be constructed within the same amount of time. For more information on (minimal) series-parallel digraphs the reader is referred to [225] and other literature pointers mentioned in that paper.

Rendl investigated the GQAP on isomorphic MVSP digraphs and showed that this problem is NP-hard. His proof for the NP-hardness of the GQAP on MVSP digraphs is valid for an even more general class of graphs as stated by the following theorem. A key element in the characterization of this class of digraphs is the so-called complete bipartite digraph $K'_{n,n}$. $K'_{n,n}$ is obtained by the complete bipartite graph $K_{n,n}$ by orienting all edges from one "part" of the vertex set to the other one. More concretely, $K'_{n,n}$ has vertex set given as union $U \cup V$, where $U = \{u_1, u_2, \ldots, u_n\}$, $V = \{v_1, v_2, \ldots, v_n\}$, and arc set $E = \{(u_i, v_j) : 1 \leq i, j \leq n\}$ (see also Figure 6.3-a,b).

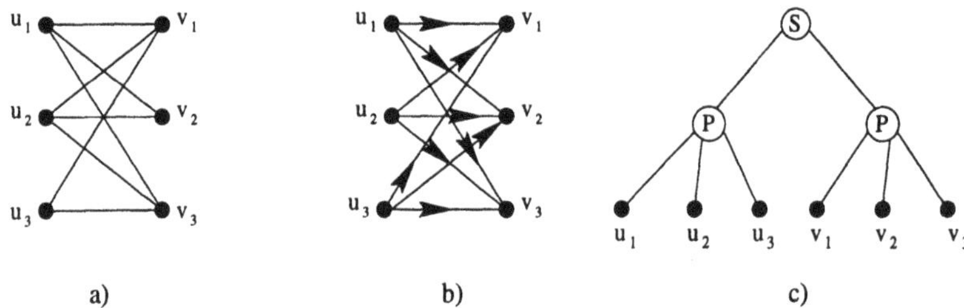

Figure 6.3 a) The complete bipartite graph $K_{3,3}$. b) The complete bipartite digraph $K'_{3,3}$. c) The canonical decomposition tree of $K'_{3,3}$ as a minimum vertex series-parallel (MVSP) digraph.

Theorem 6.22 (Rendl [192], 1986)
Consider a class $\mathcal{K}$ of digraphs such that for all $n \in \mathbb{N}$, $n \geq 2$, there exists a graph G with $2n$ vertices from $\mathcal{K}$ which contains the complete bipartite digraph $K'_{n,n}$ as a (vertex) induced subgraph. Let A and B be the weighted adjacency matrices of two isomorphic graphs G_1 and G_2 from $\mathcal{K}$. (Both graphs are part of the input.) Then, the problem $GQAP(A, B, \mathcal{I}(G_1, G_2))$ is NP-hard. □

In [192] it is shown that a QAP of size n can be polynomially reduced to a $GQAP(A, B, \mathcal{I}(G_1, G_2))$, where both G_1 and G_2 are weighted complete bipartite digraphs isomorphic to $K'_{n,n}$ and A, B are their weighted adjacency matrices, respectively. Notice that the $K'_{n,n}$ is a MVSP digraph (see also Figure 6.3)-c, where the canonical decomposition tree of this MVSP digraph is represented). Besides MVSP digraphs, another class of digraphs which fulfill the conditions of Theorem 6.22 is the class of cographs. (The complete bipartite digraphs $K_{n,n}$, $n \in \mathbb{N}$, are cographs.) Other examples of classes of digraphs for which analogous NP-hardness results can be derived are, for example, the *interval graphs*, the *chordal graphs* and the *split graphs*. (For definitions and properties of this graph classes see for example Golumbic [100].)

Although problem (6.2) is NP-hard for general MVSP digraphs, it becomes polynomially solvable when restricted on specific subclasses of MVSP digraphs.

Theorem 6.23 (Rendl [192], 1986)
Let G_1 and G_2 be two isomorphic minimal vertex series-parallel (MVSP) digraphs with arbitrary weights on the edges, and let A and B be their weighted adjacency matrices. If none of G_1 and G_2 contains the complete bipartite digraph $K'_{2,2}$ as (vertex) induced subgraph, then $GQAP(A, B, \mathcal{I}(G_1, G_2))$ can be solved in polynomial time.

Sketch of the proof. The proof of this result makes use of the *canonical decomposition trees* of G_1 and G_2. Rendl observes that an isomorphism between G_1 and G_2 leads to an isomorphism between their canonical decomposition trees, say T_1 and T_2, respectively. Moreover, this isomorphism between T_1 and T_2 preserves the labels "P", "S" on the nodes of these trees. Thus, the canonical decomposition trees of two isomorphic MVSP digraphs are isomorphic. Under these conditions, it takes some additional technicalities to transform (in a certain sense) the GQAP on the given graphs into a GQAP on the corresponding canonical decomposition trees. The main idea relies on the fact that if an isomorphism $\pi \in \mathcal{I}(G_1, G_2)$ maps a vertex v of tree T_1 to a vertex v' of tree T_2, then it maps the subtree of T_1 rooted at v to the subtree of T_2 rooted at v'. If for each vertex v, the assignment problems related to subtrees rooted at sons of v are somehow *independent*, then a polynomial time dynamic programming approach similar to that described in the proof of Theorem 6.21 can be applied. Otherwise this approach does not work. Such "nice independencies" occur if and only if tree T_1 (and T_2, as well) contains no father labeled by S having two sons labeled by P. Finally, it is not difficult to show that the canonical decomposition tree of a MVSP digraph has this property if and only if the digraph itself does not contain $K'_{2,2}$ as (vertex) induced subgraph. □

The result of Theorem 6.23 can be generalized for *edge series-parallel* (ESP) graphs or digraphs. The edge series-parallel digraphs are defined recursively as described below, and the edge series-parallel graphs are obtained from the edge series-parallel digraphs by removing the orientations of the arcs (cf. [225]).

Definition 6.5 *An ESP digraph is defined recursively by the following rules:*

(i) A digraph consisting of two vertices joined by a single arc is ESP.

(ii) If G_1 and G_2 are ESP digraphs, so are the digraphs constructed by each of the following operations:
a) Two-terminal parallel composition: Identify the source of G_1 with the source of G_2, and the sink of G_1 with the sink of G_2.
b) Two-terminal series composition: Identify the sink of G_1 with the source of G_2.

An ESP graph is obtained by an ESP digraph by ignoring edge directions.

Analogously as for MVSPs, the canonical decomposition tree of an ESP (di)graph can be introduced. Again, such a tree is unique up to the reordering of the successors of "P" vertices. Canonical decomposition trees of isomorphic ESP (di)graphs are isomorphic and vice-versa. Consequently, we can again apply the dynamic programming scheme of Theorem 6.21 in order to find the "cheapest" isomorphism between the canonical decomposition trees of the given graphs. Here, the assignments corresponding to the subtrees rooted at different sons of the same father are always independent and therefore, no additional conditions need to be put on the given graphs to guarantee the polynomial solvability of the problem.

It remains to be investigated whether similar methods can be applied for solving the GQAP concerning other classes of "well-behaving" graphs. In this context, the next classes of graphs to be considered might be planar graphs or partial k-trees.

The following result due to Christofides and Gerrard shows that relaxing the isomorphism condition in Theorem 6.21 leads to NP-hard problems. Namely, problem (6.1) is NP-hard in the case that G_1 is a tree and G_2 is a complete graph.

Theorem 6.24 (Christofides and Gerrard [53], 1976)
Let A be the weighted adjacency matrix of a tree G_1 and B be the weighted adjacency matrix of a complete graph G_2. Then, the problem

$$\min_{G\in\mathcal{G}(G_1,G_2)} \left\{ \min_{\pi\in\mathcal{I}(G_1,G)} \sum_{i=1}^{n}\sum_{j=1}^{n} a_{\pi(i)\pi(j)} b_{ij}^{G} \right\}, \tag{6.4}$$

is NP-hard. □

However, for some special trees we get polynomially solvable GQAPs.

Theorem 6.25 (Christofides and Gerrard [53], 1976)
Let $A = (a_{ij})$ be the weighted adjacency matrix of a star G_1 on n vertices and $B = (b_{ij})$ be the weighted adjacency matrix of a complete graph G_2 on n vertices. Then the generalized GQAP (6.4) is solvable in $O(n^2 \log n)$ time.

Proof. Let the vertex sets of G_1 and G_2 be given as $\{u_1, u_2, \ldots, u_n\}$ and $\{v_1, v_2, \ldots, v_n\}$, respectively. The weights on the edges of both graphs are denoted by w

for simplicity, e.g. $w(v_i, v_j)$, $w(u_i, u_j)$. W.l.o.g. denote the central vertex of the star G_1 by u_1. First, consider all permutations π which map k to 1, for some prespecified $1 \le k \le n$. (That is, fix the vertex v_k of G_2 to which u_1 is embedded.) Find π_k which minimizes the double sum $\sum_{i=1}^{n} \sum_{j=1}^{n} a_{\pi(i)\pi(j)} b_{ij}$ over all π as above. It is easily seen that this minimization problem is equivalent to the minimization of the scalar product of the vectors $\left(w(u_1, u_j)\right)^{\pi}$, $2 \le j \le n$, and $\left(w(v_k, v_j)\right)$, $j \ne k$, over all permutations π of $\{1, 2, \ldots, n-1\}$. The latter problem can be solved in $n \log n$ time as stated by Proposition 2.1. This procedure is repeated n times for $1 \le k \le n$, considering all possible embeddings of the central vertex u_1. Then a permutation is chosen which yields the minimum value of the objective function among π_k, $1 \le k \le n$, derived as above. Clearly, this is an optimal solution of the considered problem and it takes $O(n^2 \log n)$ elementary operations to compute it. □

We conclude this section with an NP-hardness result concerning GQAPs which are not related to isomorphic graphs. Namely, we show that minimizing the objective function of the quadratic assignment problem over pyramidal permutations, i.e. solving the *pyramidal QAP*, is NP-hard[2]. First, let us define a pyramidal permutation.

Definition 6.6 *A permutation $\pi \in \mathcal{S}_n$ is called pyramidal if there exists an $i_0 \in \{1, 2, \ldots, n\}$ such that either $\pi(i) < \pi(i+1)$ for $i < i_0$ and $\pi(i) < \pi(i-1)$ for $i > i_0$, or $\pi(i) > \pi(i+1)$ for $i < i_0$ and $\pi(i) > \pi(i-1)$ for $i > i_0$.*
The subset of $\mathcal{S}_n$ consisting of the pyramidal permutations of $\{1, 2, \ldots, n\}$ is denoted by $\mathcal{Y}_n$.

Theorem 6.26 *Let A and B be two arbitrary $2n \times 2n$ matrices. The following GQAP is $\mathcal{NP}$-hard:*

$$\min_{\pi \in \mathcal{Y}_{2n}} \sum_{i=1}^{2n} \sum_{j=1}^{2n} a_{\pi(i)\pi(j)} b_{ij}. \tag{6.5}$$

Proof. The proof basically consists of showing the existence of an NP-hard version of the QAP which possesses a pyramidal optimal solution. Such a problem is $QAP(A, B)$ with $A \in \textsc{Prod}(\textsc{Sym}, \textsc{Even})$ and $B = (b_{ij})$ a $2n \times 2n$ matrix defined by

$$b_{ij} = \begin{cases} 1 & \text{if } i, j \in \{1, 2, \ldots, n\} \text{ or } i, j \in \{n+1, \ldots, 2n\} \\ -1 & \text{otherwise} \end{cases} \tag{6.6}$$

[2] At this point, the QAP again seems to be "more difficult" than the TSP. It is well known that an optimal pyramidal tour on n cities can be found in $O(n^2)$ time, see e.g. [153].

The NP-hardness of this problem follows from Theorem 4.24. Indeed, $B \subseteq_\pi$ CHESS and therefore $QAP(A,B)$ is equivalent to PROD(SYM,EVEN)×CHESS. In the proof of Theorem 4.24 it has been shown that the EQUIPARTITION problem can be reduced to CHESS×PROD(SYM,EVEN). Thus, $QAP(A,B)$, with A and B as described above, is NP-hard. Further, we show that $QAP(A,B)$ possesses a pyramidal optimal solution. Let (α_i) be the generating vector of matrix A. Then, for all $\pi \in \mathcal{S}_{2n}$, the objective function of $QAP(A,B)$ can be written as

$$Z(A,B,\pi) = \sum_{i=1}^{n}\sum_{j=1}^{n} \alpha_{\pi(i)}\alpha_{\pi(j)} + \sum_{i=n+1}^{2n}\sum_{j=n+1}^{2n} \alpha_{\pi(i)}\alpha_{\pi(j)} - 2\sum_{i=1}^{n}\sum_{j=n+1}^{2n} \alpha_{\pi(i)}\alpha_{\pi(j)}.$$

The last equality implies that two permutations π_1 and π_2 which fulfill the condition "for all $1 \le i \le n$, there exists a $1 \le j \le n$ with $\pi_1(i) = \pi_2(j)$", yield the same objective function value, i.e. $Z(A,B,\pi_1) = Z(A,B,\pi_2)$. Consider now an optimal solution π_0 to $QAP(A,B)$. Let $\{i_1,i_2,\ldots,i_n\} = \{1,2,\ldots,n\}$ such that $\pi_0(i_1) < \pi_0(i_2) < \ldots < \pi_0(i_n)$ and $\{i_{n+1},i_{n+2},\ldots,i_{2n}\} = \{n+1,n+2,\ldots,2n\}$ such that $\pi_0(i_{n+1}) > \pi_0(i_{n+2}) > \ldots > \pi_0(i_{2n})$. Consider a new permutation $\pi' \in \mathcal{S}_{2n}$ defined by $\pi' = \langle \pi_0(i_1), \pi_0(i_2), \ldots, \pi_0(i_{2n})\rangle$, i.e. $\pi'(j) = \pi_0(i_j)$, for $1 \le j \le n$. Obviously, permutation π' is pyramidal and moreover, $\{\pi_0(1),\pi_0(2),\ldots,\pi_0(n)\} = \{\pi'(1),\pi'(2),\ldots,\pi'(n)\}$. Therefore, according to the above remark, $Z(A,B,\pi_0) = Z(A,B,\pi')$. Hence π' is an optimal pyramidal solution to $QAP(A,B)$. Thus, we have proven that $GQAP(A,B,\mathcal{Y}_{2n})$ is equivalent to $QAP(A,B)$, where $A \in$ PROD(SYM, EVEN) and B is given by (6.6). The NP-hardness of QAP(A,B) implies the NP-hardness of $GQAP(A,B,\mathcal{Y}_{2n})$ and this completes the proof. □

7

ON THE BIQUADRATIC ASSIGNMENT PROBLEM (BIQAP)

In this chapter we discuss a natural generalization of the QAP, namely the *biquadratic assignment problem*, shortly denoted by BiQAP. In the BiQAP a weighted sum of products of four variables x_{ij} is to be minimized subject to assignment constraints on the variables x_{ij}. Given two arrays $A = (a_{ijkl})$ and $B = (b_{mpst})$ of n^4 elements each and the variables x_{ij}, $i, j = 1, \ldots, n$, te BiQAP can be formulated as follows:

$$\min \sum_{i,j=1}^{n} \sum_{k,l=1}^{n} \sum_{m,p=1}^{n} \sum_{s,t=1}^{n} a_{ijkl} b_{mpst} x_{im} x_{jp} x_{ks} x_{lt}$$

subject to

$$\sum_{i=1}^{n} x_{ij} = 1 \qquad j = 1, 2, \ldots, n \tag{7.1}$$

$$\sum_{j=1}^{n} x_{ij} = 1 \qquad i = 1, 2, \ldots, n$$

$$x_{ij} \in \{0, 1\}.$$

Notice that the constraints on x_{ij} are the usual assignment constraints. Thus, there exists a one to one correspondence between the feasible solutions x_{ij} for $1 \le i, j \le n$ and the permutations of $\{1, 2, \ldots, n\}$, as described in Section 1.3 and the BiQAP can be formulated as follows:

$$\min_{\pi \in \mathcal{S}_n} \sum_{i=1}^{n} \sum_{j=1}^{n} \sum_{k=1}^{n} \sum_{l=1}^{n} a_{\pi(i)\pi(j)\pi(k)\pi(l)} b_{ijkl} \,, \tag{7.2}$$

where $\mathcal{S}_n$ denotes, as before, the set of all permutations of $\{1, 2, \ldots, n\}$. The formulation in (7.2) clearly shows that the BiQAP is a generalization of the quadratic assignment problem (QAP) to which it owes its name. The BiQAP was introduced by Burkard, Çela and Klinz in [32]. The investigation of this problem was motivated by an application in the field of VLSI (very large scale integrated) synthesis where an optimization problem arising along with the design of synchronous sequential VLSI circuits can be mathematically formulated as a BiQAP. This application will be discussed in detail in the next section. The terminology we use for the BiQAP is very similar to that used for the QAP. The notations BiQAP(A,B), $Z(A,B,\pi)$ and the terms *optimal solution*, *optimal value*, *size of the BiQAP*, *equivalent BiQAPs*, have a completely analogous meaning to that of the corresponding terms used for QAPs. A QAP and a BiQAP are said to be *equivalent* in the case that once we know an optimal solution to one of the problems, we may construct in *polynomial time* an optimal solution to the other one. Based on the similarity of the BiQAP with the QAP, it can be easily shown that the BiQAP is NP-complete in the strong sense (see Section 7.2). Hence, similarly as for the QAP, aspects of the BiQAP concerning MILP formulations, lower bounds, asymptotic behavior, generation of instances with known optimal solution, and heuristic approaches, are of particular relevance.

This chapter is organized as follows. In the first section we describe an optimization problem which arises in the VLSI design and can be mathematically modeled as a BiQAP. Further, in the next section, two MILP formulations for the BiQAP are given. In the third section, a Gilmore-Lawler-type bound for the BiQAP is introduced. Similarly as in the case of QAPs, this bound can be strengthened by using reduction methods. We briefly introduce two of these methods and present the outlet of a lower bounding algorithm for BiQAPs. Further, in Section 7.4, a procedure for generating BiQAP instances with known optimal solution is presented. This procedure is very much in the flavor of analogous procedures for QAPs proposed by Palubeckis [180] and Li and Pardalos [154]. In particular, it generalizes and uses one of the methods presented in [154]. In Section 7.5, the asymptotic behavior of the BiQAP is investigated. Further, several heuristics for the BiQAP are briefly introduced and discussed in Section 7.6. Finally, in a concluding section we summarize the results obtained on BiQAPs and outline some open questions for further research.

7.1 AN APPLICATION OF THE BIQAP ARISING IN THE VLSI SYNTHESIS

Almost all VLSI circuits are so-called sequential circuits. A sequential circuit is an interconnection of storage elements, so-called flip-flops, and logic gates. There are two types of sequential circuits, synchronous and asynchronous circuits. In the following we deal only with synchronous sequential circuits that employ storage elements which are allowed to change their value only at discrete instants of time.

Sequential circuits are most often modeled by using finite state machines (FSMs). An FSM is a mathematical model of a system with discrete inputs, discrete outputs and a finite number of internal configurations or states. FSMs are commonly represented by state transition tables which describe outputs and next states as a function of inputs and present states. (Note that there is no dependence on previous states.) If the state of an FSM changes from s to s', we say that a transition from s to s' takes place and write $s \rightarrow s'$ for short. The state of a sequential circuit is represented by the states of its flip-flops. A flip-flop is a storage device capable of storing one bit of information. It maintains its binary state (equal to 0 or 1) until directed by a clock pulse to switch states. The difference among various types of flip-flops is in the number of inputs they possess and in the manner in which the inputs affect the binary state. Two simple types of flip-flops with a single input are the D (data) flip-flop and the T (toggle) flip-flop . The differences between these two types are as follows: For the D flip-flop the next state is always equal to the input and does not depend on the present state. This does not hold any longer for the T flip-flop where the next state is obtained by evaluating the exclusive-OR (XOR) function with the input and the present state as arguments. (For more details about flip-flops and sequential circuits in general the reader is referred to Mano [162].) The T flip-flop turns out to be more useful in practice than the D flip-flop, but, as we will see below, it is also more difficult to deal with them.

The first step of the procedure for designing a sequential circuit consists of translating the circuit specifications obtained from the practical application into a state transition table. In the next step one tries to find an encoding of the internal symbolic states of the FSM as binary strings such that the eventual implementation after encoding and applying logic minimization is of minimum size. For instance, if the sequential circuit is to be implemented in programmable logic array (PLA) form, then one wishes to minimize the number of product terms (or the area) of the resulting PLA. The state encoding problem is an extremely challenging problem. Most algorithms from the literature focus on area minimization (see e.g. Ashar, Devadas and Newton [5] and Eschermann [73]), but the encoding has also a profound effect on the testability and performance of the resulting implementation (for some work on testability aspects see Eschermann and Wunderlich [74]). Even if the set of binary codes used to encode the n internal states is fixed in advance, a brute force enumerative approach to the state encoding problem would require $O(n!)$ calls of a logic minimization program. This method is obviously infeasible for practical sizes of n. Almost all newer state encoding algorithms for two-level (binary) logic follow a two phase strategy first proposed by De Micheli, Brayton and Sangiovanni-Vincentelli [169]. In the first phase the states of the FSM are encoded symbolically in the following way: The i-th state, $i = 1, 2, \ldots, n$, is encoded by a 0-1 string which has a 1 at position i and 0's elsewhere. Then a logic minimization program is applied to this symbolic representation. The aim of this phase is to generate conditions on the relationships between the codes for different states. The actual form of these constraints depends on the technology and the types of elements which are used to implement the sequential circuit. This issue will be discussed below. Given the coding constraints obtained in the first phase, it is the task of the second phase to find an encoding of the states which satisfies as many of the coding constraints obtained in the first phase as possible, while using a much smaller encoding length than in the symbolic encoding. (The symbolic encoding uses n bits and hence requires a large number of flip-flops to represent the state of the system.)

What remains to be discussed is the manner how the coding constraints are generated in the first step. Let us start with a simple case before proceeding to the description of the case which leads to a biquadratic assignment problem. Suppose that all employed flip-flops are D flip-flops. We divide the transitions into groups where two transitions, say $t_1 : s_1 \to s_3$ and $t_2 : s_2 \to s_4$ belong to the same group if they are induced by the same input and produce the same output. In the case of D flip-flops two such transitions can be merged, i.e. implemented by a single product term of PLA, if $s_1 \neq s_2, s_3 = s_4$ and if the states s_1 and s_2 are assigned adjacent codes, i.e. codes with Hamming distance 1. The problem of determining an encoding which satisfies as many of these coding constraints as possible leads to a quadratic assignment problem. (For details see Eschermann and Wunderlich [74].)

Now let us turn to the case where all flip-flops are T flip-flops. Let $f(s)$ denote the code of state s and $d(c, c')$ the Hamming distance of the codes c and c'. Due to the specific properties of T flip-flops, two transitions $t_1 : s_1 \to s_3$ and $t_2 : s_2 \to s_4$ which belong to the same group can be merged if the following three constraints are satisfied:

1. $s_i \neq s_j$ for all $i \neq j$, $i, j = 1, 2, 3, 4$, i.e. the four involved states are pairwise disjoint,
2. $d(f(s_1), f(s_2)) = 1$, i.e. s_1 and s_2 are assigned adjacent codes, and
3. $f(s_1) \oplus f(s_3) = f(s_2) \oplus f(s_4)$, where $\oplus$ denotes the bitwise exclusive-OR (XOR) operation.

Note that condition (ii) already occurred for D flip-flops, while condition (iii) ensures that the input to the T flip-flops is the same for both transitions. (Recall that for a T flip-flop the input is equal to the XOR of the value representing its present state and the value representing its next state.) A mathematical model of this problem can be obtained as follows: Let S be the set of states and C be the set of codes which we would like to use for the state encoding. For simplicity we assume $|S| = |C| = n$ and number the states and the codes from 1 to n, i.e. $S = \{s_1, s_2, \ldots, s_n\}$ and $C = \{c_1, c_2, \ldots, c_n\}$. (Usually C is chosen as subset of the set of all 0-1 strings of length $\lceil \log_2 n \rceil$.) Furthermore the following notational conventions turn out to be useful. We define a set $T \subseteq \{1, 2, \ldots, n\} \times \{1, 2, \ldots, n\}$, where $(i, j) \in T$ if and only if the transition table contains a transition $s_i \rightarrow s_j$ from state s_i to state s_j. Two pairs $(i, j) \in T$ and $(k, l) \in T$ are said to be mergeable, if i, j, k, l are pairwise disjoint and if the transitions $s_i \rightarrow s_j$ and $s_k \rightarrow s_l$ belong to the same group (i.e. are induced by the same input and produce the same output). For two mergeable pairs (i, j) and (k, l) we write $(i, j) \sim (k, l)$ for short. Now, two four dimensional arrays of n^4 elements each are defined. The first array $A = (a_{ijkl})$, $i, j, k, l = 1, 2, \ldots, n$, is related to the states and the transitions. The entries of A are given by

$$a_{ijkl} = \begin{cases} 1 & \text{if } (i,j) \in T,\ (k,l) \in T \text{ and } (i,j) \sim (k,l) \\ 0 & \text{otherwise.} \end{cases}$$

The second array $B = (b_{mpst})$, $m, p, s, t = 1, 2, \ldots, n$, is related to the codes. Its entries are given by

$$b_{mpst} = \begin{cases} 0 & \text{if } c_m \oplus c_p = c_s \oplus c_t \text{ and } d(c_m, c_s) = 1 \\ 1 & \text{otherwise} \end{cases}$$

where $\oplus$ again denotes the XOR operation. We note that $a_{ijkl} = 1$ exactly when the two transitions $s_i \rightarrow s_j$, $s_k \rightarrow s_l$ can potentially be merged, i.e. can be merged if the involved states are encoded appropriately. Furthermore, we have $b_{mpst} = 0$ exactly when the codes c_m, c_p, c_s and c_t satisfy the coding constraints (ii) and (iii). It is straightforward to see that the state encoding problem can be formulated as BiQAP(A,B) where the coefficient arrays A and B are as defined above. In this case we want to find a permutation $\pi_0 \in \mathcal{S}_n$ which minimizes the following objective function

$$\sum_{i=1}^{n} \sum_{j=1}^{n} \sum_{k=1}^{n} \sum_{l=1}^{n} a_{\pi(i)\pi(j)\pi(k)\pi(l)} b_{ijkl} \tag{7.3}$$

over all permutations $\pi \in \mathcal{S}_n$. For the VLSI synthesis problem, $\pi(i)$ denotes the index of the state to which code c_i is assigned, i.e. code c_i is assigned to state $s_{\pi(i)}$. Hence, the term

$$a_{\pi(i)\pi(j)\pi(k)\pi(l)} b_{ijkl}$$

is equal to 1 exactly when the two transitions $s_{\pi(i)} \rightarrow s_{\pi(j)}$ and $s_{\pi(k)} \rightarrow s_{\pi(l)}$ could be potentially merged, but this merging step cannot be performed because of the codes of the states $s_{\pi(i)}, s_{\pi(j)}, s_{\pi(k)}$ and $s_{\pi(l)}$. Therefore, the minimization of the sum (7.3) is equivalent to the minimization the number of pairs of transitions which cannot be merged because of the encoding chosen for the states, and the latter is exactly the objective of state encoding algorithms mentioned above.

7.2 DIFFERENT FORMULATIONS FOR THE BIQAP

The BiQAP is a combinatorial optimization problem which can be put in different forms. Most generally we can state it in the form

$$\min_{\pi \in \mathcal{S}_n} \sum_{i=1}^{n} \sum_{j=1}^{n} \sum_{k=1}^{n} \sum_{l=1}^{n} d_{\pi(i)i\pi(j)j\pi(k)k\pi(l)l} \tag{7.4}$$

where $\mathcal{S}_n$ denotes, as before, the set of all permutations of $\{1, 2, \ldots, n\}$, and $(d_{ijklmpst})$ is an 8-dimensional array of coefficients. Notice that the objective function in (7.2) is a special case of the objective function of problem (7.4). In (7.2) the coefficients $d_{\pi(i)i\pi(j)j\pi(k)k\pi(l)l}$ are split as product of two factors:

$$d_{\pi(i)i\pi(j)j\pi(k)k\pi(l)l} = a_{\pi(i)\pi(j)\pi(k)\pi(l)} b_{ijkl}, \quad i, j, k, l = 1, 2, \ldots, n. \tag{7.5}$$

In analogy to QAPs, we refer to BiQAPs which have property (7.5) as *Koopmans-Beckmann BiQAPs.* Without loss of generality, we assume that the coefficients $d_{\pi(i)i\pi(j)j\pi(k)k\pi(l)l}$ in (7.4) as well as the coefficients $a_{\pi(i)\pi(j)\pi(k)\pi(l)}$ and b_{ijkl} in (7.5) are nonnegative[1]. We mainly deal with BiQAPs in Koopmans-Beckmann form, but all results in this section hold also for the more general problem (7.4).

Next we derive two (mixed) integer linear programming (MILP) formulations for BiQAPs. We start with problem (7.1) and linearize its objective function by introducing $O(n^8)$ new variables y_{ijkl}, $z_{ijklmpst}$, $i, j, k, l, m, p, s, t = 1, 2, \ldots, n$.

Theorem 7.1 *The BiQAP (7.1) is equivalent to the following 0-1 linear program:*

$$\min \quad \sum_{i,j=1}^{n} \sum_{k,l=1}^{n} \sum_{m,p=1}^{n} \sum_{s,t=1}^{n} a_{ijkl} \cdot b_{mpst} \cdot z_{ijklmpst} \tag{7.6}$$

subject to

[1] Otherwise we can add a constant to all coefficients of the BiQAP such that all of them become nonnegative. This would result into adding a constant to the objective function, but would not change the optimal solution.

$$\sum_{i=1}^{n}\sum_{j=1}^{n}\sum_{m=1}^{n}\sum_{p=1}^{n} y_{ijmp} = n^2 \qquad (*)$$

$$\sum_{i,j=1}^{n}\sum_{k,l=1}^{n}\sum_{m,p=1}^{n}\sum_{s,t=1}^{n} z_{ijklmpst} = n^4 \qquad (**)$$

$$y_{ijmp} + y_{klst} - 2z_{ijklmpst} \geq 0 \qquad i,j,k,l,m,p,s,t = 1,2,\ldots,n$$

$$x_{im} + x_{jp} - 2y_{ijmp} \geq 0 \qquad i,j,m,p = 1,2,\ldots,n$$

$$\sum_{i=1}^{n} x_{ij} = 1 \qquad j = 1,2,\ldots,n$$

$$\sum_{j=1}^{n} x_{ij} = 1 \qquad i = 1,2,\ldots,n$$

$$x_{ij} \in \{0,1\} \qquad i,j = 1,2,\ldots,n$$

$$y_{ijmp} \in \{0,1\} \qquad i,j,m,p = 1,2,\ldots,n$$

$$z_{ijklmpst} \in \{0,1\} \qquad i,j,k,l,m,p,s,t = 1,2,\ldots,n$$

Proof. Let $\{x_{ij}\}_{i,j=1,2,\ldots,n}$ be a feasible solution of (7.1). We define

$$y_{ijmp} = x_{im} \cdot x_{jp} \text{ and } z_{ijklmpst} = y_{ijmp} \cdot y_{klst} \qquad (7.7)$$

for $i,j,k,l,m,p,s,t = 1,2\ldots,n$. Then, it is straightforward to see that

$$\{\{x_{ij}\}_{i,j=1,2,\ldots,n}\ ,\{y_{ijmp}\}_{i,j,m,p=1,2,\ldots,n}\ ,\{z_{ijklmpst}\}_{i,j,k,l,m,p,s,t=1,2,\ldots,n}\}$$

is a feasible solution of (7.6). Moreover these feasible solutions of (7.1) and (7.6) yield the same value for the corresponding objective functions.

Conversely let

$$\{\{x_{ij}\}_{i,j=1,2,\ldots,n}\ ,\{y_{ijmp}\}_{i,j,m,p=1,2,\ldots,n}\ ,\{z_{ijklmpst}\}_{i,j,k,l,m,p,s,t=1,2,\ldots,n}\}$$

be a feasible solution of (7.6). Since the values of x_{ij} which appear in a feasible solution of (7.6) fulfill the assignment constraints, those $\{x_{ij}\}$ constitute a feasible solution of (7.1). Let us show that these feasible solutions of (7.1) and (7.6) yield equal values of the corresponding objective functions. From $x_{im}+x_{jp}-2y_{ijmp} \geq 0$ and $x_{im} \in \{0,1\}$, $x_{jp} \in \{0,1\}$, $y_{ijmp} \in \{0,1\}$ it follows that $x_{im} = x_{jp} = 1$, whenever $y_{ijmp} = 1$. Now let i,j,m,p such that $x_{im} = x_{jp} = 1$. The variables $\{x_{ij}\}_{i,j=1,2,\ldots,n}$ fulfill the assignment constraints and therefore, there are at most n^2 quadruples of indices (i,m,j,p) for which $(x_{im},x_{jp}) = (1,1)$. As argued above, we might have $y_{ijmp} = 1$ only for such quadruples. But according to $(*)$, there are exactly n^2 quadruples (i,m,j,p) for which $y_{ijmp} = 1$. Summarizing, we must have $y_{imjp} = 1$ for all quadruples (i, j, m, p) for which $x_{im} = x_{jp} = 1$. Therefore, $y_{ijmp} = 0$ implies $x_{im} = 0$ or $x_{jp} = 0$ and hence, $y_{ijmp} = x_{im}x_{jp}$.

By using (**) instead of (*) in arguments analogous to those given above, one can quite similarly show that $z_{ijklmpst} = y_{ijmp}y_{klst}$. Then, by putting things together we get $z_{ijklmpst} = x_{im}x_{jp}x_{ks}x_{lt}$, for all i, j, k, l, m, p, s, t. Finally, we obtain the objective function of (7.1) by replacing $z_{ijklmpst}$ by $x_{im}x_{jp}x_{ks}x_{lt}$ in the objective function of (7.6), and this completes the proof. □

The BiQAP formulation introduced by Theorem 7.1 is a 0-1 linear program with $n^8+n^4+n^2$ variables and n^8+n^4+2n+2 constraints. A smaller linearization, i.e. a linearization with less variables and constraints, can be obtained by combining the technique used above with a linearization idea of Kaufman and Broeckx [138] (see Section 1.4.1). This leads to a MILP with $2n^4 + n^2$ variables and $2n^4 + 2n + 2$ constraints. Some more notations are needed in order to introduce this linearization. For each fixed 4-tuple of indices (i, m, j, p) consider an array of n^4 elements

$$D^{(i,m,j,p)} = \left(d_{kslt}^{(i,m,j,p)}\right),$$

defined by

$$d_{kslt}^{(i,m,j,p)} = a_{ijkl} \cdot b_{mpst} \text{ for } k, s, l, t = 1, 2 \ldots, n .$$

For two arrays $D = (d_{ijkl})$ and $Y = (y_{ijkl})$ of n^4 entries each define the product DY as:

$$DY := \sum_{i=1}^{n}\sum_{j=1}^{n}\sum_{k=1}^{n}\sum_{l=1}^{n} d_{ijkl}y_{ijkl}$$

Now, we can prove the following theorem.

Theorem 7.2 *The BiQAP (7.1) is equivalent to the following MILP:*

$$\min \quad \sum_{i=1}^{n}\sum_{m=1}^{n}\sum_{j=1}^{n}\sum_{p=1}^{n} w_{imjp} \tag{7.8}$$

subject to

$$\sum_{i=1}^{n} x_{ij} = 1 \quad , j = 1, 2, \ldots, n$$

$$\sum_{j=1}^{n} x_{ij} = 1 \quad , i = 1, 2, \ldots, n$$

$$\sum_{i=1}^{n}\sum_{m=1}^{n}\sum_{j=1}^{n}\sum_{p=1}^{n} y_{imjp} = n^2$$

$$\begin{aligned}
x_{im} + x_{jp} - 2y_{imjp} &\geq 0 &&, i,m,j,p = 1,2,\ldots,n \\
c_{imjp}y_{imjp} + D^{(i,m,j,p)}Y - w_{imjp} &\leq c_{imjp} &&, i,m,j,p = 1,2,\ldots,n \qquad (*) \\
w_{imjp} &\geq 0 &&, i,m,j,p = 1,2,\ldots,n \\
y_{imjp} &\in \{0,1\} &&, i,m,j,p = 1,2,\ldots,n \\
x_{im} &\in \{0,1\} &&, i,m = 1,2,\ldots,n
\end{aligned}$$

where

$$c_{imjp} = \sum_{k=1}^{n}\sum_{l=1}^{n}\sum_{s=1}^{n}\sum_{t=1}^{n} a_{ijkl}b_{mpst} \qquad i,m,j,p = 1,2,\ldots,n.$$

Proof. If $\{x_{ij}\}_{i,j=1,2,\ldots,n}$ is a feasible solution of (7.1), then we obtain a feasible solution of (7.8) by defining

$$y_{imjp} = x_{im}x_{jp} \qquad i,m,j,p = 1,2,\ldots,n, \qquad Y = (y_{ijmp})$$

$$w_{imjp} = y_{imjp}(D^{(i,m,j,p)}Y) \qquad i,m,j,p = 1,2,\ldots,n.$$

It is easily seen that

$$\{\{x_{ij}\}_{i,j=1,2,\ldots,n}, \{y_{imjp}\}_{i,j,m,p=1,2,\ldots,n}, \{w_{imjp}\}_{i,j,m,p=1,2,\ldots,n}\}$$

is a feasible solution of (7.8), and that these solutions yield the same value for the corresponding objective functions.

Conversely let us assume that

$$\{\{x_{ij}\}_{i,j=1,2,\ldots,n}, \{y_{imjp}\}_{i,m,j,p=1,2,\ldots,n}, \{w_{imjp}\}_{i,m,j,p=1,2,\ldots,n}\}$$

is a feasible solution of (7.8). The variables $\{x_{ij}\}_{i,j=1,2,\ldots,n}$ fulfill the assignment constraints and therefore form a feasible solution of (7.1). We show moreover that these solutions of (7.8) and (7.1) yield the same value of the corresponding objective functions and this completes the proof. To this end, it suffices to prove that

$$y_{imjp} = x_{im}x_{jp} \qquad i,m,j,p = 1,2,\ldots,n, \tag{7.9}$$

$$w_{imjp} = y_{imjp}(D^{(i,m,j,p)}Y) \qquad i,m,j,p = 1,2,\ldots,n. \tag{7.10}$$

Then, simple calculations show the equality of the above mentioned objective function values. The proof of equality (7.9) is done in the same way as in the preceding theorem. In order to prove the correctness of equality (7.10), notice that under the basic assumption that the coefficients of a BiQAP are nonnegative, i.e. the entries of the arrays A, B are nonnegative (see the beginning of this section), the inequality $D^{(i,m,j,p)}Y \leq c_{imjp}$ always holds. Therefore, whenever $y_{imjp} = 0$, the nonnegativity constraints are the only constraints on the variables w_{imjp}. Since we are minimizing the sum of the w_{imjp}, it follows that $w_{imjp} = 0$, whenever $y_{imjp} = 0$. In the case that $y_{imjp} = 1$ holds, constraint $(*)$ reads

$$w_{imjp} \geq D^{(i,m,j,p)}Y$$

and this is the only constraint which has to be fulfilled by variables w_{imjp}. Again, the fact that we are minimizing the sum of the variables w_{imjp} yields

$$w_{imjp} = D^{(i,m,j,p)}Y = y_{imjp}(D^{(i,m,j,p)}Y)$$

as required. □

Next, let us say a few words about the computational complexity of the BiQAP which is almost evident. An arbitrary QAP(A,B) of size n, with $A = (a_{ij})$, $B = (b_{ij})$, is equivalent to a BiQAP(A′,B′) of size n with $A' = (a'_{ijkl})$, $B' = (b'_{ijkl})$, where:

$$a'_{ijkl} = a_{ij} \text{ and } b'_{ijkl} = b_{ij}\ , \quad \text{for all } i, j, k, l = 1, 2, \ldots, n$$

The following equality holds for each $\pi \in \mathcal{S}_n$ and shows that QAP(A,B) and BiQAP(A′,B′) have a common set of optimal solutions:

$$\begin{aligned} Z(A', B', \pi) &= \sum_{i=1}^{n}\sum_{j=1}^{n}\sum_{k=1}^{n}\sum_{l=1}^{n} a'_{\pi(i)\pi(j)\pi(k)\pi(l)} b'_{ijkl} \\ &= n^2 \sum_{i=1}^{n}\sum_{j=1}^{n} a_{\pi(i)\pi(j)} b_{ij} = n^2 Z(A, B, \pi). \end{aligned}$$

This relationship between the QAP and the BiQAP shows that the computational complexity results derived for the QAP hold also for the BiQAP.

7.3 BOUNDS FOR THE BIQAP

In this section we derive a Gilmore-Lawler-type lower bound for BiQAPs and briefly describe two *reduction methods* which can be used in order to improve the quality of this bound, similarly as in the case of QAPs. The reduction methods for the BiQAP are obtained by generalizing well known reduction ideas for QAPs (see e.g. Burkard [28], Roucairol [200]). For more information on these topics and a layout of the related algorithms the reader is referred to [32].

Consider a BiQAP(A,B) of size n with $A = (a_{ijkl})$ and $B = (b_{ijkl})$, $i,j,k,l = 1,2\dots,n$. For fixed indices i,j,m,p, let f_{ijmp} be a lower bound for the following QAP:

$$\min_{\varphi\in\widetilde{\mathcal{S}}_n} \sum_{k=1}^{n}\sum_{l=1}^{n} a_{ij\pi(k)\pi(l)} b_{mpkl} \,, \tag{7.11}$$

where $\widetilde{\mathcal{S}}_n = \{\pi \in \mathcal{S}_n : \pi(m) = i,\, \pi(p) = j\}$. (Notice that what we actually have here is a GQAP, but this GQAP can be easily rewritten as a generalized Koopmans-Beckmann QAP with an additional linear term in the objective function as in (1.3). If the size of the original BiQAP is n, the size of this generalized Koopmans-Beckman QAP will be $n-2$.) Then, for any permutation π of $\{1,2,\dots,n\}$, the following inequality holds:

$$f_{\pi(i)\pi(j)ij} \le \sum_{k=1}^{n}\sum_{l=1}^{n} a_{\pi(i)\pi(j)\pi(k)\pi(l)} b_{ijkl} \tag{7.12}$$

Let $\underline{z}$ be a lower bound for the QAP

$$\min_{\pi\in\mathcal{S}_n} \sum_{i=1}^{n}\sum_{j=1}^{n} f_{\pi(i)\pi(j)ij} \tag{7.13}$$

By summing up over i and j in both sides of (7.12), and by taking then the minimum of each side over $\pi \in \mathcal{S}_n$, we get

$$\underline{z} \le \min_{\pi\in\mathcal{S}_n} \sum_{i=1}^{n}\sum_{j=1}^{n} f_{\pi(i)\pi(j)ij} \le \min_{\pi\in\mathcal{S}_n} \sum_{i=1}^{n}\sum_{j=1}^{n}\sum_{k=1}^{n}\sum_{l=1}^{n} a_{\pi(i)\pi(j)\pi(k)\pi(l)} b_{ijkl}. \tag{7.14}$$

Thus, $\underline{z}$ is a lower bound for BiQAP(A,B) which will be termed Gilmore-Lawler-like bound for the BiQAP. In order to compute the bound $\underline{z}$ for the QAP in (7.13) we have to compute n^4 lower bounds f_{ijmp} for n^4 Koopmans-Beckmann QAPs and then, a lower bound for the QAP in (7.13). Whereas for Koopmans-Beckmann QAPs a variety of bounding techniques can be applied, basically only Gilmore-Lawler-like approaches can be applied to the general QAP in (7.13). Since a large number (n^4) of lower bound computations for Koopmans-Beckmann QAPs have to be performed, it is reasonable to apply some cheap bounding procedure, e.g. the Gilmore-Lawler bound. If the Gilmore-Lawler bound is applied, the bound $\underline{z}$ can be computed in $O(n^7)$ time.

Similarly as for the QAP, the above proposed bound can be improved by applying *reduction* methods. The basic idea of reduction methods for the BiQAP consists of splitting the objective function of the given BiQAP instance as a sum of different summands: the objective function of a new BiQAP instance with *smaller* coefficients than the original BiQAP instance, and the objective functions of *lower-degree* assignment problems (QAPs and LAPs), respectively. This idea is justified by the hope that tighter bounds can be computed for lower degree assignment problems. In a reduction method for BiQAPs introduced in [32] new coefficients $\overline{a}_{ijkl}$ and $\overline{b}_{mpst}$ are derived, in terms of the coefficients a_{ijkl}, b_{mpst}, $i,j,k,l,m,p,s,t = 1,2,\ldots,n$, of the given BiQAP(A,B):

$$a_{ijkl} = \overline{a}_{ijkl} + a^{(1)}_{ij} + a^{(2)}_{ik} + a^{(3)}_{il} + a^{(4)}_{jk} + a^{(5)}_{jl} + a^{(6)}_{kl}\,, \tag{7.15}$$

$$b_{mpst} = \overline{b}_{mpst} + b^{(1)}_{mp} + b^{(2)}_{ms} + b^{(3)}_{mt} + b^{(4)}_{ps} + b^{(5)}_{pt} + b^{(6)}_{st}\,, \tag{7.16}$$

The adjusting coefficients $a^{(1)}_{ij}$, $a^{(2)}_{ij}$, $b^{(1)}_{ij}$, $b^{(2)}_{ij}$, $1 \le i,j \le n$, are appropriately chosen so as to produce as many zeros as possible in the 4-dimensional arrays $\bar{A} = (\bar{a}_{ijkl})$ and $\bar{B} = (\bar{b}_{ijkl})$. Moreover, the entries of both arrays $\bar{A}$, $\bar{B}$ should remain nonnegative. For example, the adjusting coefficients can be chosen as minima over the coefficients of the corresponding array having two of their indices equal to prespecified values, e.g. $a^{(1)}_{ij}$ would be the minimum of a_{ijkl} over all $1 \le k,l \le n$, where n is the size of the problem. Such a choice produces at least n^2 zeros in each of the arrays $\bar{A}$ and $\bar{B}$, whereas the time-complexity of the corresponding procedure is $O(n^4 \log(n))$. After plugging the new coefficients $\bar{a}$, $\bar{b}$ in the objective function of the given BiQAP(A,B), the latter can be written as a sum of the objective functions of a new problem BiQAP($\overline{\mathrm{A}},\overline{\mathrm{B}}$), 8 QAPs, and 24 linear assignment problems with a constant. By summing up the coefficients of the resulting QAPs (LAPs), a new QAP (LAP) is obtained. Thus, the objective function of BiQAP(A,B) is then presented as a sum of four summands: the objective function of BiQAP($\overline{\mathrm{A}},\overline{\mathrm{B}}$), the objective function of a QAP, the objective function of an LAP, and a constant. A lower bound for this decomposed problem can be obtained as a sum of lower bounds for the corresponding summands. (For the LAP, the optimal solution can be computed, instead of a lower bound.) As the coefficients of BiQAP($\bar{A},\bar{B}$) are smaller then those of BiQAP(A,B), and moreover, at least n^2 of them are equal to zero, one hopes that a lower bound for BiQAP($\bar{\mathrm{A}},\bar{\mathrm{B}}$) computed as described previously in this section, will be tighter than the corresponding bound for BiQAP(A,B).

In the second reduction method, the largest entry of each of the coefficients arrays A and B is iteratively decreased. This reduction process is applied a fixed number of times, say M, to both arrays A and B. Computational experiments show that for BiQAP instances of size up to 30 the control parameter M does not need to be larger than 5, since further reductions do not lead to any improvement of the bound. (A similar idea for QAPs has been suggested by Roucairol [200].) After applying this reduction method, the given problem BiQAP(A,B) splits into a new problem BiQAP($\bar{A}$, $\bar{B}$) and a QAP (not necessarily of Koopmans-Beckmann form), where the largest entries of arrays $\bar{A}$, $\bar{B}$ are smaller than those of A, B, respectively. The time cost of this reduction amounts to $O(n^4)$ per iteration.

The Gilmore-Lawler-like bound $\underline{z}$ together with reduction methods have been tested on instances of the BiQAP with known optimal solution (generated as described in the next section). Both reduction methods described above were involved in the experiments. It turned out that none of them is more advantageous than the other concerning the quality of the produced bounds. The quality of the bounds is not good and becomes poor when the size of the problem increases, a phenomenon already known for the QAP. This is not surprising, since linearizing a power-four function yields already in simpler models a large error. As one would expect, the bounding procedure produces better bounds when applied to instances with small coefficients. For more information on the numerical tests the reader is referred to [32].

7.4 BIQAPS WITH KNOWN OPTIMAL SOLUTION

A possible way to test the quality of a bounding procedure (for a certain problem) is to apply it to test instances with known optimal value. Based on the analogous ideas for QAPs, a generator of BiQAP instances with known optimal solution has been derived in [32]. The authors use as a starting point QAP instances generated by the generator of Li and Pardalos (see Section 3.7). The coefficients a_{ijkl}, b_{ijkl} of the generated BiQAP are obtained from the coefficients $a_{ij}^{(1)}$, $a_{ij}^{(2)}$, $b_{ij}^{(1)}$, $b_{ij}^{(2)}$, of two problems $QAP(A^{(1)}, B^{(1)})$ and $QAP(A^{(2)}, B^{(2)})$ with a common optimal solution π_0, generated by the generator of Li and Pardalos:

$$a_{ijkl} = a_{ij}^{(1)} a_{kl}^{(2)} \,, \quad b_{mpst} = b_{mp}^{(1)} b_{st}^{(2)} \,, \quad i,j,k,l,m,p,s,t = 1,2,\ldots,n \,, \tag{7.17}$$

where $A^{(t)} = \left(a_{ij}^{(t)}\right)$ and $B^{(t)} = \left(b_{ij}^{(t)}\right)$, $t = 1,2$, are $n \times n$ matrices. Clearly, the objective function of BiQAP(A,B) can then be written as

$$\begin{aligned} Z(A,B,\pi) &= \sum_{i=1}^{n}\sum_{j=1}^{n}\sum_{k=1}^{n}\sum_{l=1}^{n} a_{\pi(i)\pi(j)}^{(1)} a_{\pi(k)\pi(l)}^{(2)} b_{ij}^{(1)} b_{kl}^{(2)} \\ &= Z(A^{(1)},B^{(1)},\pi) Z(A^{(2)},B^{(2)},\pi), \end{aligned} \tag{7.18}$$

for each $\pi \in \mathcal{S}_n$ Thus, the objective function of BiQAP(A,B) is given as a product of the objective functions of the problems $QAP(A^{(1)}, B^{(1)})$ and $QAP(A^{(2)}, B^{(2)})$. As argued many times by now, without loss of generality we may consider only QAPs and BiQAPs with nonnegative coefficients, and therefore, nonnegative values of the objective function. Under this assumption, if $Z(A^{(1)}, B^{(1)}, \pi)$, $Z(A^{(2)}, B^{(2)}, \pi)$ attain their minima at $\pi = \pi_0$, so does $Z(A, B, \pi)$, and hence, π_0 is an optimal solution for BiQAP(A,B). This simple idea is summarized by the following algorithm. The input of the algorithm consists of the size n of the BiQAP instance to be generated and a permutation π_0 of $\{1, 2, \ldots, n\}$ which will be the optimal solution of the generated BiQAP instance. The variables involved in this algorithm are the following: the coefficients a_{ijkl}, b_{ijkl} of the BiQAP instance to be generated, the optimal value Z_{opt} of this BiQAP, the coefficients $a_{ij}^{(1)}$, $a_{ij}^{(2)}$, $b_{ij}^{(1)}$, $b_{ij}^{(2)}$ of the auxiliary problems $QAP(A^{(1)}, B^{(1)})$, $QAP(A^{(2)}, B^{(2)})$ together with their respective optimal values Z_1, Z_2, and the auxiliary variables i, j, k, l. The algorithm makes use of the routine *Random* which generates some positive integer at random. Moreover, the algorithm calls Li&Pardalos' Generator (see Section 3.7) twice to generate the coefficient matrices $A^{(i)}$, $B^{(i)}$ of $QAP(A^{(i)}, B^{(i)})$ with optimal solution π_0 and optimal value Z_i, $i = 1, 2$, respectively. The output consists of the coefficient arrays $A = (a_{ijkl})$, $B = (b_{ijkl})$ and the optimal value Z_{opt} of $BiQAP(A, B)$.

```
Generator_BIQAP(n, π0)
    Δ1 := Random(); Δ2 := Random();
    Δ3 := Random(); Δ4 = Random(); /* generate Δ1, Δ2, Δ3, Δ4 */
    (A^(1), B^(1), Z1) := Li&Pardalos' Generator(n, Δ1, Δ3, π0);
                 /* Generate QAP(A^(1), B^(1)) with opt. value Z1 */
    (A^(2), B^(2), Z2) := Li&Pardalos' Generator(n, Δ2, Δ4, π0);
                 /* Generate QAP(A^(2), B^(2)) with opt. value Z2 */
    Z_opt := Z1 Z2;
    for i = 1 to n do
        for j = 1 to n do
            for k = 1 to n do
                for l = 1 to n do
                    a_ijkl := a_ij^(1) a_kl^(2); /* generate array A */
                    b_ijkl := b_ij^(1) b_kl^(2); /* generate array B */
                endfor
            endfor
        endfor
    endfor
    return A = (a_ijkl), B = (b_ij kl), Z_opt;
```

The correctness of this algorithm follows from the correctness of the generator of Li and Pardalos for QAPs with known optimal solution (see Theorem 3.3). According to Theorem 3.3 the generation of matrices $A^{(1)}$, $B^{(1)}$ and $A^{(2)}$, $B^{(2)}$ by Li&Pardalos' Generator takes $O(n^2 \log n)$ time. The generation of the 4-dimensional arrays A, B takes $O(n^4)$ elementary operations. Hence, the overall time-complexity of this algorithm amounts to $O(n^4)$. Notice that this is the best possible time-complexity for a generator of BiQAPs as the output itself consists of $O(n^4)$ integers. Next, let us generate a BiQAP instance of size 3 with known optimal solution by applying *Generator_BiQAP*.

Example 7.1 Generating a BiQAP instance of size 3 with an optimal solution $\pi_0 = \langle 3, 1, 2 \rangle$.

Let $\Delta_1 = 3$, $\Delta_2 = 2$, $\Delta_3 = 3$, $\Delta_4 = 3$. After the initialization steps (lines 2–13 in Li&Pardalos' Generator, Section 3.7) the matrices $A^{(1)}$, $A^{(2)}$, $B^{(1)}$, $B^{(2)}$ look as follows

$$A^{(1)} = \begin{pmatrix} 0 & 3 & 3 \\ 3 & 0 & 3 \\ 3 & 3 & 0 \end{pmatrix} \qquad A^{(2)} = \begin{pmatrix} 0 & 2 & 2 \\ 2 & 0 & 2 \\ 2 & 2 & 0 \end{pmatrix}.$$

$$B^{(1)} = \begin{pmatrix} 0 & 1 & 1 \\ 3 & 0 & 2 \\ 1 & 3 & 0 \end{pmatrix} \qquad B^{(2)} = \begin{pmatrix} 0 & 3 & 1 \\ 3 & 0 & 2 \\ 2 & 2 & 0 \end{pmatrix}.$$

Suppose that the vectors X, Y, generated at random and sorted non-decreasingly (lines 15–18 of Li&Pardalos' Generator) are given by $X^t = (0, 0, 2, 2, 2, 2)$ and $Y^t = (0, 0, 0, 1, 1, 1)$. Then, the transformation of matrices $A^{(1)}$, $A^{(2)}$, as described in lines 19–25 of Li&Pardalos' Generator yields:

$$A^{(1)} = \begin{pmatrix} 0 & 3 & 3 \\ 1 & 0 & 1 \\ 1 & 1 & 0 \end{pmatrix} \qquad A^{(2)} = \begin{pmatrix} 0 & 1 & 2 \\ 1 & 0 & 2 \\ 2 & 1 & 0 \end{pmatrix}.$$

We have $Z_1 = \sum_{i=1}^{3} \sum_{j=1}^{3} a_{ij}^{(1)} b_{ij}^{(1)} = 15$ and $Z_2 = \sum_{i=1}^{3} \sum_{j=1}^{3} a_{ij}^{(2)} b_{ij}^{(2)} = 18$ Next, the rows and the columns of both matrices $B^{(1)}$ and $B^{(2)}$ are permuted by permutation π_0. We get

$$B^{(1)} = \begin{pmatrix} 1 & 1 & 3 \\ 1 & 2 & 1 \\ 2 & 3 & 2 \end{pmatrix} \qquad B^{(2)} = \begin{pmatrix} 2 & 2 & 2 \\ 1 & 2 & 3 \\ 2 & 3 & 3 \end{pmatrix}.$$

Finally, by applying (7.17) with matrices $A^{(1)}, A^{(2)}$ and $B^{(1)}, B^{(2)}$ as above, we get a BiQAP instance of size 3, for which π_0 is an optimal solution. The corresponding optimal value is $Z_{opt} = Z_1 Z_2 = 270$.

7.5 THE ASYMPTOTIC BEHAVIOR OF THE BIQAP

In Section 1.5 it is shown that the approximation problem for the QAP is NP-hard. Since the QAP is a special case of the BiQAP, as shown in Section 7.2, even the approximation problem for the latter is NP-hard. Nevertheless, similarly as for the QAP, the approximation problem for the BiQAP becomes in a certain sense simple, when the size of the problem increases. Namely, it can be shown that, under natural probabilistic constraints on the problem data, the ratio of the best and the worst value of the objective function gets arbitrarily close to one with probability tending to one, as the size of the problem tends to infinity. Similarly as for QAPs, Theorems 3.5 and 3.6 can be applied to BiQAPs, with coefficients fulfilling appropriate probabilistic constraints. One would get for BiQAPs a corollary analogous to Corollary 3.7 for QAPs. Here, we apply Theorem 3.5 to get a slightly more general result. First, let us introduce some notational conventions. For a BiQAP instance of size n, BiQAP($A^{(n)}, B^{(n)}$), where $A^{(n)} = \left(a_{ijkl}^{(n)}\right)$ and $B^{(n)} = \left(b_{ijkl}^{(n)}\right)$, let us denote by $\pi_{\text{opt}}^{(n)}$ and $\pi_{\text{wor}}^{(n)}$ an optimal and a worst solution, respectively, i.e.

$$Z(A^{(n)}, B^{(n)}, \pi_{\text{opt}}^{(n)}) = \min_{\pi \in \mathcal{S}_n} Z(A^{(n)}, B^{(n)}, \pi)$$

$$Z(A^{(n)}, B^{(n)}, \pi_{\text{wor}}^{(n)}) = \max_{\pi \in \mathcal{S}_n} Z(A^{(n)}, B^{(n)}, \pi)$$

With these notations, the above mentioned result on the asymptotic behavior of the BiQAP is formulated below.

Corollary 7.3 *Consider a sequence of problems BiQAP($A^{(n)}, B^{(n)}$), $n \in \mathbb{N}$, with coefficient arrays $A^{(n)} = \left(a_{ijkl}^{(n)}\right)$, $B^{(n)} = \left(b_{ijkl}^{(n)}\right)$ of n^4 elements each. Consider moreover the sets I_n, $n \in \mathbb{N}$, such that $I_n \subseteq \{1, 2, \ldots, n\}^4$. Assume that for all $n \in \mathbb{N}$ the random variables $a_{ijkl}^{(n)}$, $(i, j, k, l) \in I_n$, are identically, independently distributed on $[0, M]$, where M is some positive constant. Moreover, these variables have finite expected value and variance. The other coefficients $a_{ijkl}^{(n)}$ with $(i, j, k, l) \notin I_n$, are equal to zero. Furthermore, assume that for all $n \in \mathbb{N}$ the random variables $b_{ijkl}^{(n)}$, $i, j, k, l = 1, 2, \ldots, n$, are identically, independently distributed on $[0, M]$ with finite expected value and variance. Finally, assume that for each $n \in \mathbb{N}$ the variables $a_{ijkl}^{(n)}$ and $b_{ijkl}^{(n)}$, $i, j, k, l = 1, 2, \ldots, n$, are independently distributed. If, additionally, the following holds*

$$\lim_{n \to \infty} \frac{|I_n|}{n \ln n} = \infty, \tag{7.19}$$

then

$$\forall \epsilon > 0, \quad \lim_{n\to\infty} P\left\{ \frac{Z(A^{(n)}, B^{(n)}, \pi^{(n)}_{\text{wor}})}{Z(A^{(n)}, B^{(n)}, \pi^{(n)}_{\text{opt}})} < 1+\epsilon \right\} = 1\,. \tag{7.20}$$

Proof. The proof is a straightforward application of Theorem 3.5. First, notice that Theorem 3.5 remains true when considering problems with coefficients on $[0, M]$ instead of $[0, 1]$. Next, notice that the above claim holds trivially in the case that the expected value of the coefficients $a^{(n)}_{ijkl}$, denoted by $E(a)$, equals zero, or the variance of the coefficients $b^{(n)}_{ijkl}$, denoted by $\sigma^2(b)$, is equal to zero. In the first case, all coefficients a_{ijkl} would be equal to zero, for all $i, j, k, l = 1, 2, \ldots, n$. This would imply $Z(A^{(n)}, B^{(n)}, \pi) = 0$, for all $\pi \in \mathcal{S}_n$ and for all $n \in \mathbb{N}$. In the second case, all coefficients b_{ijkl} would have the same value and therefore the objective function $Z(A^{(n)}, B^{(n)}, \pi)$ would be constant over $\pi \in \mathcal{S}_n$ for all $n \in \mathbb{N}$. Consequently, we may assume that $E(a) > 0$ and $\sigma^2(b) > 0$ throughout the rest of the proof.

We show that the conditions (BF1), (BF2) and (BF3) of Theorem 3.5 are fulfilled. Consider the formulation of the BiQAP as a general combinatorial optimization problem (see Section 1.5.2 for a generic formulation of combinatorial optimization problems and Section 3.8.1 for a formulation of the QAP in that generic formalism). For each $n \in \mathbb{N}$, the ground sets $\mathcal{E}_n$, the feasible solution $X^{(\pi)}_n$ corresponding to $\pi \in \mathcal{S}_n$, and the set of feasible solutions $\mathcal{F}_n$, are given as follows:

$$\begin{aligned}
\mathcal{E}_n &= \Big\{ (i,j,k,l,m,p,s,t) \in \{1,2,\ldots,n\}^8 : (i,j,k,l) \in I_n \Big\} \\
X^{(\pi)}_n &= \Big\{ (\pi(i), \pi(j), \pi(k), \pi(l), i, j, k, l) \colon (\pi(i), \pi(j), \pi(k), \pi(l)) \in I_n \Big\} \\
\mathcal{F}_n &= \Big\{ X^{(\pi)}_n \colon \pi \in \mathcal{S}_n \Big\}.
\end{aligned}$$

Notice that the feasible solution $X^{(\pi)}_n$ consists only of those 8-tuples $(\pi(i), \pi(j), \pi(k), \pi(l), i, j, k, l)$ for which the corresponding coefficient $a_{\pi(i)\pi(j)\pi(k)\pi(l)}$ is not fixed to zero. Hence, we clearly have $|X^{(\pi)}_n| = |I_n|$, for each $\pi \in \mathcal{S}_n$ and any $n \in \mathbb{N}$. Thus, condition (BF1) is fulfilled. Since some of the sets $X^{(\pi)}_n$, $\pi \in \mathcal{S}_n$, might be identical, we have $1 \le |\mathcal{F}_n| \le n!$. Thus, under the conditions of the theorem, the following holds for each $\lambda_0 > 0$:

$$\lambda_0 |X^{(\pi)}_n| - \ln(|\mathcal{F}_n|) \ge \lambda_0 |I_n| - \ln(n!) \to \infty \quad \text{as} \quad n \to \infty.$$

Thus, condition (BF3) of Theorem 3.5 is also fulfilled.

Next, we consider the random variables $c_n(x)$, $x \in \mathcal{E}_n$, where c_n is the cost function related to BiQAP($A^{(n)}, B^{(n)}$), and show that condition (BF2) of Theorem 3.5 is fulfilled. The cost function $c_n(x)$ maps the elements $x = (i,j,k,l,m,p,s,t)$ of the ground set $\mathcal{E}$ to the reals

$$c_n(x) = a_{ijkl}^{(n)} b_{mpst}^{(n)} .$$

Under the assumptions of the theorem $c_n(x)$, $n \in \mathbb{N}$, $x \in \mathcal{E}$ are identically distributed random variables on $[0, M^2]$. Moreover for each $n \in \mathbb{N}$ the variables $c_n(x)$, $x \in \mathcal{E}$ are independently distributed. Further, it can be easily shown that the common distribution of $c_n(x)$ has finite expected value and a *positive* finite variance. Indeed, for independently distributed random variables a and b we have:

$$\sigma^2(ab) = E((ab - E(ab))^2) = E(a^2)\sigma^2(b) + E^2(b)\sigma^2(a)$$

Now, the claim follows by plugging in the above equality $a_{\pi(i)\pi(j)\pi(k)\pi(l)}^{(n)}$ and $b_{ijkl}^{(n)}$ instead of a and b, respectively, and taking into account that $E(a^2) > 0$ and $\sigma^2(b) > 0$.

Summarizing, we have shown that all conditions of Theorem 3.5 are fulfilled. Hence, this theorem applies and (7.20) follows. □

Under probabilistic conditions which are a bit more restrictive than those of Corollary 7.3 a similar asymptotic result is obtained as a straightforward corollary of Theorem 3.6.

Theorem 7.4 *Consider a sequence of problems BiQAP($A^{(n)}, B^{(n)}$), $n \in \mathbb{N}$, with coefficient arrays $A^{(n)} = (a_{ijkl}^{(n)})$, $B^{(n)} = (b_{ijkl}^{(n)})$ of n^4 elements each. Assume that the coefficients $a_{ijkl}^{(n)}$, $n \in \mathbb{N}$, $1 \le i,j,k,l \le n$, are random variables, identically distributed on $[0, M]$, where M is some positive real number. The expected value $E(a)$, the variance and the third moment of the common distribution are finite. Analogously, the coefficients $b_{ijkl}^{(n)}$, $n \in \mathbb{N}$, $1 \le i,j,k,l \le n$ are also random variables, identically distributed on $[0, M]$. The expected value $E(b)$, the variance, and the third moment of the corresponding distribution are finite. Moreover, assume that for each $n \in \mathbb{N}$, the random variables $a_{ijkl}^{(n)}$, $b_{ijkl}^{(n)}$, $1 \le i,j,k,l \le n$, are independently distributed. Finally, assume that the worst values $Z(A^{(n)}, B^{(n)}, \pi_{\text{wor}}^{(n)})$ of the objective functions form a non-decreasing sequence for increasing n. Under these conditions, the following equalities hold almost surely:*

$$\begin{aligned} Z(A^{(n)}, B^{(n)}, \pi_{\text{opt}}^{(n)}) &= n^4 E(a)E(b) - o(n^4) \\ Z(A^{(n)}, B^{(n)}, \pi_{\text{wor}}^{(n)}) &= n^4 E(a)E(b) + o(n^4) \end{aligned}$$

□

7.6 HEURISTICS FOR BIQAPS

The relationship between QAPs and BiQAPs described in Section 7.2 suggests that the BiQAP is at least as hard as the QAP, from both a theoretical and a practical point of view. Moreover, the quality of Gilmore-Lawler-type bounds for BiQAPs, introduced in Section 7.3, is unsatisfactory, especially when the size of the problem increases. Under these conditions, heuristics are a reasonable, and perhaps, currently the only realistic approach to cope with BiQAPs. The goal is to design heuristics with a good trade-off between the quality of the produced solutions and the requirements on computational resources (e.g. memory and time requirements). Some local search algorithms and some metaheuristics (simulated annealing, tabu search) for the BiQAP have been proposed by Burkard and Çela [30]. A GRASP heuristic for the BiQAP has been proposed by Mavridou, Pardalos, Pitsoulis and Resende [163]. In this section we give a brief description of these heuristics and discuss upon the comparison of their performance. For more information and details the reader is referred to the above mentioned papers.

7.6.1 Deterministic improvement methods

In Section 3.2, we gave a general description of deterministic improvement methods for QAPs. Analogously as for the QAP, three deterministic improvement approaches have been tested on BiQAPs [30]: the *first* improvement method, the *best* improvement method, and *Heider's* improvement method, abbreviated by FIRST, BEST and HEID, respectively. These approaches and all heuristic algorithms described in this section make use of the *pair-exchange* neighborhood (introduced in Section 3.2). For an arbitrary permutation $\pi_0 \in \mathcal{S}_n$, its neighborhood $\mathcal{N}_2(\pi_0) \subset \mathcal{S}_n$ is given by

$$\mathcal{N}_2(\pi_0) = \Big\{(i,j) \circ \pi_0 \colon (i,j) \in I\Big\}, \tag{7.21}$$

where I is the set of transpositions on $\{1,2,\ldots,n\}$ (and the operation $\circ$ is defined by: $\phi \circ \psi(i) = \phi(\psi(i))$, for all i). The *stop criterion* involved in BEST, FIRST and HEID, is the standard one: If $Z(A,B,\pi) \geq Z(A,B,\pi_0)$ for each $\pi \in \mathcal{N}_2(\pi_0)$, where π_0 is the current permutation, then stop. In the case of BEST we must compute the value of the objective function $Z(A,B,\pi)$ for all $n(n-1)/2$ neighbors π of the current permutation π_0, in each iteration. As for FIRST and HEID, $O(n^2)$ is just an upper bound on the number of computations of $Z(A,B,\pi)$ per iteration. Given a neighbor $\pi = (k_0,l_0) \circ \pi_0$ of π_0, where $(k_0,l_0) \in I$, the corresponding $\Delta Z = Z(A,B,\pi_0) - Z(A,B,\pi)$ can be in $O(n^3)$ time, as shown in [30]. Consequently, in the worst case, $O(n^5)$ elementary operations have to be performed in each iteration of BEST, (FIRST, HEID). However, in practice the number of ΔZ-computations per iteration of FIRST and HEID is much smaller than $n(n-1)/2$. As for BEST, analogously as in the case of the QAP, it can be shown that except for the first iteration, the computation can be speeded up. Namely, in any iteration but the first one the evaluation of the neighborhood, i.e. the computation of all differences ΔZ, can be performed in $O(n^4)$ time, where n is the size of the problem. Clearly, this is the lowest possible time cost, because the computation of value the objective function itself amounts to $O(n^4)$ elementary operations. The proof of this result is elementary and can be found in [30].

Theorem 7.5 (Burkard and Çela [30], 1995)
The best improvement method (BEST) for BiQAPs can be implemented such that each iteration except for the first one performs $O(n^4)$ elementary operations, where n is the size of the BiQAP instance. The first iteration performs $O(n^5)$ elementary operations. □

As for the number of iterations these improvement methods (FIRST, BEST, HEID) run through, $n!$ is a trivial upper bound and no better bound is known, to the best of our knowledge. In numerical experiments the number of iterations of FIRST, BEST and HEID amounts to $O(n)$, as reported in [30],

7.6.2 Simulated annealing algorithms

Different variants of simulated annealing have been applied to the BiQAP. The main differences between these variants concern the temperature schedule and the so-called "equilibrium criterion", i.e. the criterion which decides when to stop the search at a given value of the temperature and drop to the next temperature in the schedule.

In the algorithm ANNEAL in [30] the cooling process drops from the current temperature value to the next one in the schedule either if the equilibrium is reached, i.e. the quality of the solutions found during the "last" steps of the search at the current temperature value does not vary a lot, or a "large" number of permutations is already considered at the current temperature value. In this case the maximum number of the considered neighboring permutations at a fixed value of the temperature is a control parameter. The meaning of "last" above is determined by another control parameter. The temperature schedule involved in this version of simulated annealing is defined by $t_k = q^k t_0$, where t_k is the k-th value of the temperature in the schedule. The number r of the cooling steps, i.e. the number of the values of the temperature in the schedule, and $q \in (0,1)$ are two other control parameters.

In another version of simulated annealing for BiQAPs, denoted by SIMANN2, the equilibrium criterion has been eliminated. The amount of attempted transpositions at a given temperature value does not exçeed $Kn(n-1)$, where $0 < K \leq 1/2$ is a control parameter. The temperature drops from the current state to the next one in the schedule if, either a solution is found which yields a decrease in the value of the objective function and hence is accepted, or $Kn(n-1)$ transpositions have already been attempted. The temperature schedule is generated by the following formula

$$t_{k+1} := \frac{t_k}{1+\beta \cdot t_k}, \tag{7.22}$$

where β is a control parameter, $0 < \beta \ll t_0$. The parameter β is chosen much smaller than t_0 in order to ensure a slow cooling process.

Another simulated annealing approach for the BiQAP, introduced and tested in [30] and denoted by SIMANN3, involves a so-called *"optimal" value of the temperature*. The "optimal" value of the temperature was originally used by Connolly [59] in a simulated annealing approach for QAPs. The idea of the "optimal" temperature is based on the following arguments. Experiments with simulated annealing algorithms applied to different combinatorial optimization problems have shown that the range of the temperature schedule has a strong impact on the performance of these algorithms. If the system is kept too "hot", neighboring permutations which yield a *large* increase in the value of the objective function are accepted. Therefore, the search may become chaotic and it may be impossible to get to a local minimum. If the system is kept too "cold", only neighboring permutations which yield very *small* increases or decreases in the value of the objective function are accepted. In this case, the search may easily get stuck to a bad local minimum. This leads to the idea that somewhere between these two extremes there must be an "optimal" value of the temperature. Several efforts have been made to elaborate an appropriate concept of the "optimal" value of the temperature in simulated annealing algorithms for different problems (see Kirkpatrick, Gelatt and Vecchi [140], Vanderbilt and Louie [226]). In the case of BiQAPs Burkard and Çela [30] derive an "optimal" value of the temperature out of a given temperature schedule. This is done by considering a fixed amount of transpositions in each step of the cooling process and by computing the percentage of accepted transpositions in each cooling step. Then, the arithmetical mean of these percentages, the so-called *acceptance average*, is computed and the deviations of all percentages from this mean are evaluated. As "optimal" value of the temperature is chosen that value in the schedule which yields the minimum deviation from the acceptance average. After the cooling process an appropriate number of search iterations are performed with value of the temperature fixed at the optimal one.

For a detailed description and an outline of ANNEAL, SIMANN2, and SIMANN3, the reader is referred to [30].

7.6.3 Tabu search approaches

A description of tabu search techniques in general, and tabu search schemes for QAPs in particular, is given in Section 3.2. Some versions of tabu search for BiQAPs have been introduced, discussed and tested in [30]. In all these algorithms the movement in the neighborhood of the current permutation is done by means of transpositions. The tabu list is a FIFO list with fixed maximum length and consists of transpositions that correspond to forbidden moves. One of the algorithms (denoted by TAB in [30]) accepts a forbidden move, i.e. applies this move to the current permutation, only if this moves yields an improvement of the best value of the objective function known so far. Thus, the improvement of the currently best value of the objective function is an *aspiration criterion*. The algorithm starts with an empty tabu list. In each iteration, all non-tabu transpositions (moves) and those tabu transpositions which fulfill the aspiration criterion are considered. One transposition among the considered transpositions is em accepted, i.e. is applied apply to the current permutation. As transposition to be accepted is chosen the one which yields the smallest objective function value. Then, the tabu list is updated by adding the last accepted transposition to the end of the list, if this is a non-tabu transposition, or by moving this transposition to the end of the list, if it is a tabu transposition. If a transposition is to be added to the tabu list and the length of the list has already reached the prespecified maximum, the first tabu transposition in the list becomes a non-tabu one and the transposition to become "tabu" is added at the end of the list. This procedure is applied a fixed number of times.

The performance of TAB strongly depends on the initial permutation and on the length of the tabu list. In order to improve the robustness of the algorithm Burkard et al. perform more than one run of the algorithm with different values for the length of the tabu list and/or different initial permutations. Moreover, as different sets of accepted transpositions at the beginning of a run of TAB may lead to suboptimal solution of different quality at the end of the algorithm, all runs of TAB but the first one are started with a number of forbidden (tabu) transpositions. In the i-th consecutive run of TAB the tabu transpositions are selected among the permutations accepted at the beginning of the previous runs. This is done by starting the i-th run with a full tabu list obtained as a concatenation of the *first* full tabu lists generated during the previous runs of TAB. The algorithm which results by repeating TAB a certain number of times as described above, is called *long term* tabu search and is denoted by LTMTAB.

The reader is referred to [30] for more information on TAB, LTMTAB, other tabu search approaches to the BiQAP, and computational experiments.

7.6.4 Greedy randomized adaptive search for the BiQAP

In Section 3.6 we introduced the greedy randomized adaptive search procedure, so-called GRASP, which is a heuristic approach to solve hard combinatorial optimization problems. Mavridou, Pardalos, Pitsoulis and Resende [163] have applied GRASP to the BiQAP obtaining very good experimental results on BiQAP instances with known optimal solution generated as described in Section 7.4. The construction phase of GRASP for the BiQAP starts by selecting a *partial permutation* which consists of the assignment of four indices i, j, k, l from $\{1, 2, \ldots, n\}$, where n is the size of the BiQAP to be solved. This partial permutation is not selected among all partial permutations of four indices (the number of such partial permutations amounts to $O(n^4)$) but among Λ randomly chosen partial permutations of four indices, where Λ is a control parameter. Mavridou et al. report that $\Lambda = 1000$ is a good choice for this control parameter. After having generated these Λ partial permutations the selection of one of them is made by combining greedy criteria with random elements as described in Section 3.6 with $\alpha = 0.25$ and $\beta = 0.3$. Then the construction phase completes the partial permutation into a permutation of $\{1, 2, \ldots, n\}$ by assigning one index at a time in the usual GRASP manner. In the local improvement phase the pair-exchange neighborhood in the set $\mathcal{S}_n$ of permutations is used. The stop criterion is a maximum number of iterations which in the experiments of Mavridou et al. is chosen to be 150.

7.7 ON COMPUTATIONAL RESULTS

In this section we briefly report on computational experience with the heuristic approaches described in the previous section. For a more complete discussion the reader is referred to [30, 163]. The heuristics have been tested on BiQAP instances with known optimal solution generated by the Generator_BiQAP. The sizes of the test instances are even integers between 10 and 40 and the coefficients a_{ijkl}, b_{ijkl} are integers between 1 and 9. The initialization of the improvement methods, simulated annealing, and tabu search algorithms involves the generation of some initial permutation. For each even n between 10 and 40, 40 permutations of $\{1, 2, \ldots, n\}$ are generated at random and serve as initial permutations for the above mentioned heuristics.

Deterministic improvement methods. According to Burkard et al. [30], among the three improvement methods FIRST, BEST and HEID, HEID offers the best trade-off between of computation time requirements and solution quality. Each heuristic was applied 40 times to test instances from the benchmark starting with 40 different initial permutations. For each instance-heuristic pair the average relative error, the average CPU time, the average iteration number, and the frequency of finding an optimal solution over the corresponding 40 runs were computed. The computation time tends to become very large as the size of the test instances increases. Testing with instances of size larger than 36 was considered as fruitless, as already for sizes larger than 30 the CPU time required by each improvement method exceeded 1 hour!

Simulated annealing approaches. Computational experiments confirm the high sensitivity of simulated annealing algorithms to the number of cooling steps and the annealing scheme. The choices for these control parameters were made based on numerous numerical tests. Among simulated annealing approaches tested in [30], the third approach, i.e. the one involving the so-called "optimal value" of the temperature, produces solutions of a better quality whereas it exhibits computational time requirements which are accordingly high. However, based on their numerical results, the authors conclude that this approach yields the best trade-off between computation time and solution quality.

Tabu search approaches. The experiments made in [30] show that the tested heuristics produce good solutions after a number of iterations which is large enough. The problem is that the application of such a large number of iterations requires too much computation time and makes the algorithms infeasible. As reported by Burkard et al., among the tested tabu search algorithms, the one-phase-tabu search TAB produces the best trade-off between computation time and solution quality. After numerous experiments, 40 iterations and a tabu list of length equal to 25 were found to be relatively good settings for the corresponding control parameters. With these settings, the quality of the suboptimal solutions yielded by TAB is better than the quality of the solutions yielded by HEID when applied to BiQAP instances of size $n \leq 20$, whereas the corresponding CPU costs are about 5 times higher. For larger problems the quality of the produced solution is not satisfactory. It seems that the large relative errors are due to the fact that 40 iterations are to few and cannot lead to solutions of good quality, especially for the larger test instances.

GRASP and comparison. GRASP seems to produce very good results when applied to BiQAP instances with known optimal solution from the benchmark described above. As reported in [163] 10 runs of the algorithm (with different seeds for the generation of the pseudo-random stream of numbers) were performed for each of the test instances. GRASP has found the optimal solution in each run and for each test instance. Mavridou et al. report on the average number of iterations (in the local improvement phase) and the average CPU times of GRASP in these experiments. It seems that the very good performance of GRASP in terms of solution quality requires computation times which are higher than those required by HEID and SIMANN3. We say "it seems" because the experiments with the above mentioned algorithms are made in different machines: HEID, SIMANN3 and TAB were implemented in a 40 MHz DIGITAL DECstation 2000/240 and those with GRASP in a 150MHz Silicon Graphics Challenge computer. Some results on the comparison of these algorithms are given in Table 7.1 below. The table presents average results obtained on test instances from our benchmark. For each instance and for each algorithm the average relative error, the average running time, and the amount of the instances solved to optimality are given in the third, fourth and fifth column, respectively. The average values are obtained over all runs of the given algorithm on the given test instance. (As described above these runs differ on the starting permutation or on the seed for the generator of the stream of pseudo-random numbers.)

Finally, we would like to notice that it is not clear whether the specific structure of these test instances is representative for the BiQAP and whether GRASP and the other algorithms can be applied with the same success to other instances of the BiQAP, e.g. to instances generated at random. To author's knowledge there exist no investigations concerning this question.

7.8 CONCLUSIONS AND OPEN QUESTIONS

In this chapter a generalization of the quadratic assignment problem, the BiQAP, was considered. The BiQAP is a problem of practical relevance which arises in VLSI design. There exists a strong relationship between the QAP and the BiQAP, in the sense that each QAP can be equivalently formulated as a BiQAP. This fact implies that the BiQAP is at least as hard as the QAP, from a theoretical and from a practical point of view.

From a theoretical point of view, two aspects of the BiQAP have been investigated. First, as a generalization of analogous results for the QAP, a 0-1 and a mixed integer

Table 7.1 Performance of HEID, SIMANN3, TAB and GRASP

Size	Algorithm	Rel. Error (in %)	CPU Time (in seconds)	Opt. solved (in %)
10	HEID	6.08	4.1	73.7
	SIMANN3	0.09	14.0	72.1
	TAB	0.00	69.5	100.0
	GRASP	0.00	2.3	100.0
14	HEID	9.49	23.6	7.8
	SIMANN3	0.43	86.3	98.4
	TAB	0.81	384.9	88.9
	GRASP	0.00	23.9	100.0
16	HEID	5.25	65.2	55.9
	SIMANN3	0.67	177.2	95.6
	TAB	4.71	487.2	82.4
	GRASP	0.00	289.6	100.0
20	HEID	11.53	212.4	48.2
	SIMANN3	1.05	579.1	93.2
	TAB	11.15	1360.5	42.3
	GRASP	0.00	184.2	100.0
24	HEID	6.98	520.9	2.9
	SIMANN3	5.37	1489.5	23.5
	TAB	3.81	3028.0	21.5
	GRASP	0.00	8092.2	100.0
28	HEID	4.49	1325.2	76.8
	SIMANN3	8.65	3289.2	8.9
	TAB	7.86	6087.5	13.7
	GRASP	0.00	4130.4	100.0
32	HEID	18.52	4459.0	6.7
	SIMANN3	4.61	6978.6	78.9
	TAB	17.32	11482.4	4.8
	GRASP	0.00	12585.1	100.0

linear programming (MILP) formulation for the BiQAP have been derived. It is not surprising that the size of these linearizations explodes as the size of the BiQAP increases. Secondly, the asymptotic behavior of the BiQAP has been analyzed. It turns out that, under natural probabilistic constraints, the problem becomes "easy" as its size tends to infinity, i.e. the best and worst values of the objective function get in some sense closer as the size of the problem increases. From a practical point of view this means that for problems of large size - where the meaning of "large" is not clear yet - each feasible solution is close enough to an optimal solution. It remains an interesting and apparently difficult open problem to derive a lower bound for the size of instances which show such an asymptotic behavior.

From a practical point of view, the results presented in this chapter concern three aspects of the problem. First, as a generalization of the Gilmore-Lawler bound for the QAP, Gilmore-Lawler-like bounds for the BiQAP have been derived. Secondly, a generator of BiQAP instances with known optimal solution has been presented. The BiQAP instances produced by this generator are combinatorially related to QAPs with known optimal solution generated as proposed by Li and Pardalos [154] (see also Section 3.7). These BiQAP instances can be used for the evaluation of heuristics and bounding procedures. At this point an interesting open question arises: How difficult are the generated test instances from a theoretical point of view, i.e. in terms of computational complexity, and from a more practical point of view, i.e. in terms of the so-called *average complexity*? Obviously, the answer to this question has an immediate impact on the relevance of the above mentioned test instances in evaluation of heuristics and bounding procedures. If the generated test instances are in a certain sense "easy", they are not representative for the BiQAP.

Finally, we mentioned and briefly described some heuristics for BiQAPs like deterministic improvement methods, simulated annealing algorithms, tabu search approaches, and GRASP. Numerical results concerning the evaluation of these heuristics have been obtained by testing them on the above mentioned BiQAP instances with known optimal solution. However a lot remains to be done on this matter. Another probably interesting research direction is the investigation of BiQAPs whose coefficient arrays have some special structure and the exploitation of this structure in designing specific efficient algorithms for these special problems. (Recall that the coefficient arrays of BiQAP instances occurring in VLSI design as described in Section 7.1 are very sparse.)

REFERENCES

[1] E. H. L. Aarts and J. Korst, *Simulated Annealing and Boltzmann Machines: A Stochastic Approach to Combinatorial Optimization and Neural Computing*, Wiley, Chichester, 1989.

[2] W. P. Adams and T. Johnson, Improved linear programming based lower bounds for the quadratic assignment problem, in *Quadratic Assignment and Related Problems*, P. Pardalos and H. Wolkowicz, eds., *DIMACS Series in Discrete Mathematics and Theoretical Computer Science* **16**, 1994, pp. 43–77, AMS, Providence, Rhode Island.

[3] D. Adolphson and T. C. Hu, Optimal linear ordering, *SIAM Journal on Applied Mathematics* **25**, 1973, 403–423.

[4] R. K. Ahuja, J. B. Orlin, and A. Tivari, A greedy genetic algorithm for the quadratic assignment problem, Working paper 3826-95, Sloan School of Management, MIT, 1995.

[5] P. Ashar, S. Devadas, and R. A. Newton, *Sequential Logic Synthesis*, Kluwer Academic Publishers, Boston, 1992.

[6] S. Arora, A. Frieze, and H. Kaplan, A new rounding procedure for the assignment problem with applications to dense graph arrangement problems, *Proceedings of the 37-th Annual IEEE Symposium on Foundations of Computer Science (FOCS)*, 1996, 21–30.

[7] A. A. Assad, and W. Xu, On lower bounds for a class of quadratic 0-1 programs, *Operations Research Letters* **4**, 1985, 175–180.

[8] A. S. Azarionok, Packing graphs of width two, *Vesti Akad. Navuk BSSR, Physics and Math. Sci. Minsk* **3**, 1987, 9–16, (in Russian).

[9] E. Balas and J. B. Mazzola, Quadratic 0-1 programming by a new linearization, presented at *the Joint ORSA/TIMS National Meeting*, 1980, Washington D.C.

[10] E. Balas and J. B. Mazzola, Nonlinear programming: I. Linearization techniques, *Mathematical Programming* **30**, 1984, 1–21.

[11] E. Balas and J. B. Mazzola, Nonlinear programming: II. Dominance relations and algorithms, *Mathematical Programming* **30**, 1984, 22–45.

[12] H. J. Bandelt and A. W. M. Dress, A canonical decomposition theory for matrices of a finite set, *Advances in Mathematics* **92**, 1992, 47–105.

[13] F. Barahona and A. R. Mahjoub, On the cut polytope, *Mathematical Programming* **36**, 1986, 157–173.

[14] F. Barahona, J. Fonlupt, and A. R. Mahjoub, Composition of graphs and polyhedra iv: acyclic spanning subgraphs, *SIAM Journal on Discrete Mathematics* **7**, 1994, 390–402.

[15] R. Battiti and G. Tecchiolli, The reactive tabu search, *ORSA Journal on Computing* **6**, 1994, 126–140.

[16] R. Battiti and G. Tecchiolli, Simulated annealing and tabu search in the long run: A comparison on QAP tasks, *Computers and Mathematics with Applications* **28**, 1994, 1–8.

[17] M. S. Bazaraa and O. Kirca, Branch and bound based heuristics for solving the quadratic assignment problem, *Naval Research Logistics Quarterly* **30**, 1983, 287–304.

[18] M. S. Bazaraa and H. D. Sherali, Benders' partitioning scheme applied to a new formulation of the quadratic assignment problem, *Naval Research Logistics Quarterly* **27**, 1980, 29–41.

[19] M. S. Bazaraa and H. D. Sherali, On the use of exact and cutting plane methods for the quadratic assignment problem, *Journal of the Operational Research Society* **33**, 1982, 991–1003.

[20] J. F. Benders, Partitioning procedures for solving mixed variable programming problems, *Numerische Mathematik* **4**, 1962, 238–252.

[21] B. Boffey and J. Karkazis, Location of transfer centers of a communication network with proportional traffic, *Journal of the Operational Research Society* **40**, 1989, 729–734.

[22] S. H. Bokhari, On the mapping problem, *IEEE Transactions on Computers* **C-30**, 1981, 207–214.

[23] B. Bollobás, *Extremal Graph Theory*, Academic Press, London, 1978.

[24] B. Bollobás and S. E. Eldrige, Packing of graphs and applications to computational complexity, *Journal of Combinatorial Theory B* **25**, 1978, 105–124.

[25] P. A. Bruijs, On the quality of heuristic solutions to a 19×19 quadratic assignment problem, *European Journal of Operational Research* **17**, 1984, 21–30.

[26] A. Bruenegger, J. Clausen, A. Marzetta, and M. Perregaard, Joining forces in solving large-scale quadratic assignment problems in parallel, Technical Report DIKU TR-96-23, Department of Computer Science, University of Copenhagen, Denmark, 1996. To be presented at IPPS 97, the IEEE International Parallel Processing Symposium.

[27] R. E. Burkard, Die Störungsmethode zur Lösung quadratischer Zuordnungsprobleme, *Operations Research Verfahren* **16**, 1973, 84-108, (in German).

[28] R. E. Burkard, Locations with Spatial Interactions: The Quadratic Assignment Problem, in *Discrete Location Theory*, P.B. Mirchandani and R.L. Francis, eds., Wiley, New York, 1991, pp. 387–437.

[29] R. E. Burkard and T. Bönniger, A heuristic for quadratic Boolean programs with applications to quadratic assignment problems, *European Journal of Operational Research* **13**, 1983, 374–386.

[30] R. E. Burkard and E. Çela, Heuristics for biquadratic assignment problems and their computational comparison, *European Journal of Operational Research* **83**, 1995, 283–300.

[31] R. E. Burkard, E. Çela, V. M. Demidenko, N. N. Metelski, and G. J. Woeginger, Easy and hard cases of the quadratic assignment problem: A survey, SFB Report 104, Institute of Mathematics, Technical University Graz, Austria, 1997.

[32] R. E. Burkard, E. Çela, and B. Klinz, On the biquadratic assignment problem, in *Quadratic Assignment and Related Problems*, P. Pardalos and H. Wolkowicz, eds., *DIMACS Series in Discrete Mathematics and Theoretical Computer Science* **16**, 1994, pp. 117–146, AMS, Providence, Rhode Island.

[33] R. E. Burkard, E. Çela, G. Rote, and G. J. Woeginger, The quadratic assignment problem with an Anti-Monge and a Toeplitz matrix: Easy and hard cases, SFB Report 34, Institute of Mathematics, Technical University Graz, Austria, 1995. To appear in *Mathematical Programming*.

[34] R. E. Burkard, V. G. Deĭneko, R. van Dal, J. A. A. van der Veen, and G. J. Woeginger, Well solvable cases of the TSP: A survey, SFB Report 52, Institute of Mathematics, Technical University Graz, Austria, 1995.

[35] R. E. Burkard and U. Derigs, *Assignment and Matching Problems: Solution Methods with FORTRAN-Programs*, *Lecture Notes in Economics and Mathematical Systems* **184**, Springer-Verlag, Berlin, 1980.

[36] R. E. Burkard and U. Fincke, Probabilistic asymptotic properties of some combinatorial optimization problems, *Discrete Applied Mathematics* **12**, 1985, 21–29.

[37] R. E. Burkard, S. E. Karisch, and F. Rendl, QAPLIB - a quadratic assignment problem library, *Journal of Global Optimization* **10**, 1997, 391–403. An on-line version is available via World Wide Web at the following URL: `http://opt.math.tu-graz.ac.at/~karisch/qaplib/`

[38] R. E. Burkard, B. Klinz, and R. Rudolf, Perspectives of Monge properties in optimization, *Discrete Applied Mathematics* **70**, 1996, 95–161.

[39] R. E. Burkard and J. Offermann, Entwurf von Schreibmaschinentastaturen mittels quadratischer Zuordnungsprobleme, *Zeitschrift für Operations Research* **21**, 1977, B121–B132, (in German).

[40] R. E. Burkard and F. Rendl, A thermodynamically motivated simulation procedure for combinatorial optimization problems, *European Journal of Operational Research* **17**, 1983, 169–174.

[41] V. N. Burkov, M. I. Rubinstein, and V. B. Sokolov, Some problems in optimal allocation of large-volume memories, *Avtomatika i Telemekhanika* **9**, 1969, 83–91, (in Russian).

[42] R. E. Burkard, R. Rudolf, and G. J. Woeginger, Three-dimensional axial assignment problems with decomposable cost coefficients, *Discrete Applied Mathematics* **65**, 1996, 123–139.

[43] P. Carraresi and F. Malucelli, A new lower bound for the quadratic assignment problem, *Operations Research* **40**, 1992, Suppl. No. **1**, S22–S27.

[44] P. Carraresi and F. Malucelli, A reformulation scheme and new lower bounds for the QAP, in *Quadratic Assignment and Related Problems*, P. Pardalos and H. Wolkowicz, eds., *DIMACS Series in Discrete Mathematics and Theoretical Computer Science* **16**, 1994, pp. 147–160, AMS, Providence, Rhode Island.

[45] E. Çela, The Quadratic Assignment Problem: Special Cases and Relatives, *Ph.D. Thesis*, Technical University Graz, Austria, 1995.

[46] E. Çela and G. J. Woeginger, A note on the maximum of a certain bilinear form, SFB Report 8, Institute of Mathematics, Technical University Graz, Austria, 1994.

[47] V. Černý, Thermodynamical approach to the traveling salesman problem: An efficient simulation algorithm, *Journal of Optimization Theory and Applications* **45**, 1985, 41–51.

[48] J. Chakrapani and J. Skorin-Kapov, Massively parallel tabu search for the quadratic assignment problem, *Annals of Operations Research* **41**, 1993, 327–342.

[49] J. Chakrapani and J. Skorin-Kapov, A constructive method to improve lower bounds for the quadratic assignment problem, in *Quadratic Assignment and Related Problems*, P. Pardalos and H. Wolkowicz, eds., *DIMACS Series in Discrete Mathematics and Theoretical Computer Science* **16**, 1994, pp. 161–171, AMS, Providence, Rhode Island.

[50] G. Christofer, M. Farach, and M. Trick, The structure of circular decomposable metrics, Proceedings of the 4-th European Symposium on Algorithms (ESA), *Lecture Notes in Computer Science* **1136**, Springer-Verlag, 1996, 406–418.

[51] N. Christofides, Worst case analysis of a new heuristic for the traveling salesman problem, Technical Report 338, Graduate School of Industrial Administration, Carnegie-Mellon University, Pittsburgh, PA, 1976.

[52] N. Christofides and E. Benavent, An exact algorithm for the quadratic assignment problem on a tree, *Operations Research* **37**, 1989, 760–768.

[53] N. Christofides and M. Gerrard, Special cases of the quadratic assignment problem, *Management Science Research Report* 391, Carnegie-Mellon University, Pittsburgh, PA, 1976.

[54] N. Christofides and M. Gerrard, A graph theoretic analysis of bounds for the quadratic assignment problem, in *Studies on Graphs and Discrete Programming*, P.Hansen, ed., North Holland, 1981, pp. 61–68.

[55] F. R. K. Chung, On optimal linear arrangements of trees, *Computers and Mathematics with Applications* **10**, 1984, 43–60.

[56] D. Cyganski, R. F. Vaz, and V. G. Virball, manuscript, 1993.

[57] J. Clausen, S. E. Karisch, M. Perregaard, and F. Rendl, On the applicability of lower bounds for solving rectilinear quadratic assignment problems in parallel, Technical Report DIKU TR-96-24, Department of Computer Science,

University of Copenhagen, Denmark, 1996. To appear in *Computational Optimization and Applications*.

[58] J. Clausen and M. Perregaard, Solving large quadratic assignment problems in parallel, Technical Report DIKU TR-94-22, Department of Computer Science, University of Copenhagen, Denmark, 1994. To appear in *Computational Optimization and Applications*.

[59] D. T. Connolly, An improved annealing scheme for the QAP, *European Journal of Operational Research* **46**, 1990, 93–100.

[60] K. Conrad, *Das quadratische Zuordnungsproblem und zwei seiner Spezialfälle*, Mohr Siebeck Publishers, Tübingen, Germany, 1971, (in German).

[61] D. G. Corneil, Y. Perl, and L. K. Stewart, A linear recognition algorithm for cographs, *SIAM Journal on Computing* **14**, 1985, 926–934.

[62] L. Davis, *Genetic Algorithms and Simulated Annealing*, Pitman, London, 1987.

[63] V. G. Deĭneko and V. L. Filonenko, On the reconstruction of specially structured matrices, in *Aktualnyje Problemy EVM, Programmirovanije*, DGU, Dnepropetrovsk, 1979, (in Russian).

[64] V. G. Deĭneko, R. Rudolf, and G. J. Woeginger, Sometimes traveling is easy, Proceedings of the 3-d European Symposium on Algorithms (ESA), *Lecture Notes in Computer Science* **979**, Springer-Verlag, Berlin, 1995, 128–141.

[65] V. G. Deĭneko and G. J. Woeginger, A solvable case of the quadratic assignment problem, SFB Report 88, Institute of Mathematics, Technical University Graz, Austria, 1996.

[66] M. Dell'Amico and S. Martello, Linear assignment, to appear in *Annotated Bibliographies in Combinatorial Optimization*, M. Dell'Amico, F. Maffioli, S. Martello, eds., Wiley, Chichester, 1997.

[67] V. M. Demidenko and R. Rudolf, A note on Kalmanson matrices, *Optimization* **40**, 1997, 285–294.

[68] J. W. Dickey and J. W. Hopkins, Campus building arrangement using TOPAZ, *Transportation Research* **6**, 1972, 59–68.

[69] J. Edmonds and R. Giles, A min-max relation for submodular functions on graphs, *Annals of Discrete Mathematics* **1**, 1977, 185–204.

[70] C. S. Edwards, The derivation of a greedy approximator for the Koopmans-Beckmann quadratic assignment problem, *Proceedings of the 77-th Combinatorial Programming Conference (CP77)*, 1977, 55-86.

[71] C. S. Edwards, A branch and bound algorithm for the Koopmans-Beckmann quadratic assignment problem, *Mathematical Programming Study* **13**, 1980, 35–52.

[72] A. N. Elshafei, Hospital layout as a quadratic assignment problem, *Operations Research Quarterly* **28**, 1977, 167–179.

[73] B. Eschermann, State assignment methods for synchronous sequential circuits, in *Progress in Computer Aided VLSI Design*, G.Zobrist, ed., Ablex, Norwood, 1990.

[74] B. Eschermann and H.-J. Wunderlich, Optimized synthesis of self-testable finite state machines, presented at the *20th International Symposium on Fault-Tolerant Computing (FTCS 20)*, Newcastle upon Tyne, 1990.

[75] S. Evan and Y. Shiloach, NP-Completeness of several arrangement problems, Technical Report 43, Computer Science Department, Technion, Haifa, Israel, 1975.

[76] U. Faigle and W. Kern, Note on the convergence of simulated annealing algorithms, *SIAM Journal on Control and Optimization* **29**, 1991, 153–159.

[77] T. Feo and M. G. C. Resende, Greedy randomized adaptive search procedures, *Journal of Global Optimization* **6**, 1995, 109–133.

[78] G. Finke, R. E. Burkard, and F. Rendl, Quadratic assignment problems, *Annals of Discrete Mathematics* **31**, 1987, 61–82.

[79] C. Fleurent and J. Ferland, Genetic hybrids for the quadratic assignment problem, in *Quadratic Assignment and Related Problems*, P. Pardalos and H. Wolkowicz, eds., *DIMACS Series in Discrete Mathematics and Theoretical Computer Science* **16**, 1994, pp. 173–187, AMS, Providence, Rhode Island.

[80] J. H. Forsberg, R. M. Delaney, Q. Zhao, G. Harakas, and R. Chandran, Analyzing lanthanide-included shifts in the NMR spectra of lanthanide(III) complexes derived from 1,4,7,10-tetrakis(N,N-diethylacetamido)-1,4,7,10-tetraazacyclododecane, *Inorganic Chemistry* **34**, 1994, 3705–3715.

[81] A. Frank, How to make a digraph strongly connected, *Combinatorica* **1**, 1981, 145–153.

[82] J. B. G. Frenk, M. van Houweninge, and A. H. G. Rinnooy Kan, Asymptotic properties of the quadratic assignment problem, *Mathematics of Operations Research* **10**, 1985, 100–116.

[83] A. M. Frieze and J. Yadegar, On the quadratic assignment problem, *Discrete Applied Mathematics* **5**, 1983, 89–98.

[84] A. M. Frieze, J. Yadegar, S. El-Horbaty, and D. Parkinson, Algorithms for assignment problems on an array processor, *Parallel Computing* **11**, 1989, 151–162.

[85] H. N. Gabow, A representation for crossing set families with applications to submodular flow problems, *Proceedings of the 4-th Annual ACM-SIAM Symposium on Discrete Algorithms (SODA)*, 1993, 202–211.

[86] H. N. Gabow and R. E. Tarjan, A linear time algorithm for a special case of disjoint set union, *Proceedings of the 15-th Annual ACM Symposium on Theory of Computing (STOC)*, 1983, 246–251.

[87] G. Gallo, M. D. Grigoriadis, and R. E. Tarjan, A fast parametric maximum flow algorithm and applications, *SIAM Journal on Computing* **18**, 1989, 30-55.

[88] M. R. Garey and D. S. Johnson, *Computers and Intractability: A Guide to the Theory of NP-Completeness*, W.H. Freeman and Company, New York, 1979.

[89] J. W. Gavett and N. V. Plyter, The optimal assignment of facilities to locations by branch and bound, *Operations Research* **14**, 1966, 210–232.

[90] F. Gavril, Some NP-complete problems on graphs, *Proceedings of the 11-th Conference on Information Sciences and Systems*, 1977, 91–95.

[91] A. M. Geoffrion and G. W. Graves, Scheduling parallel production lines with changeover costs: Practical applications of a quadratic assignment/LP approach, *Operations Research* **24**, 1976, 595–610.

[92] P. C. Gilmore, Optimal and suboptimal algorithms for the quadratic assignment problem, *SIAM Journal on Applied Mathematics* **10**, 1962, 305–313.

[93] F. Glover, Tabu search - Part I, *ORSA Journal on Computing* **1**, 1989, 190–206.

[94] F. Glover, Tabu search - Part II, *ORSA Journal on Computing* **2**, 1989, 4–32.

[95] F. Glover, M. Laguna, E. Taillard, and D. de Werra, eds., Tabu search, *Annals of Operations Research* **41**, 1993.

[96] M. X. Goemans and D. P. Williamson, Improved approximation algorithms for maximum cut and satisfiability problems using semidefinite programming, *Journal of the ACM* **42**, 1995, 1115–1145.

[97] D. E. Goldberg, *Genetic Algorithms in Search, Optimization and Machine Learning*, Addison-Wesley, Wokingham, England, 1989.

[98] M. K. Goldberg and I. A. Klipker, Minimal placing of trees on a line, Technical Report, Physico-Technical Institute of Low Temperatures, Academy of Sciences of the Ukrainian SSR, USSR, 1976, (in Russian).

[99] G. H. Golub and C. F. Van Loan, *Matrix Computations*, North Oxford Academic, London, 1986.

[100] M. C. Golumbic, *Algorithmic Graph Theory and Perfect Graphs*, Academic Press, New York, 1980.

[101] R. E. Gomory and T. C. Hu, Multi-terminal network flows, *SIAM Journal on Applied Mathematics* **9**, 1961, 551–570.

[102] G. W. Graves and A. B. Whinston, An algorithm for the quadratic assignment problem, *Management Science* **17**, 1970, 453–471.

[103] M. Grötschel, Approaches to hard combinatorial optimization problems, in *Modern Applied Mathematics: Optimization and Operations Research*, B. Korte, ed., North Holland, 1982, pp. 437–515.

[104] M. Grötschel, Discrete Mathematics in Manufacturing, in *Proceedings of the Second International Conference on Industrial and Applied Mathematics (ICIAM 1991)*, Robert E. O'Malley, ed., SIAM, 1992, pp. 119–145.

[105] M. Grötschel, M. Jünger, and G. Reinelt, On the acyclic subgraph polytope, *Mathematical Programming* **33**, 1985, 28–42.

[106] M. Grötschel, L. Lovász, and A. Schrijver, *Geometric Algorithms and Combinatorial Optimization*, Springer-Verlag, Berlin, 1989.

[107] M. Grötschel and Y. Wakabayashi, A cutting plane algorithm for a clustering problem, *Mathematical Programming* **45**, 1989, 59–96.

[108] P. Hahn and T. Grant, Lower bounds for the quadratic assignment problem based upon a dual formulation, to appear in *Operations Research*.

[109] P. Hahn, T. Grant, and N. Hall, Solution of the quadratic assignment problem using the Hungarian method, to appear in *European Journal of Operational Research*.

[110] M. Hanan and J. M. Kurtzberg, A review of the placement and quadratic assignment problems, *SIAM Review* **14**, 1972, 324–342.

[111] S. W. Hadley, F. Rendl, and H. Wolkowicz, Bounds for the quadratic assignment problem using continuous optimization techniques, *Proceedings of the 1-st International Integer Programming and Combinatorial Optimization Conference (IPCO)*, 1990, 237–248.

[112] S. W. Hadley, F. Rendl, and H. Wolkowicz, Symmetrization of nonsymmetric quadratic assignment problems and the Hoffman-Wielandt inequality, *Linear Algebra and its Applications* **167**, 1992, 53–64.

[113] S. W. Hadley, F. Rendl, and H. Wolkowicz, A new lower bound via projection for the quadratic assignment problem, *Mathematics of Operations Research* **17**, 1993, 727–739.

[114] G. H. Hardy, J. E. Littlewood, and G. Pólya, The maximum of a certain bilinear form, *Proceedings of the London Mathematical Society* **25**, 1926, 265–282.

[115] G. H. Hardy, J. E. Littlewood, and G. Pólya, *Inequalities*, Cambridge University Press, Cambridge, 1952.

[116] S. M. Hedetniemi, S. T. Hedetniemi, and P. J. Slater, A linear algorithm for packing trees into K_N, Technical Report CS-TR-79-15, University of Oregon, 1979.

[117] S. M. Hedetniemi, S. T. Hedetniemi, and P. J. Slater, A note on packing trees into K_N, *Ars Combinatorica* **11**, 1981, 149–153.

[118] D. R. Heffley, Assigning runners to a relay team, in *Optimal Strategies in Sports*, S.P. Ladany and R.E. Machol, eds., North-Holland, Amsterdam, 1977, pp. 169–171.

[119] C. H. Heider, A computationally simplified pair exchange algorithm for the quadratic assignment problem, Paper No. **101**, Center for Naval Analysis, Arlington, Virginia, 1972.

[120] K. Hoffman and M. W. Padberg, LP-based combinatorial problem solving, *Annals of Operations Research* **4**, 1985/1986, 145–195.

[121] J. H. Holland, *Adaptation in Natural and Artificial Systems*, University of Michigan Press, Ann Arbor, 1975.

[122] J. E. Hopcroft and R. E. Tarjan, Dividing a graph into triconnected components, *SIAM Journal on Computing* **2**, 1973, 135–158.

[123] J. E. Hopcroft and R. E. Tarjan, Efficient planarity testing, *Journal of the ACM* **21**, 1974, 549–568.

[124] L. J. Hubert, *Assignment methods in combinatorial data analysis*, Marcel Dekker Inc., New York, 1987.

[125] D. L. Isaacson and R. W. Madsen, *Markov Chains: Theory and Applications*, Wiley, New York, 1976.

[126] M. Jünger, *Polyhedral Combinatorics and the Acyclic Subdigraph Problem*, Heldermann Verlag, Berlin, Germany, 1985.

[127] M. Jünger, A. Martin, G. Reinelt, and R. Weissmantel, Quadratic 0/1 optimization and a decomposition approach for the placement of electronic circuits, *Mathematical Programming* **63**, 1994, 257–279.

[128] M. Jünger and V. Kaibel, A basic study of the QAP polytope, Technical Report 96.215, Institut für Informatik, Universität zu Köln, Germany, 1996.

[129] M. Jünger and V. Kaibel, On the SQAP polytope, Technical Report 96.241, Institut für Informatik, Universität zu Köln, Germany, 1996.

[130] D. S. Johnson, C. H. Papadimitriou, and M. Yannakakis, How easy is local search, *Journal of Computer and System Sciences* **37**, 1988, 79–100.

[131] B. K. Kaku and G. L. Thompson, An exact algorithm for the general quadratic assignment problem, *European Journal of Operational Research* **2**, 1986, 382–390.

[132] S. E. Karisch, Nonlinear Approaches for Quadratic Assignment and Graph Partition Problems, *Ph.D. Thesis*, Technical University Graz, Austria, 1995.

[133] S. E. Karisch and F. Rendl, Lower bounds for the quadratic assignment problem via triangle decompositions, *Mathematical Programming* **71**, 1995, 137–151.

[134] S. E. Karisch, F. Rendl, and H. Wolkowicz, Trust regions and relaxations for the quadratic assignment problem, in *Quadratic Assignment and Related Problems*, P. Pardalos and H. Wolkowicz, eds., *DIMACS Series in Discrete Mathematics and Theoretical Computer Science* **16**, 1994, pp. 199–220, AMS, Providence, Rhode Island.

[135] N. K. Karmarkar and K. G. Ramakrishnan, Computational results of an interior point algorithm for large scale linear programming, *Mathematical Programming* **52**, 1991, 555–586.

[136] R. M. Karp, Reducibility among combinatorial problems, in *Complexity of Computer Computations*, R. E. Miller and J. W. Thatcher, eds., Plenum Press, New York, 1972, pp. 85–103.

[137] A. V. Karzanov, On the minimal number of arcs of a digraph meeting all its directed cut sets (abstract), *Graph Theory Newsletters* **8**, No. **4**, 1979.

[138] L. Kaufman and F. Broeckx, An algorithm for the quadratic assignment problem using Benders' decomposition, *European Journal of Operational Research* **2**, 1978, 204–211.

[139] B. Kernighan and S. Lin, An efficient heuristic procedure for partitioning graphs, *Bell Systems Journal* **49**, 1972, 291–307.

[140] S. Kirkpatrick, C. D. Gelatt, and M. P. Vecchi, Optimization by simulated annealing, *Science* **220**, 1983, 671–680.

[141] T. C. Koopmans and M. J. Beckmann, Assignment problems and the location of economic activities, *Econometrica* **25**, 1957, 53–76.

[142] N. M. Korneenko, On the complexity of computing a distance between graphs, *Vesti of Belarus Acad. Sci.* **1**, 1981, 10–13, (in Russian).

[143] J. Krarup and P. M. Pruzan, Computer-aided layout design, *Mathematical Programming Study* **9**, 1978, 75–94.

[144] A. V. Krushevski, The linear programming problem on a permutation group, *Proceedings of the seminar on methods of mathematical modeling and theory of electrical circuits*, Institute of Cybernetics of the Academy of Sciences of Ukraine, No. **3**, 1964, 364–371, (in Russian).

[145] A. V. Krushevski, Extremal problems for linear forms on permutations and applications, *Proceedings of the seminar on methods of mathematical modeling and theory of electrical circuits*, Institute of Cybernetics of the Academy of Sciences of Ukraine, No. **3**, 1965, 262–269, (in Russian).

[146] L. Lovász and A. Schrijver, Cones of matrices and set functions and 0-1 optimization, *SIAM Journal on Optimization* **1**, 1991, 166–190.

[147] P. J. M. van Laarhoven and E. H. L. Aarts, *Simulated Annealing: Theory and Applications*, D. Reidel Publishing Company, Dordrecht, 1988.

[148] A. M. Land, A problem of assignment with interrelated costs, *Operations Research Quarterly* **14**, 1963, 185–198.

[149] G. Laporte and H. Mercure, Balancing hydraulic turbine runners: A quadratic assignment problem, *European Journal of Operational Research* **35**, 1988, 378–382.

[150] P. S. Laursen, Simple approaches to parallel branch and bound, *Parallel Computing* **19**, 1993, 143–152.

[151] P. S. Laursen, Simulated annealing for the QAP - Optimal tradeoff between simulation time and solution quality, *European Journal of Operational Research* **69**, 1993, 238–243.

[152] E. L. Lawler, The quadratic assignment problem, *Management Science* **9**, 1963, 586–599.

[153] E. L. Lawler, J. K. Lenstra, A. H. G. Rinnooy Kan, and D. B. Shmoys, *The Traveling Salesman Problem*, Wiley, Chichester, 1985.

[154] Y. Li and P. M. Pardalos, Generating quadratic assignment test problems with known optimal permutations, *Computational Optimization and Applications* **1**, 1992, 163–184.

[155] P. M. Pardalos, K. G. Ramakrishnan, M. G. C. Resende, and Y. Li, Implementation of a variance reduction based lower bound in a branch and bound algorithm for the quadratic assignment problem, *SIAM Journal on Optimization* **7**, 1997, 280–294.

[156] Y. Li, P. M. Pardalos, K. G. Ramakrishnan, and M. Resende, Lower bounds for the quadratic assignment problem, *Annals of Operations Research* **50**, 1994, 387–410.

[157] Y. Li, P. M. Pardalos, and M. Resende, A greedy randomized adaptive search procedure for the quadratic assignment problem, in *Quadratic Assignment and Related Problems*, P. Pardalos and H. Wolkowicz, eds., *DIMACS Series in Discrete Mathematics and Theoretical Computer Science* **16**, 1994, 237–261, AMS, Providence, Rhode Island.

[158] L. Lovász, On two minimax theorems in graph, *Journal of Combinatorial Theory B* **21**, 1976, 96–103.

[159] C. L. Lucchesi, A Minimax Inequality for Directed Graphs, *Ph.D. Thesis*, University of Waterloo, Ontario, Canada, 1976.

[160] C. L. Lucchesi and D. H. Younger, A minimax theorem for directed graphs, *Journal of the London Mathematical Society* **17**, 1978, 369–374.

[161] F. Malucelli, Quadratic Assignment Problems: Solution Methods and Applications, *Ph.D. Thesis* **TD-9/93**, University of Pisa, Italy, 1993.

[162] M. M. Mano, *Computer System Architecture*, 2nd edition, Prentice Hall, Englewood Cliffs, NJ, 1982.

[163] T. Mavridou, P. M. Pardalos, L. Pitsoulis, and M. G. C. Resende, A GRASP for the biquadratic assignment problem, to appear in *European Journal of Operational Research*, 1997.

[164] T. Mautor and C. Roucairol, A new exact algorithm for the solution of quadratic assignment problems, *Discrete Applied Mathematics* **55**, 1992, 281–293.

[165] N. N. Metelski, On extremal values of quadratic forms on symmetric groups, *Vesti Akad. Navuk BSSR, Ser. Fiz.-Mat. Navuk* **6**, 1972, 107–110, (in Russian).

[166] N. N. Metelski, Some special problems in placements and packings of graphs (abstract), *Vesti Akad. Navuk BSSR, Physics and Math. Sci. Minsk* **5**, 1984, p. 110, (in Russian).

[167] N. Metropolis, A. Rosenbluth, M. Rosenbluth, A. Teller, and E. Teller, Equations of state calculations by fast computing machines, *Journal of Chemical Physics* **21**, 1953, 1087–1092.

[168] P. B. Mirchandani and T. Obata, Algorithms for a class of quadratic assignment problems, presented at *the Joint ORSA/TIMS National Meeting*, 1979, New Orleans.

[169] G. De Micheli, R. K. Brayton, and A. Sangiovanni-Vincentelli, Symbolic design of combinational and sequential logic circuits implemented by two-level logic macros, *IEEE Transactions on Computer Aided Design* **5**, 1986, 597–616.

[170] J. Mosevich, Balancing hydraulic turbine runners — a discrete combinatorial optimization problem, *European Journal of Operational Research* **26**, 1986, 202–204.

[171] K. A. Murthy, P. Pardalos, and Y. Li, A local search algorithm for the quadratic assignment problem, *Informatica* **3**, 1992, 524–538.

[172] H. Müller-Merbach, *Optimale Reihenfolgen,* Springer-Verlag, Berlin, Heidelberg, New York, 1970, pp. 158–171.

[173] G. L. Nemhauser and L. A. Wolsey, *Integer and Combinatorial Optimization*, Wiley, Chichester, 1988.

[174] C. E. Nugent, T. E. Vollmann, and J. Ruml, An experimental comparison of techniques for the assignment of facilities to locations, *Journal of Operations Research* **16**, 1969, 150–173.

[175] Z. Nutov and M. Penn, Minimum feedback arc set and maximum integral dicycle packing algorithms in $K_{3,3}$-free digraphs, Technical Report, Technion, Faculty of Industrial Engineering and Management, Haifa, Israel, June 1994.

[176] M. W. Padberg, *Linear Optimization and Extensions*, Springer-Verlag, Berlin, 1995.

[177] M. W. Padberg and M. P. Rijal, *Location, Scheduling, Design and Integer Programming*, Kluwer Academic Publishers, Boston, 1996.

[178] M. W. Padberg and G. Rinaldi, A branch and cut algorithm for the resolution of large-scale symmetric traveling salesman problems, *SIAM Review* **33**, 1991, 60–100.

[179] M. W. Padberg and G. Rinaldi, Optimization of a 532-city symmetric traveling salesman problem by a branch and cut algorithm, *Operations Research Letters* **6**, 1987, 1–7.

[180] G. S. Palubeckis, Generation of quadratic assignment test problems with known optimal solutions, *U.S.S.R. Comput. Maths. Math. Phys.* **28**, 1988, 97–98, (in Russian).

[181] C. H. Papadimitriou and K. Steiglitz, Local Search, in *Combinatorial Optimization: Algorithms and complexity*, Prentice Hall, Englewood Cliffs, NJ, pp. 454–481.

[182] C. H. Papadimitriou and D. Wolfe, The complexity of facets resolved, *Proceedings of the 25-th Annual IEEE Symposium on Foundations of Computer Science (FOCS)*, 1985, 74–78.

[183] P. Pardalos and J. Crouse, A parallel algorithm for the quadratic assignment problem, *Proceedings of the Supercomputing Conference 1989*, ACM Press, 1989, 351–360.

[184] P. Pardalos, F. Rendl, and H. Wolkowicz, The quadratic assignment problem: A survey and recent developments, in *Quadratic Assignment and Related Problems*, P. Pardalos and H. Wolkowicz, eds., *DIMACS Series in Discrete Mathematics and Theoretical Computer Science* **16**, 1994, 1–42, AMS, Providence, Rhode Island.

[185] U. Pferschy, R. Rudolf, and G. J. Woeginger, Monge matrices make maximization manageable, *Operations Research Letters* **16**, 1994, 245–254.

[186] M. A. Pollatschek, N. Gershoni, and Y. T. Radday, Optimization of the typewriter keyboard by computer simulation, *Angewandte Informatik* **10**, 1976, 438–439.

[187] V. R. Pratt, An $N \log N$ algorithm to distribute N records optimally in a sequential access file, in *Complexity of Computer Computations*, R. E. Miller and J. W. Thatcher, eds., Plenum Press, New York, 1972, pp. 111–118.

[188] M. Queyranne, Performance ratio of polynomial heuristics for triangle inequality quadratic assignment problems, *Operations Research Letters* **4**, 1986, 231–234.

[189] V. Ramachandran, Finding a minimum feedback arc set in reducible flow graphs, *Journal of Algorithms* **9**, 1988, 299–313.

[190] G. Reinelt, *The Linear Ordering Problem: Algorithms and Applications*, Heldermann Verlag, Berlin, Germany, 1985.

[191] F. Rendl, Ranking scalar products to improve bounds for the quadratic assignment problem, *European Journal of Operational Research* **20**, 1985, 363–372.

[192] F. Rendl, Quadratic assignment problems on series parallel digraphs, *Zeitschrift für Operations Research* **30**, 1986, 161–173.

[193] F. Rendl, Das Quadratische Zuordnungsproblem: Spezialfälle, Näherungsverfahren und Untere Schranken, *Habilitationsschrift*, Institute of Mathematics, Technical University Graz, Austria, 1988, (in German).

[194] F. Rendl and H. Wolkowicz, Applications of parametric programming and eigenvalue maximization to the quadratic assignment problem, *Mathematical Programming* **53**, 1992, 63–78.

[195] A. Rényi, *Probability Theory*, North Holland, Amsterdam, 1970.

[196] M. G. C. Resende, K. G. Ramakrishnan, and Z. Drezner, Computing lower bounds for the quadratic assignment problem with an interior point algorithm for linear programming, *Operations Research* **43**, 1995, 781–791.

[197] W. T. Rhee, A note on asymptotic properties of the quadratic assignment problem, *Operations Research Letters* **7**, 1988, 197–200.

[198] W. T. Rhee, Stochastic analysis of the quadratic assignment problem, *Mathematics of Operations Research* **16**, 1991, 223–239.

[199] C. Roucairol, A reduction method for quadratic assignment problems, *Operations Research Verfahren* **32**, 1979, 183–187.

[200] C. Roucairol, A parallel branch and bound algorithm for the quadratic assignment problem, *Discrete Applied Mathematics* **18**, 1987, 221–225.

[201] M. I. Rubinstein, Problems and methods in combinatorial programming, Technical Report, Institute of Control Sciences, Moscow, 1976, (in Russian).

[202] M. I. Rubinstein, *The Optimal Grouping of Related Objects*, NAUKA Publishers, Moscow, 1989, (in Russian).

[203] M. I. Rubinstein, The two and four points conditions in the quadratic assignment problems, manuscript, 1994.

[204] R. Rudolf, Recognition of d-dimensional Monge arrays, *Discrete Applied Mathematics* **52**, 1994, 71–82.

[205] R. Rudolf and G. J. Woeginger, The cone of Monge matrices: Extremal rays and applications, *ZOR – Mathematical Methods of Operations Research* **42**, 1995, 161–168.

[206] S. Sahni and T. Gonzalez, P-complete approximation problems, *Journal of the ACM* **23**, 1976, 555–565.

[207] N. Sauer and J. Spencer, Edge disjoint placement of graphs, *Journal of Combinatorial Theory B* **25**, 1978, 295–302.

[208] S. L. Savage, Some theoretical implications of local search, *Mathematical Programming* **10**, 1976, 354–366.

[209] A. Schäffer and M. Yannakakis, Simple local search problems that are hard to solve, *SIAM Journal on Computing* **20**, 1991, 56–87.

[210] D. Schlegel, Die Unwucht-optimale Verteilung von Turbinenschaufeln als quadratisches Zuordnungsproblem, *Ph.D. Thesis*, ETH Zürich, Switzerland, 1987, (in German).

[211] A. Schrijver, *Theory of Linear and Integer Programming*, Wiley, Chichester, 1986.

[212] Y. Shiloach, A minimum linear arrangement algorithm for undirected trees, *SIAM Journal on Computing* **8**, 1979, 15–32.

[213] J. Skorin-Kapov, Tabu search applied to the quadratic assignment problem, *ORSA Journal on Computing* **2**, 1990, 33–45.

[214] W. Stallings, *Local Networks: An introduction*, MacMillan, New York, 1984.

[215] L. Steinberg, The backboard wiring problem: A placement algorithm, *SIAM Review* **3**, 1961, 37–50.

[216] F. Supnick, Extreme Hamiltonian lines, *Annals of Mathematics* **66**, 1957, 179–201.

[217] W. Szpankowski, Combinatorial optimization problems for which almost every algorithm is asymptotically optimal!, *Optimization* **33**, 1995, 359–367.

[218] E. Taillard, Robust taboo search for the quadratic assignment problem, *Parallel Computing* **17**, 1991, 443–455.

[219] E. Taillard, Comparison of iterative searches for the quadratic assignment problem, *Location Science* **3**, 1995, 87–105.

[220] R. E. Tarjan, Testing flow graph reducibility, *Journal of Computer System Sciences* **9**, 1974, 355–365.

[221] R. E. Tarjan, *Data Structures and Network Algorithms*, SIAM, Philadelphia, 1983.

[222] D. M. Tate and A. E. Smith, A genetic approach to the quadratic assignment problem, *Computers and Operations Research* **22**, 1995, 73–83.

[223] B. B. Timofeev and V. A. Litvinov, On the extremal value of a quadratic form, *Kibernetica* **4**, 1969, 56–61, (in Russian).

[224] I. Ugi, J. Bauer, J. Friedrich, J. Gasteiger, C. Jochum, and W. Schubert, Neue Anwendungsgebiete für Computer in der Chemie, *Angewandte Chemie* **91**, 1979, 99–111, (in German).

[225] J. Valdes, R. E. Tarjan, and E. L. Lawler, The recognition of series parallel digraphs, *SIAM Journal on Computing* **11**, 1982, 471–506.

[226] D. Vanderbilt and S. G. Louie, A Monte Carlo simulated annealing approach to optimization over continuous variables, *Journal of Computational Physics* **56**, 1984, 259–271.

[227] J. A. A. van der Veen, Solvable Cases of the Traveling Salesman Problem with Various Objective Functions, *Ph.D. Thesis*, University of Groningen, The Netherlands, 1992.

[228] D. H. West, Algorithm 608: Approximate solution of the quadratic assignment problem, *ACM Transcations on Mathematical Software* **9**, 1983, 461–466.

[229] M. R. Wilhelm and T. L. Ward, Solving quadratic assignment problems by "simulated annealing", *IIE Transcations* **1**, 1987, 107–119.

[230] Q. F. Yang, R. E. Burkard, E. Çela, and G. J. Woeginger, Hamiltonian cycles in circulant digraphs with two stripes, SFB Report 20, Institute of Mathematics, Technical University Graz, Austria, 1995. To appear in *Discrete Mathematics*.

[231] Q. Zhao, Semidefinite Programming for Assignment and Partitioning Problems, *Ph.D. Thesis*, University of Waterloo, Ontario, Canada, 1996.

[232] Q. Zhao, S. E. Karisch, F. Rendl, and H. Wolkowicz, Semidefinite relaxations for the quadratic assignment problem, Technical Report DIKU TR-96-32, Department of Computer Science, University of Copenhagen, Denmark, 1996. To appear in *Journal of Combinatorial Optimization*.

NOTATION INDEX

The following is a list of commonly used notations. For each symbol its meaning (or name) is given together with the page number where the symbol occurs for the first time in the text. Most of the symbols are listed in the order of their first occurrence. There are also some standard symbols which are defined nowhere in the text. For these symbols which are listed at first we give an $*$ instead of a page number. There is also some overloaded notation, e.g. (i,j), which has more than one meaning.

$\subset$	proper subset	*
$\subseteq$	subset	*
$\setminus$	set difference	*
$\Rightarrow$	implies	*
$\cap$	matrix intersection	*
$\cup$	matrix union	*
$f\colon A \to B$	function which maps set A to set B	*
$\mathbb{R}$	the set of the real numbers	*
$\mathbb{R}^+$	the set of the nonnegative real numbers	*
$\mathbb{R}^n$	the set of the n-dimensional real vectors	*
$\mathbb{N}$	the set of the natural numbers	*
$\lfloor x \rfloor$	lower integer part of x	*
$\lceil x \rceil$	upper integer part of x	*
A^t	transpose of matrix A	*
π^{-1}	inverse of permutation π	*
$\mathcal{P}$	class of problems solvable in polynomial time	*
$\mathcal{NP}$	class of problems solvable in nondeterministic polynomial time	*

$\mathcal{PLS}$	class of local search problems solvable in polynomial time	*
$o(f(x))$	the little-oh notation	*
$O(f(x))$	the big-oh notation	*
$\Omega(f(x))$	the big-omega notation	*
$G = (V, E)$	graph with vertex set V and edge set E	*
$K_{n,m}$	complete bipartite graph with n vertices on one part and m on the other one	*
K_n	complete graph on n vertices	*
QAP	the quadratic assignment problem	1
MILP	mixed integer linear programming	1
QAP(A,B)	the Koopmans-Beckmann quadratic assignment problem with coefficient matrices A and B	2
$\mathcal{S}_n$	the set of permutations of the numbers $1, 2, \ldots, n$	2
$Z(A, B, \phi)$	the value of the objective function of QAP(A,B) yielded by permutation ϕ	2
$A = (a_{ij})$	matrix with entries a_{ij}	2
$V = (v_i)$	vector with components v_i	2
V^π	vector permuted by permutation π, $V^\pi = (v_{\pi(i)})$	2
$D = (d_{ijkl})$	four-dimensional array with coefficients d_{ijkl}	3
$X = (x_{ij})$	permutation matrix	5
Π_n	the set of $n \times n$ permutation matrices	5
π_X	permutation corresponding to the permutation matrix X	5
$f_{A,B}$	the objective function of QAP(A,B) in the trace formulation	6
$tr(A)$	trace of matrix A	7
$conv(S)$	convex hull of set S	11
QAP_n	the n-dimensional QAP polytope	11
$SQAP_n$	the n-dimensional symmetric QAP polytope	14
HC	the Hamiltonian cycle problem	17

π_{opt}	optimal solution of a QAP (or a BiQAP)	18 (240)
TSP	the traveling salesman problem	20
$R_\Upsilon(A,B)$	performane ratio of algorithm Υ applied to the instance QAP(A,B)	20
R_Υ	performance ratio of algorithm Υ	20
R_Υ^∞	asymoptotic performance ratio of algorithm Υ	20
$\mathcal{E}$	ground set of a combinatorial optimization problem	22
$\mathcal{F}$	set of the feasible solutions of a combinatorial optimization problem	22
$\mathcal{N}(S)$	neighborhood of solution S	22
$(P,\mathcal{N})$	local search problem with neighborhood structure $\mathcal{N}$	23
(i,j)	ordered pair of indices	24
(i,j)	transposition which maps i to j and j to i	24
$\circ$	composition of permutations	24
$\mathcal{N}_{\text{K-L}}$	Kernighan-Lin-type neighborhood	24
$\mathcal{N}_2$	pair-exchange neighborhood	25
$A_{(i,.)}$ $(A_{(.,i)})$	i-th row (column) of matrix A considered as a vector	28
$\langle U,V\rangle$	scalar product of vectors U and V	29
$\langle U,V\rangle_+$	maximum value of the scalar product $\langle U,V^\pi\rangle$, where the maximum is taken over all permutations π	29
$\langle U,V\rangle_-$	minimum value of the scalar product $\langle U,V^\pi\rangle$, where the minimum is taken over all permutations π	29
GL	Gilmore-Lawler bound	29
FY1	first bound of Frieze and Yadegar	31
FY2	second bound of Frieze and Yadegar	31
CG	bound of Christofides and Gerrard	32
$\gamma(M)$	average of matrix M	34
$V(M)$	variance of matrix M	34
$T(M,\lambda)$	total variance of matrix M with parameter λ	34

SUBJECT INDEX

GPSR Compliance
The European Union's (EU) General Product Safety Regulation (GPSR) is a set of rules that requires consumer products to be safe and our obligations to ensure this.

If you have any concerns about our products, you can contact us on

ProductSafety@springernature.com

In case Publisher is established outside the EU, the EU authorized representative is:

Springer Nature Customer Service Center GmbH
Europaplatz 3
69115 Heidelberg, Germany

www.ingramcontent.com/pod-product-compliance
Ingram Content Group UK Ltd.
Pitfield, Milton Keynes, MK11 3LW, UK
UKHW051127260726
13967UKWH00010B/2902

* 9 7 8 1 4 7 5 7 2 7 8 8 3 *